Changing Things – Moving People

Strategies for Promoting Sustainable Development at the Local Level

Ruth Kaufmann-Hayoz
Heinz Gutscher
(Editors)

Birkhäuser Verlag
Basel · Boston · Berlin

Prof. Dr. Ruth Kaufmann-Hayoz
Universität Bern
Interfakultäre Koordinationstelle
für Allgemeine Ökologie (IKAÖ)
Falkenplatz 16
CH-3012 Bern

Prof. Dr. Heinz Gutscher
Psychologisches Institut
Abteilung Sozialpsychologie
Universität Zürich
Plattenstrasse 14
CH-8032 Zürich

A CIP catalogue record for this book is available from the Library of Congress, Washington D.C., USA

Deutsche Bibliothek Cataloging-in-Publication Data
Changing things – moving people : strategies for promoting sustainable development at the local level / Ruth Kaufmann-Hayoz ; Heinz Gutscher (Hrsg.) - Basel ; Boston ; Berlin : Birkhäuser, 2001
ISBN 3-7643-6252-9

Member of the BertelsmannSpringer Publishing Group
Camera-ready copy prepared by U. Wittwer
Cover illustration (large picture): Luftbild Schweiz
Printed on acid-free paper produced from chlorine-free pulp. TFC ∞

ISBN 3-7643-6252-9

9 8 7 6 5 4 3 2 1

www.birkhauser.ch

Table of Contents

Empirical Studies on Selected Instruments and Instrument Combinations

Conclusions

Annex

Foreword

This book results from a pioneering effort to organize a productive interdisciplinary research program on sustainable development policy in a small country not previously recognized as a world leader in environmental social science. The results are very promising, considering the short time frame and the high barriers to success for such an enterprise – differences in concepts and terminology, disciplinary myopia, and the inherent difficulty of the problem. In the USA, where I work, these barriers continue to pose major challenges after some 30 years of effort. Switzerland has made noteworthy progress in only five. I hope this book represents the beginning of a long-term effort at problem-oriented interdisciplinary collaboration among Swiss researchers and practitioners.

The Swiss group has succeeded in developing a unifying framework that makes a major contribution to environmental policy analysis. The framework broadens policy thinking by giving serious treatment to underutilized strategies that rely on communication and informal influence as well as to well-studied ones that rely on technological change, regulation, and economic forces. This broad typology makes it easier for an analyst to escape the tendency to presume that the policy instrument currently in fashion, whether it be market-based instruments, voluntary measures, or whatever, is the right strategy for all problems. It also encourages discipline-based analysts to consider how their favored strategies might be combined with other strategies less familiar to them, and thus to craft strategies that can take advantage of the strengths of various policy instruments. The concluding chapter offers some reasonable suggestions about how best to combine the policy instruments.

The substantive chapters in the book examine most of the main categories in the typology, thus illustrating ways that a full set of policy options might be implemented in the Swiss context. The case examples indicate highly varied degrees of success, as might be expected with what are essentially pilot projects. In addition, the coordination of policy instruments that the conceptual framework suggests has only begun to be evidenced in the pilot projects. This sort of development inevitably takes more than a few years' time. It depends on collaboration involving researchers expert in different intervention types and willing to work together on particular environment-behavior problems in particular settings. It also depends on establishing good working partnerships among researchers, policy analysts, and decision makers in the affected communities.

This initial report shows the potential for these collaborations and partnerships and the growing realization of what is needed to develop them. Progress toward the Swiss sustainable development goals depends on continued sustenance for the new relationships and continued development, elaboration, and field testing of the kind of conceptual framework that this book sets forth.

Paul C. Stern
U.S. National Research Council
Washington, D.C., U.S.A.

Reader's Guide and Acknowledgements

Ruth Kaufmann-Hayoz and Heinz Gutscher

This book is the product of inter and transdisciplinary co-operation among academic scientists, researchers in private consultancy firms, and potential beneficiaries of the research, i.e., decision makers in politics, public administration, and non-governmental organisations. Its target audience is equally heterogeneous, as we expect scholars from different disciplines as well as educated lay persons interested in policy making toward sustainability to be among its readers. As different readers will bring in very different backgrounds of knowledge, experience, and interest, not all parts of the book may appeal equally to all of them. As an aid, we will explain the structure of the book and what is to be found where.

Due to the review and decision processes within the Swiss Priority Programme "Environment", which set the institutional framework for the research co-operation, the specific goals and research questions of the individual projects were worked out more or less independently. However, the terms of this programme demanded proposals of project groups, in which the individual projects also were to make a contribution to the objectives of the project group. During a process of lively exchange and communication among the research participants a number of common goals and questions that go beyond the goals and questions of the individual projects and to which the researchers of all projects were able to contribute were developed. These contributions were based not only on the results of the researcher's ongoing projects, but also on earlier research and their entire intellectual and disciplinary backgrounds. Therefore, the resulting synthesis should not primarily be viewed as an integration of the individual projects' results with respect to the resolution of a common problem. It is instead the joint development of a theoretical framework that incorporates more than one disciplinary perspective, relates the sub-projects' results to each other, and allows common conclusions to be drawn.

The book presents those results of the research co-operation that pertain to the questions of how the still quite vague category of "new" policy instruments can be differentiated, described, and explained more precisely; how they relate to traditional instruments of environmental policy; and how multiple instruments can best be combined for designing strategies toward sustainability. The material is organised in the following way:

The **introductory chapter** by KAUFMANN-HAYOZ AND GUTSCHER[1] sets the stage by introducing the topic, goals, and questions to be treated. It also outlines the general model of human action that we use as the basic theoretical approach. To all those who are familiar with sophisticated disciplinary theories of human action (for example, psychologists, philosophers, or sociologists), this model will seem very crude. In fact, it purposely does without elaborate features because its aim is to make visible the basic structure of human-environment interaction.

The short chapter by KAUFMANN-HAYOZ AND MAUCH provides some elementary information on the **Swiss political system** and its environmental and sustainability policies. Although we believe that the relevance of our findings is not restricted to this country, the information given may help the reader unfamiliar with Switzerland to better understand some details of the empirical studies.

Readers who are interested in the praxis of inter- and transdisciplinary research should take a close look on the chapter by DEFILA AND DI GIULIO in the Conclusions section. It describes and discusses the methods that have been used for supporting the **inter and transdiciplinary processes** in the integrated project. These are the methods and procedures applied for developing integrated research results as well as for organising co-operation and communication with non-academic project partners. The chapter reports the experience gained and makes recommendations for the management of comparable integrated projects.

The results of the **synthesis work** are to be found in the Theoretical Framework section and in the Conclusions section (chapters by KAUFMANN-HAYOZ ET AL.). A comprehensive **typology of instruments** available to a variety of actors to promote sustainable development is presented. The typology contains – in addition to the well-known conventional command-and-control and economic instruments – service and infrastructure instruments, collaborative agreements, and communication and diffusion instruments. The latter three types are an attempt to order and elaborate the heterogeneous class of "new" policy instruments that are often mentioned in this context. Special emphasis is given to instruments that are based on principles of social and behavioural psychology, because these approaches are relatively new and have barely been discussed in relation to conventional instruments. The different types of instruments are described in detail from an actor-oriented perspective. Readers familiar with a specific disciplinary approach to some instruments will perhaps find the typology crude or inadequate as compared to what they know about these particular instruments. However, they will be rewarded by finding detailed descriptions of less familiar instruments.

The synthesis chapter that presents the typology is complemented by a **theoretical paper** by FLURY-KLEUBLER AND GUTSCHER. These authors describe the basic possibilities of changing the behaviour of others from a psychological point of view, and they discuss psychological mechanisms that help us to understand and clarify the acceptance and effectiveness of different types of environmental policy instruments.

[1] Throughout the book, we will refer to chapters within this volume by printing the authors' names in the font SMALL CAPITALS.

The chapter by KAUFMANN-HAYOZ ET AL. in the **Conclusions** section represents another aspect of the synthesis. Here we make suggestions for designing policy strategies for sustainable development that are based on the theoretical work as well as the empirical studies. We point out the main lessons learnt with respect to combining different types of instruments within a process of policy strategy formation. Particularly promising combinations of instruments and some caveats are discussed.

Readers who are more interested in **empirical studies** than in theoretical considerations might want to concentrate on the contributions from the individual projects. These nine chapters are like searchlights that illuminate selected aspects of the application of certain types of instruments and combinations of instruments from the perspective and with the methods of different scientific disciplines. Readers should not expect, however, to find a uniform empirical approach. They might miss methodological details in papers stemming from their own disciplines, while they might not understand every detail or be surprised by the approaches taken by studies from other disciplines.

The following questions are addressed more or less explicitly in all the studies:

- *Instruments*: How does the study relate to the typology presented as theoretical framework? What instruments are studied?
- *Actors*: Who has implemented or would implement the instruments? What are the direct target groups and who is affected indirectly?
- *Objectives*: Why have the instruments been implemented? What were the objectives?
- *Implementation*: What are the specific implementation problems of the instruments studied? What is needed in terms of costs and effort to implement them?
- *Effectiveness*: What can be said about the impacts and outcomes of applying the instruments?
- *Combinations*: What advantages and disadvantages of combining different types of instruments have become evident? What particular advantages (or disadvantages) of combining classic instruments with communication and diffusion instruments and collaborative agreements have been found or can be expected?

Two chapters deal with the problem of how **more sustainable transport** could be achieved. MAUCH, NORTH, AND PULLI discuss the question of an optimal combination of instruments in the transport sector within the context of two fundamental strategic approaches to sustainability, often called the "sufficiency revolution" and "efficiency revolution". The focus lies on the assessment of communication and diffusion as well as service and infrastructure instruments that incorporate the characteristics of a sufficiency approach. GUTSCHER, MOSLER, AND ARTHO describe a project aimed at voluntary speed reductions in residential areas of a small town, using a set of communication instruments. The goal of the intervention was to stimulate enough co-operative behaviour in the 4,000 registered car owners to clearly reduce average driving speeds in the town. They point out that the outcome of such a campaign depends not only on finding an appropriate selection of instruments to change individual behaviour, but also on careful use of mechanisms of social diffusion in order to reach a sufficient number of people.

Four studies pertain to the **energy domain**. GARBELY AND MCFARLANE present an economic analysis and model simulation of different – existing and innovative – economic instruments aimed at improving the energy efficiency of the residential building sector. They give a justification of why investment in energy conservation of buildings is essential on the way towards sustainable consumption, and they analyse the functioning and the particular implementation problems of the proposed policy instruments. They also discuss some shortcomings of economic instruments and point out the advantages of combining them with communication and diffusion instruments. GEHRIG AND NORTH analyse the action strategies of solar power suppliers in Switzerland in view of the forthcoming liberalisation of the electricity market, and they point to the mix of instruments that can be applied when targeting ecologically sounder power production and consumption. The central questions of JEGEN'S study are how key players of Swiss energy policy evaluate different energy policy instruments, and how the political structure and power relations in this policy domain can be described. Several concrete instruments have been selected for each type, including the tax on the energy quality of buildings that is the subject of GARBELY AND MACFARLANE'S study. The results reported give some important indications as to the political realities that are encountered in the implementation phase of any policy programme. The contribution by BÄTTIG AND BALTHASAR focuses upon the Swiss Confederation's network management strategy within the framework of the energy action programme "Energy 2000 for Municipalities". This is an approach that emphasises optimal co-operation among the various actors in implementing the measures. Communication and diffusion instruments play a key role, and they are implemented on both the operations and the strategic levels. Super-ordinate, strategy-oriented communication offerings are designed to prepare the ground for active energy policy. For example, they may be used to build and strengthen a network of municipalities and other actors in energy policy. Measures-oriented elements, in contrast, directly target the planning and realisation of measures in specific domains.

MÜLLER AND KRAMER evaluate two **economic instruments for water conservation** in Switzerland: The wastewater charge as an instrument in the municipality and the wastewater tax as an instrument in the canton. They examine issues in designing these economic instruments and in orchestrating their relationship to other instruments (prescriptions, prohibitions, subsidies, and communication and diffusion instruments). They take a look at obstacles to implementation, ecological effects and obstacles to effectiveness, and how these might be overcome.

The two final studies focus on the **household level**. BRUPPACHER AND ULLI-BEER outline the strategy taken by the NGO "Global Action Plan" (GAP) to "spread a sustainable lifestyle" by combining communication instruments. Potentials and effects of the applied instruments are discussed with regard to co-operation between municipalities and NGOs, such as GAP. The results reported indicate that a combination of communication instruments with traditional policy instruments (economic and command and control instruments) is a promising way to lead a community towards sustainability. The chapter by VATTER, GESSNER, AND WITTWER looks at key psychological structures and processes such as basic values, implicit assumptions, and specific action intentions of GAP participants. Their analysis helps to anticipate specific opportunities and difficulties associated with the implementation of different types of policy instruments.

A special feature of the presentation of the empirical studies are two short **lawyers' comments**. They are not research reports in the true sense, but rather comments on some questions that particular studies and the instruments they highlight raise in a lawyer's mind. The comments are provided by FRIEDERICH (on the study by MÜLLER AND KRAMER) and by WICHTERMANN (on the study by BRUPPACHER AND ULLI-BEER).

* * *

For making realisation of this book possible, we are grateful to many persons and organisations for their participation and support. We thank the Swiss National Science Foundation and those responsible for the Swiss Priority Programme "Environment" for funding the Integrated Project "Strategies and Instruments for Sustainable Development", particularly for providing the means not only for the sub-projects, but for management and thus the synthesis work and this publication as well. In memory of the many stimulating hours we shared, we give hearty thanks to our colleagues who conducted the research, worked tirelessly – especially within the Synthesis Group – on joint efforts, and wrote the contributions to this volume. We wish to express our gratitude to the members of the Monitoring Group and our partners in the municipalities and private organisations for their interest in our work and for their generosity in giving us their time and in sharing their experiences and thoughts with us. Thanks are also due to the many persons who participated in the surveys and interviews in our projects; without them and their knowledge, many of the results reported in this volume would never have been produced. We are grateful to Heidi Schelbert, Paul C. Stern, and Helmut Weidner for their careful readings of the book, their critiques, and valuable suggestions. Our great appreciation is extended to Ellen Russon for language editing of the texts and translation of large parts of the book and to Urs Wittwer for meticulous attention to the layout work. Finally, for their constant support throughout the project, even at the most hectic of times, our colleagues, friends, and families deserve our warm thanks.

Bern and Zürich, February 2001

Ruth Kaufmann-Hayoz
Heinz Gutscher

Introduction

Changing Things – Moving People.
Strategies for Promoting Sustainable
Development at the Local Level
R. Kaufmann-Hayoz/H. Gutscher (eds.)

Transformation toward Sustainability: An Interdisciplinary, Actor-Oriented Perspective

Ruth Kaufmann-Hayoz and Heinz Gutscher

1. Environmental policy as part of an integrated policy for sustainable development

1.1. Sustainable development

The notion of "sustainable development" has become very popular through the report "Our Common Future" published in 1987 by the World Commission on Environment and Development in preparation of the 1992 United Nations Conference on Environment and Development in Rio de Janeiro (WCED, 1987). The report states that sustainable development "integrates economics and ecology in decision making and law making to protect the environment and to promote development" (p. 37). It "aims at a type of development which integrates production with resource conservation and enhancement and links both to providing an adequate livelihood base and equitable access to resources" (p. 39). It tries to "reorient international relations to achieve trade, capital and technology flows that are more equitable and consistent with environmental imperatives" (p. 40). It "meets the needs of the present without compromising the ability of future generations to meet their own needs" (p. 43).

In this vision of desirable global development, environmental protection was for the first time linked explicitly to socio-economic development. "Sustainable development" is to be understood as a three-dimensional normative concept. The three dimensions are the following:

The ecological dimension: Sustainable development aims at preserving natural life-supporting systems, by improving environmental quality, by preventing residual pollution for future generations, by sustainable use of natural resources, and other measures.

The economic dimension: Sustainable development aims at ensuring economic prosperity through efficient use of resources, by providing sufficient numbers of jobs and sufficient income, by pursuing technological progress, and so on.

The socio-cultural dimension: Sustainable development aims at realising social justice and solidarity, for example by promoting a just distribution of wealth and income, by ensuring education, legal rights, cultural identity and diversity, and so on.

Within the past decade numerous attempts have been made to define the objectives of sustainable development in any of the three dimensions more precisely and in more concrete ways. It has become clear that it is impossible to define a universally applicable target state for all societies, and that instead, societal visions of sustainable development vary with time, situation, culture, and knowledge. Criteria and indicators of sustainability will also be different depending on the domain examined (such as transport, agriculture, energy, and so on), and on the perspective taken: global, regional, or local.

From a global perspective, the western industrialised nations have to improve first of all in the ecological dimension. Therefore, instruments that will promote environmentally responsible action in western societies are indispensable for sustainable development and will automatically contribute to it. But in the long range, sustainable development will only be possible if socio-cultural and economic conditions are also taken into consideration. For example, employment and social security issues must not be neglected as we strive for ecological progress. Environmental policy must become embedded into an integrated policy for sustainable development. One step on the way to this goal is a careful analysis and evaluation of environmental policy instruments. This book contributes to this by looking at "classic" and "new" instruments from an actor-oriented perspective.

1.2. Classic and new environmental policy instruments

Until recently, environmental policy in most countries has based heavily on traditional "command and control" (or "regulatory") instruments and – to a lesser extent – on the more modern "economic" instruments. Both instrument types are usually treated as "classic" policy instruments. They are based on the assumption that the state can promote or force environmentally responsible behaviour in various target groups (individual citizens and organisations) by modifying some aspects of their economic and legal or administrative frameworks of action.

The rationale, the advantages and disadvantages, and the limitations of these "classic" instruments have been discussed widely, mainly in the political and economic sciences literature. When discussing disadvantages and limitations, the following three points are often mentioned:

1. The globalisation of both the environmental impacts of human activity and of the economic activities themselves limit the influence and effectiveness of traditional national policies.
2. A great number of actors other than public authorities (private sector, non-governmental organisations – in fact, the entire civil society) are under pressure and often also motivated to contrib-

ute to sustainable development. This raises the question of what specific instruments might be available to them and how their activities interact with public policy.

3. Classic instruments of public policy often encounter considerable problems of implementation and enforcement, and thus lack efficiency and efficacy.

For these reasons, an intensive search and discussion of new instruments is underway, both in practice and in theory (see, for example, Dente, 1995), and several attempts to evaluate environmental policy have been made (e.g., Jänicke and Weidner, 1995; Weidner, 1996; Jänicke and Weidner, 1997). There seems to be a need for complementing and sometimes even replacing traditional policy instruments with a set of new instruments that are available not only to public authorities, but to other actors as well. These rely mainly on voluntary and co-operative action, thus activating the potential of self-regulation of social systems towards sustainability. In fact, various kinds of "soft" policy instruments are being discussed in the literature, such as "information" (Bemelmans-Videc et al., 1998), "mediation and information" or "instruments of dialogue and co-operation" (Jänicke and Weidner, 1995), "persuasive instruments" (Dahme and Grunow, 1983; Dente, 1995), "supporting measures" (Bartel and Hackl, 1994), or "voluntary agreements" (Albrecht, 1995; Kirchgässner, 1997). The instruments subsumed under these and similar terms are typically available not only to public authorities, but also to a variety of actors, such as environmental and consumer organisations, production industries, media, distributors, and others. However, this is a rather undifferentiated and heterogeneous class of instruments, and their rationale has often not been clearly elaborated. Dente states:

> "What we do point to is the need for a careful, wide and interdisciplinary research program on the promise of self-regulation, especially aimed at investigating the circumstances under which, and the processes through which, it can emerge, and where and when it works best in securing the preservation of the natural resources"(Dente, 1995: 16).

2. A transdisciplinary research programme on environmental policy instruments

The research on which this book is based may be regarded as a step in the direction suggested by (Dente, 1995) in the statement cited above. Within the Swiss Priority Programme "Environment" (SPPE[1]) of the Swiss National Science Foundation, an integrated transdisciplinary project was carried out from 1997 through 2000.[2] Its research objects were strategies, instruments, and measures available to various societal actors to promote sustainable development. Its aim was to provide

[1] For information on the programme, see <http://www.snf.ch/SPP_Umwelt/overview.htm>.

[2] Integrated Project "Strategies and instruments for sustainable development: Bases and evaluation of applications, with special regard to the municipality level" (Nr. 5001-48826).

knowledge about characteristics and effects of different instruments, especially to actors at local levels (such as individuals, local authorities, enterprises, citizens' organisations, or schools), which would enable municipalities to adopt the measures that would most probably help them to reach their goals. Thus, the **objectives** of the project were

- to conceive, analyse, evaluate, and optimise, in a transdisciplinary process, the interlinkages of strategies, measures, and instruments aimed at mobilising resources or at overcoming obstacles to sustainable development on the different levels of human action;
- to estimate their ecological effectiveness in selected fields;
- to improve the scientific knowledge base required for the conception of particular instruments;
- to test empirically the application of interventions and instruments and their effects in selected Swiss municipalities with their private and public actors.

In accordance with the objectives of the project the following questions were asked:

- What individual, socio-cultural, economic, political, legal, and administrative obstacles and opportunities exist for private and public actors of Swiss municipalities to contribute to sustainable development?
- How can the obstacles that are identified be overcome?
- How appropriate and how effective are particular instruments and measures?
- How do different measures and instruments interact, and how can this linkage be optimised?

Nine sub-projects uniting researchers from different disciplines (mainly psychology and social psychology, economics, political science, jurisprudence, philosophy, geography, and architecture) participated in the integrated project.[3] Each of these disciplines has its own tradition of looking at human action and societal development. Working together to synthesise their respective disciplinary theories and empirical findings in order to find integrated answers for the above questions required in part that the researchers free themselves of the traditional perspectives of their disciplines. They adopted – as a conceptual framework for their integrative work – the model of human action that is outlined in the following section.

2.1. *A model of human action*

The flow of human action is regarded as a result of the constant interplay between the internal structure of the actor and the actor's environment, or external structure (see Figure 1). If the actor is an individual person, his or her **internal structure** consists of the ensemble of psychophysiological factors and processes; if the actor is an organisation (e.g., a company), its internal structure consists of factors such as the organisational structure and goals, communication processes, and

[3] See appendix for complete titles of the sub-projects and names and affiliations of the participating researchers.

general culture within the organisation. However, there is no such thing as an isolated actor. Each and every actor is embedded in a network of social relations, and personal identity is in part defined by the quality of these relations and the way the actor is integrated in the social body.

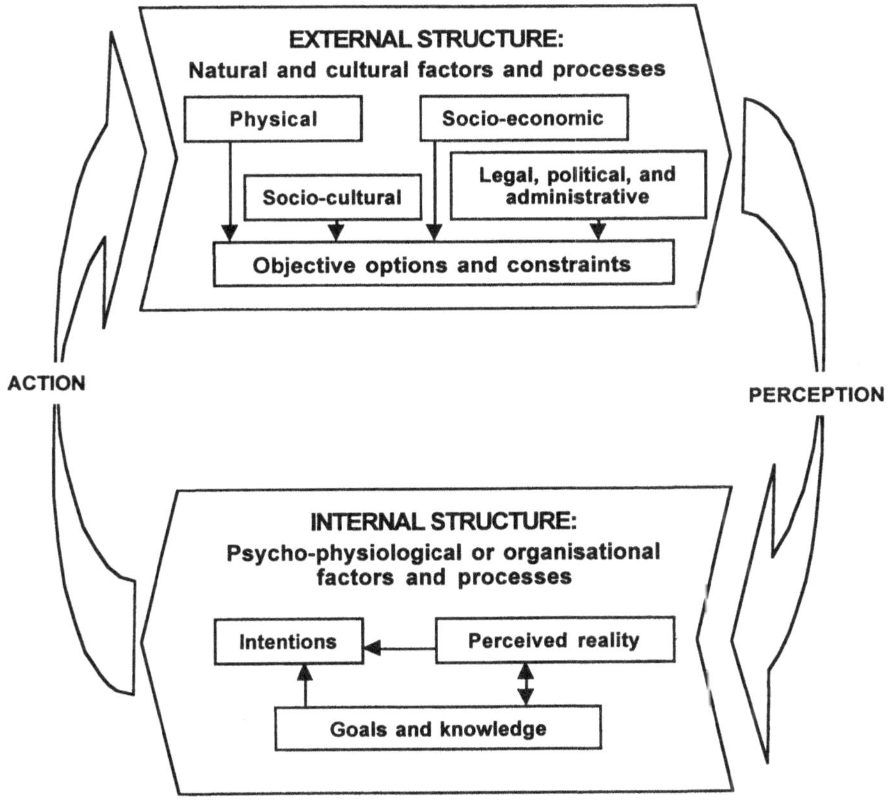

Figure 1: A simplified model of human action.

In the **external structure**, or the actor's environment, four aspects may be distinguished: (1) the physical environment, which consists of the natural environment and man-made objects and infrastructures; (2) the socio-cultural environment, which provides the culturally embodied world-views, value systems, power relations, modes of knowledge production and reproduction, and the actor's social relations; (3) the socio-economic environment, which defines the frame for economic activities and comprises factors such as welfare, economic growth, international competitiveness, employment, and so on; (4) the legal, political, and administrative environment, which is the ensemble of legal norms and the political-administrative system and institutions. Of course, these are mere analytical distinctions, a heuristic clustering of the multitude of factors that interact with the internal structures of the actors in determining the concrete actions they perform. The external structure can also be regarded from an institutional perspective. Institutions are systems of rules that give rise to social practices and guide interactions between humans and their physical environment (for example, as environmental or resource regimes). Thus, institutions and the respective organisations that administer the institutional arrangements also belong to the external structure.

Both the internal and the external structures are viewed as evolving dynamically as a combined result of inherent processes of self-organisation and of interactions between the actor and the actor's environment. According to this view, the intention to perform a specific action emerges within the actor as a result of individual goals and knowledge and the perception, interpretation, and appraisal of the actual situation, that is, the reality as perceived by the actor. When the actor actually carries out the intended **action**, its exact course and its outcome – its success – depend not only on the actor's skill, but also on the objective situation. Actors constantly monitor the situation and tune their behaviour accordingly. The outcome or result of actions is on the one hand **perceived** by the actors and appraised with respect to their goals, and it thus affects their internal structure; it has on the other hand an effect on the objective situation, it affects the external structure, the actors' (and other people's) surroundings.[4]

Today, many characteristics of the actors' internal structure as well as of the external structure constrain behaviour in non-sustainable directions (Gessner and Kaufmann-Hayoz, 1995; Gessner, 1996; Gessner and Bruppacher, 1999; Tanner, 1999). This is true for individual actors as well as for organisations (such as industrial companies or public authorities). All actors have only limited possibilities to alter their own frameworks of action, because they are determined by other actors' decisions. However, collective action or social practices stabilise and reproduce the mutual framework conditions, or, alternatively, they contribute to their change. Over time, there is a "co-evolution" of individual and collective patterns of behaviour and its framework.

In order to understand and stimulate transformation processes towards sustainability, we need to analyse the complex web of agents in a society and the mutual influences they exert on each other. No single-factor or one-dimensional intervention can be successful in stimulating changes in individual behaviour or corporate action. Examples from history show that major changes occur when many factors together create a constellation strong enough to transgress the inertia of the existing systems (Loessøe, 1996). Such constellations can neither be planned entirely nor predicted

[4] The view that is outlined here is consistent with the proposition of semiotic ecology as worked out by Lang (1992) and similar approaches found in cultural psychology (Valsiner, 1996; Valsiner and Lawrence, 1996; Boesch, 1980). It also bears similarities to the theory of structuration as proposed by Giddens (1984). Stern (2000) has recently published a conceptual framework for advancing psychological theories of environmentally significant individual behaviour, which elaborates further on what we call here the "internal structure" of individual actors.

exactly, because the social reality is too complex and is influenced simultaneously by a multitude of self-reflecting, creative agents. However, a better understanding of the dynamics of such constellations and of the instruments available to different actors to stimulate change in the desired direction is crucial for adequate policymaking for sustainable development.

References

Albrecht, C. (1995) Vorschriften und Anreize in der schweizerischen Umweltpolitik. Buwal, Bern.

Bartel, R. and Hackl, F. (1994) Einführung in die Umweltpolitik. Franz Vahlen, München.

Bemelmans-Videc, M.-L., Rist, R.C. and Vedung E. (eds.) (1998) Carrots, sticks and sermons. Policy instruments and their evaluation. Transaction Publishers, New Brunswick, New Jersey.

Boesch E.E. (1980) Kultur und Handlung. Einführung in die Kulturpsychologie. Hans Huber, Bern, Stuttgart, Wien.

Dahme, J. and Grunow, D. (1983) Implementation persuasiver Programme. *In*: Implementation politischer Programme II, Mayntz, R. (ed.), Westdeutscher Verlag GmbH, Opladen.

Dente, B. (ed.) (1995) Environmental policy in search of new instruments. Kluwer Academic Publishers, Dordrecht, Boston, London.

Gessner, W. (1996) Der lange Arm des Fortschritts. Versuch über die umweltrelevanten Strukturen der Lebenswelt. *In*: Umweltproblem Mensch. Humanwissenschaftliche Zugänge zu umweltverantwortlichem Handeln, Kaufmann-Hayoz, R. and Di Giulio, A. (eds.), Haupt, Bern.

Gessner, W. and Bruppacher, S. (1999) Restriktionen individuellen umweltverantwortlichen Handelns. *In*: Umweltgerechtes handeln. Barrieren und Brücken, Linneweber, V. and Kals, E. (eds.), Springer, Berlin.

Gessner, W. and Kaufmann-Hayoz, R. (1995) Die Kluft zwischen Wollen und Können. *In*: Ökologisches Handeln als sozialer Prozess, Fuhrer, U. (ed.), Birkhäuser, Basel.

Giddens, A. (1984) The constitution of society. Outline of the theory of structuration. Polity Press, Cambridge.

Jänicke, M. and Weidner, H. (eds.) (1995) Successful environmental policy. Rainer Bohn Verlag, Berlin.

Jänicke, M. and Weidner, H. (eds.) (1997) National environmental policies. Springer, Berlin.

Kirchgässner, G. (1997) Environmental policy in Switzerland: Methods, results, problems and challenges. *In*: Economic policy in Switzerland, Baccheta, P. and Wasserfallen, W. (eds.), Macmillan Press, Houndmills.

Lang, A. (1992) Kultur als "externe Seele". Eine semiotisch-ökologische Perspektive. *In*: Psychologische Aspekte des kulturellen Wandels, Allesch, Ch., Billmann-Mahecha, E. and Lang, A. (eds.), Asanger, Heidelberg.

Loessøe, J. (1996) Consumption, lifestyles and constellatoric strategies. Nytenkning omkring effektiv energibruk og bërekraftig forbrul i husholdinger, Norges forskingsgråd.

Stern, P.C. (2000) Toward a coherent theory of environmentally significant behavior. *Journal of Social Issues, 56/3*: 407-424.

Tanner, C. (1999) Constraints on environmental behavior. *Journal of Environmental Psychology, 19*: 145-157.

Valsiner, J. (1996) Cultural organization of cognitive functions. *In*: Cognitive ecology. The evolutionary ecology of information processing and decision making, Dukas, R. (ed.), Academic Press.

Valsiner, J. and Lawrence, J.A. (1996) Human development in culture across the life span. *In*: Handbook of cross-cultural psychology, Berry, J.W., Dasen, P.R. and Saraswathi, T.S. (eds.), Allyn and Bacon, Boston.

WCED/World Commission on Environment and Development (1987) Our common future. Oxford University Press, Oxford.

Weidner, H. (1996) Basiselemente einer erfolgreichen Umweltpolitik. Rainer Bohn Verlag, Berlin.

Changing Things – Moving People.
Strategies for Promoting Sustainable
Development at the Local Level
R. Kaufmann-Hayoz/H. Gutscher (eds.)

Switzerland's Environmental and Sustainability Policies

Ruth Kaufmann-Hayoz and Ursula Mauch

The studies presented in this book relate to the situation in Switzerland. For readers unfamiliar with the Swiss political system and Swiss environmental policy, we give a brief outline in order to facilitate comprehension of the studies. A more in-depth evaluation of Switzerland's environmental policy is found in Knoepfel (1997) or Kirchgässner (1997). Extensive information is also available in the latest Swiss environmental report (SFSO and SAEFL, 1997) and online in the WWW at <http://www.admin.ch>.

1. The political structure of Switzerland

The Swiss Confederation consists of 26 (semi-autonomous) member states, the cantons, with roughly 3000 municipalities (45% of them counting less than 500 inhabitants). The correct application of national environmental legislation is to a very large extent the responsibility of the cantons and municipalities. They may pass their own environmental laws and regulations, as long as these do not contradict federal law. For instance, although waste disposal is regulated in principle in the federal Environmental Protection Law (EPL), the individual cantons have widely differing interpretations and forms of implementation, which take local conditions into consideration. The actual collection of municipal refuse, the running of recycling stations, and so on is mainly the responsibility of the municipalities. One of the problems for local authorities is that they perceive very little room for manoeuvre, because they are busy implementing and enforcing the decrees from the upper political levels. Moreover, enforcement can only be successful if the public authorities co-operate with various private actors. This situation clearly calls for instruments that encourage voluntary action of private actors and enhance co-operation between private actors and the authorities.

2. Environmental policy

Switzerland has a federal Environmental Protection Law (EPL), which was introduced in 1983 (latest amendment in 1995), and a number of specific laws pertaining to the environment. Some of them were introduced several decades earlier than the comprehensive EPL (such as the Law on the Preservation of Nature and the Landscape, Law on Water Pollution Control, Law on Forests, Law on Spatial Planning, and others).

The legislative and monitoring system has worked quite well in past decades. Cases of severe local pollution of water, air, or soil are rare; close to 100 % of households and industries are connected to waste water treatment plants; waste is collected and disposed of in a controlled way. The majority of these achievements has been made possible by the application of "end-of-pipe" technical improvements, such as wastewater treatment, dust filters and the cleaning of waste gases in waste incinerators. However, there are three types of problems that can be identified today:

1. The outcome of environmental policy is not always as it should be. In the case of air pollution, for example, the impact levels of nitrogen dioxide and ozone are too high, and the impact thresholds set by law are frequently exceeded. Motor traffic, which is a major source of these pollutants, appears to be especially difficult to control.
2. There is a growing insight that "end-of-pipe" solutions alone do not lead to long-term solutions. The benefits that they bring are often overridden by the sheer quantitative increase of resource consumption and emissions. Also, this type of solution is becoming increasingly difficult and expensive, and it seems desirable to preserve natural ecosystems and values rather than to repair damages and to restore losses. The principle of action should be to avoid emissions and prevent damages in the first place. But this principle is not yet well realised.
3. Switzerland's overall consumption of environmental resources is – despite the local success of its environmental policy during the past decades – by no means sustainable. For example, Switzerland – as all industrialised countries of the northern hemisphere – ought to cut down its output of carbon dioxide by far more than half to reach a sustainable basis (Infras, 1995).

Therefore, the integration of sustainability goals into all policy areas as opposed to environmental policy as a separate policy sector has become the primary objective of a modern concept of environmental policy. In 1997, the Federal Council issued a Strategy and an Action Plan for Sustainable Development (SAEFL, 1997) that defined the following eight areas as priority fields for action: international commitment, energy, trade and industry, consumer behaviour, security policy, ecological tax reform, federal expenditure, and implementation and monitoring. According to this programme, more importance than hitherto is given to economic instruments and voluntary steps.

3. The implementation of sustainable development

3.1. Follow up of Rio

In March 1993, the Swiss government established the Interdepartmental Committee on Rio (IDARio). Meanwhile, almost 30 agencies of all ministries on the federal level participate in this committee. Its responsibility is to implement Agenda 21 as well as the other Rio agreements of the Earth Summit of 1992, which were signed by the Swiss government. The "Strategy for Sustainable Development" (SAEFL, 1997) has so far been the basic document for the work within IDARio.

As suggested by the international community, an independent "Council on Sustainable Development" was also established by the Swiss government. Its main task was to be the counselling of the government on matters of sustainable development, especially the co-ordination of major political decisions with implications for sustainable development issues. For various reasons, the Council did never really get "off the ground" and was suspended by the government by the end of the year 2000. Its tasks will be delegated to the "Council of Spatial Planning".

On 18 April 1999, the Swiss people approved the new Federal Constitution, which explicitly declares its support for sustainable development. No special "law on sustainable development" has been planned, but rather, issues concerning sustainable development should be included in all relevant laws.

On the international level, Switzerland plays an active role in the UN Commission on Sustainable Development (CSD), the United Nations Environment Programme (UNEP), and the Organisation for Economic Co-operation and Development (OECD).

3.2. Looking back and ahead

With a view to the year 2002, when the world will look back over the *10 years since Rio*, the IDARio has started a project that will focus on two aspects:

1. What has been achieved with regard to implementation of sustainable development in Switzerland so far?
2. What are the prospects of future development and what further actions towards sustainable development are needed?

The main result of the project will be an agenda of priority fields of action for federal policies as well as suggestions for further activities for the attainment of a sustainable society. The report will present a policy-related reflection on past and future ways of achieving sustainable development in Switzerland. It bases on a normative framework that offers guidance on an ethically conscious policy for sustainability. The political-normative framework is set by the new federal constitution.

The characteristics and standards for sustainable development are based on the three-dimensional model according to Agenda 21: Environment, Economy, and Society. It also takes into account the needs of present and future generations.

Within the past years since Rio, much knowledge has been acquired, and many strategies have dealt with the dimension of the environment. Yet, according to Rio declarations, all three dimensions of sustainable development should have equal weight within governmental policies. In order to establish "more equality among the three sustainability dimensions", the report will focus particularly on the social and economic dimensions. Moreover, it will emphasise clearly that the three-dimensional model of sustainability should not be understood as a harmonious whole. There are numerous conflicts among the dimensions for which political solutions must be found. What is required is a coherent policy for sustainable development on the governmental level. The Swiss government is aware of the fact that this ambitious goal has not yet been reached.

References

INFRAS (1995) Quantitative Aspekte einer zukunftsfähigen Schweiz. Infras, Zürich.

Kirchgässner, G. (1997) Environmental policy in Switzerland: Methods, results, problems and challenges. *In*: Economic policy in Switzerland, Baccheta, P. and Wasserfallen, W. (eds.), Macmillan Press, Houndmills.

Knoepfel, P. (1997) Switzerland. *In*: Jänicke, M. and Weidner, H. (eds.), National environmental policies. Springer, Berlin u.a.

SAEFL/Swiss Agency for the Environment, Forests and Landscape (1997) Sustainable development in Switzerland: Strategy of the Federal Council. SAEFL, Bern.

SFSO/Swiss Federal Statistical Office and SAEFL/Swiss Agency for the Environment, Forests and Landscape (1997) The environment in Switzerland 1997 – Facts, figures, perspectives. SFSO and SAEFL, Bern.

Theoretical Framework

Changing Things – Moving People.
Strategies for Promoting Sustainable
Development at the Local Level
R. Kaufmann-Hayoz/H. Gutscher (eds.)

A Typology of Tools for Building Sustainability Strategies

Ruth Kaufmann-Hayoz, Christoph Bättig, Susanne Bruppacher, Rico Defila, Antonietta Di Giulio, Peter Flury-Kleubler, Ueli Friederich, Myriam Garbely, Heinz Gutscher, Christian Jäggi, Maya Jegen, Hans-Joachim Mosler, André Müller, Nicole North, Silvia Ulli-Beer, and Jürg Wichtermann[1]

This chapter introduces and describes a typology of policy instruments that has a dual purpose. It serves (1) as a conceptual tool for integrating the findings of the different studies that were part of our inter- and transdisciplinary research and (2) as a useful frame of reference for political actors when choosing appropriate sets of instruments for policy strategies. The instruments we have included in the typology focus on the ecological dimension of sustainability; they are applied primarily to promote environmentally responsible action. However, we believe that the five basic types of instruments – command and control instruments, economic instruments, service and infrastructure instruments, collaborative agreements, and communication and diffusion instruments – are of a general character that would allow for the inclusion of specific groups of instruments for promoting economic and social sustainability. Future work will tackle the issue of complementing the typology so that it covers the full range of instruments available for promoting all three dimensions of sustainable development. The chapter consists of five sections.

After a brief introduction, we present an overview of the typology by characterising the five types of instruments and relating them to each other (section 2, including the trilingual survey Tables 1A-C on pages 40-45). In section 3, we describe the specific instruments of each type in detail and give illustrative examples. A preliminary comparison and evaluation of the different instrument types is given in section 4, while section 5 relates the typology to other classifications and typologies developed in several disciplines.

---

1 The responsibility for this chapter lies with the Synthesis Group (compare the chapter by DEFILA AND DI GIULIO in the Conclusions section). The Synthesis Group is grateful to Peter Flury-Kleubler, Hans-Joachim Mosler, Silvia Ulli-Beer, and Jürg Wichtermann for joining it at certain times, for offering critical comments, and for writing parts of the chapter.

1. Introduction

1.1. Aims of the typology

The typology of instruments that we present should help us to look at classic policy instruments as well as new instruments from an *actor-oriented perspective,* and it should allow instruments to be evaluated with respect to issues of *practical interest* (effectiveness, costs, and acceptance).

Classification has always been an important methodological element in the development of the sciences. The classical procedure uses the criteria of *genus proximum* and *differentia specifica* to produce a logical system of distinctive categories. Within the social sciences, the notion of "type" has played an important role in methodological discussions, certainly since Max Weber's (1904) publications. Many of the typologies developed in the social sciences do not have the strict logic of a system of mutually exclusive classes. They are more similar to the spontaneous and fuzzy everyday classifications people use in mastering their lives. "Types" in the social sciences are usually described multidimensionally, and they serve as complex perspectives that order the social reality. Because they pertain to "natural" events (rather than to artificially made, well defined objects), there are always cases that fall between classes or that could justifiably be classified as belonging to more than one class or type. By applying the logical procedure described by Max Weber as analytical accentuation of elements of reality *[pointierend hervorhebende Abstraktion]*, types can be constructed as "ideal types" or as "real" or "average types". Alternatively, statistical procedures, such as the discriminant function or cluster analysis can be applied to quantitative data (for a more extensive discussion of typologies of policy instruments see, for example, Kaufmann and Rosewitz, 1983, or Bemelmans-Videc et al., 1998).

Typologies or classifications differ from each other, even if they pertain to the same objects. They depend on the scientific purpose that they serve, on the goals and questions of the researcher. Therefore, no typology or classification is any more "true" than another is, but one of them might be more useful for a particular purpose. This becomes quite obvious, if we look at existing typologies of environmental policy instruments. Many were developed for the purpose of integrating a number of heterogeneous case studies in order to work out general conclusions. Others are based on specific disciplinary theories and serve the purpose of generating and testing scientific hypotheses. Still others were developed for the practical purpose of facilitating communication among policy makers. When researchers come up with a "new" typology, this does not usually mean that they consider existing ones to be wrong, but rather that they do not serve their exact purposes of study. "New" typologies often have some similarities with "old" ones, but they may make distinctions where others group things together and vice versa, or they may point out specific properties that were unimportant for earlier study purposes.[2]

---

[2] In section 5 of this chapter we review some typologies and classifications that have been developed in different disciplines, and we discuss their relation to the typology that we propose.

The dual purpose of our interdisciplinary resarch has been to study a number of different instruments available to different actors (mainly at the local level) to stimulate changes in one or more of the various factors that influence environmentally responsible behaviour and to communicate our results to those actors. We therefore needed a clearly arranged and comprehensive typology of instruments that would include all principal and politically acceptable possibilities known to influence human environmentally relevant behaviour and that would be...

- ...actor-oriented: It should reveal what instruments are available to what actors and show by what "mechanisms" the various instruments are supposed to affect the behaviour of the target groups,
- ...not restricted to specific disciplinary theories and approaches,
- ...practice-oriented: It should allow us to evaluate the instruments with respect to all practitioners' main questions: How effective are they? How much do they cost? How accepted are they?

1.2. A word on terminology

The notions of "strategy", "instrument", "measure", and "campaign" are not used coherently in the literature. In this volume we use them as follows:

- By "**instrument**" we mean the basic way in which an actor chooses to influence the behaviour of another actor (such as implementing an energy tax or presenting facts and options). We should more precisely speak of "intervention instruments", because they specify the type of intervention – the goal-directed influence of an actor upon the conditions that determine a target group's action. As the aim is not to produce effects in single individuals or organisations, but rather to affect entire populations or very large target groups, the mechanisms of the intervention instruments have to be "transported" to as many members of the target group as possible. This means that for every intervention instrument there is a specific diffusion problem that has to be resolved through the aid of special **diffusion instruments**. The diffusion instrument "transports", or carries, the intervention to the target group. For an intervention that bases on communication instruments in particular, the issue of how a target group can be reached effectively, quickly, and surely is central. For this reason, the diffusion problem in connection with that type of instrument will be treated in detail (see sections 2.5, 3.5, and the case study by GUTSCHER ET AL.).
- The term "**measure**" will be used to indicate the concrete realisation of an instrument (for example, announcing a tax of 0.3 cents per kWh non-renewable energy for a maximum duration of 15 years, or providing an information brochure on municipal waste collection to every household). Here again, intervention measures have to be complemented by diffusion measures.
- By "**strategy**" we mean the purposeful application of a set of instruments and measures over time in a given political-administrative context and embedded in an appropriate policy style, in order to reach a defined policy goal (such as the Strategy for Sustainable Development of the

Swiss Federal Council (SAEFL, 1997); for an overview of other national action plans see Jänicke et al., 1999: 114).

- "**Campaign**" also refers to the intentional application of a whole set of instruments and measures, but the goals tend to be more specific and narrower in focus than the goals of strategies. Campaigns are usually implemented in order to achieve specific, concrete changes in behaviour, or aggregated results of behaviour changes. Campaigns that are even more specifically defined in terms of content and the temporal framework we have called "**action campaigns**".

2. Overview of the typology

We propose to distinguish five types of instruments. They are understood as "types" in the sense of complex ordering perspectives as described above, and they should not be regarded as mutually completely exclusive classes. The main dimension used to define the types is their *rationale* – that is, the basic psychological mechanism by which the instruments are supposed to affect the target group's behaviour.

In this section, each type is briefly described with respect to its *rationale* (the "mechanisms" according to which the instruments are supposed to work), the main *actors* (to whom the instruments are available), and the *target groups* (i.e., whose behaviour is intended to be influenced by the application of the instruments). In addition, some remarks are made on specific aspects of *implementation and enforcement* of the instruments. The entire typology is shown in the trilingual Tables 1A-C on pages 40-45.

2.1. Command and control instruments

2.1.1. General description and rationale

We use the term command and control instruments for regulatory instruments that are direct and mandatory (IEA, 1989). Command and control instruments in this sense are legal prescriptions that have a direct impact on the range of options open to specified social actors, because they constrain certain ways of acting or exclude some forms of conduct. Specific actions and/or specific outcomes of actions are prescribed or prohibited by way of mandatory orders. Thus, the rationale of this instrument type is based exclusively on command, control, and sanction. It is assumed that actors behave according to the prescription or norm in order to avoid penalties that take effect if they violate the norm. Therefore, such enforcement mechanisms are fundamental to the viability and effectiveness of command and control instruments. Enforcement policies rely on a variety of (conditional or secondary) instruments, ranging from license withdrawal to criminal prosecution. Enforcement measures often involve an economic component in addition to the pure regulatory aspect (e.g., fines to be paid when admissible levels of emission are exceeded; see also IEA, 1989).

2.1.2. *Actors*

In western democracies, only the legislative authorities are legitimised to use command and control instruments. In Switzerland, federal, cantonal, or municipal authorities share this authority, the division of labour being organised according to the policy domain at issue. Although the implementation of command and control instruments may be (partly) delegated to private bodies, the final decision on their application (and eventual enforcement measures) remains with the public authorities.

2.1.3. *Target groups*

Command and control instruments can be used to influence the behaviour of any target group – individuals as well as corporate actors or private companies. Basically, command and control instruments apply in the same way to every actor or group of actors specified in the legal norm.

2.1.4. *Implementation and enforcement*

The implementation and enforcement of command and control instruments can be troublesome in many respects. Often, implementation places heavy demands on technical competence as well as on the amount of available human and financial resources (Mayntz, 1980; 1983). In practice, the granting of exceptions recurrently compromises the implementation of efficient command and control instruments. In Switzerland, appealing against such granting of exceptions by the executive authorities is an uncommon practice, and the number of admitted appellants is rather limited. In some cases, moreover, legal prescriptions directed toward barely accessible target groups (such as "consumers") are very difficult to enforce, because control is almost impossible. Therefore the motive to avoid sanctions disappears (such as for illegal waste disposal by burning or dumping, or spreading manure on frozen soil).

2.2. *Economic instruments*

2.2.1. *General description and rationale*

Economic instruments are based on the assumption that environmental degradation and resource depletion occur because a substantial part of the costs of economic activities is not being paid by the actors responsible but by the general public (in the form of environmental damage, security and health risks, or long-term climatic risks). In order to correct this situation, three principal forms of intervention are conceivable:

(1) *Raising the costs of polluting behaviour:* This approach is based on the principle of true costs. The costs of environmentally harmful behaviour, that is, of the respective production, purchase, and investment decisions, are raised by means of incentive taxes and charges. The actors' range of options is not influenced directly; all options (production and consumption patterns, choice of technologies) remain open insofar as they are not constrained by regulations. The actors' behaviour is supposed to be steered toward environmentally sound actions, because these are more

cost efficient. Moreover, the market mechanisms guarantee that, in order to achieve environmental goals, the least expensive measures are realised first.

(2) *Reducing the costs of environmentally sound behaviour*: Instead of raising the costs of environmentally undesirable behaviour, environmentally beneficial behaviour can be rewarded. By means of subsidies, prices for environmentally responsible behaviour are lowered, to the effect that such behaviour becomes more likely.

(3) *Establishing markets for polluting rights:* A market for the trading of resource use permits (such as the permit to pollute the air by nitrogen oxides) is created. Whoever wants to use environmental resources has to buy the respective permit on these markets.

2.2.2. *Actors*

Economic instruments are usually implemented by public authorities. They require a legal basis and can be implemented at all political levels (federal, cantonal, municipal), depending on the domain and respective authority.

2.2.3. *Target groups*

Direct target groups are firms and investors as well as private persons and households. In households, economic instruments target consumption behaviour. For companies, economic instruments make environmentally sound production more competitive. The price relations between different options (higher or lower emission levels, more or less energy intensity, more or less transport) are modified in favour of the environment. *Indirectly*, economic instruments usually affect all actors. For example, a tax on certain emissions will alter the price of products in the affected industrial sectors as well as their demand for other production factors. This can have effects on all other industrial sectors, on the prices of consumer and investment goods, and, of course, on foreign trade. Besides these price effects, the effects on the social distribution of wealth have to be taken into consideration, especially from a political point of view. Economic instruments affect the incomes of different actors not only through the modification of prices, but also through the way in which the government uses the taxes or finances the subsidies.

2.2.4. *Implementation and enforcement*

The implementation of economic instruments first and foremost raises the issue of their legal basis. If they are supported by existing constitutional bases, the questions of who is to be taxed, what is to be taxed, how the tax is to be calculated, and how tax revenue will be used will be defined at the level of the law. For successful implementation, it is particularly important that regulations be simple and flexible. Experience in many countries has shown that simple and uniform calculation bases that allow for little administrative interpretation are less vulnerable to manipulation and dispute (De Clercq, 1996).

Monitoring and enforcement are crucial to successful implementation of the instruments. The effort involved can vary widely. For example, a country like Switzerland, which imports all of its fossil fuels, can add a carbon dioxide tax to existing taxes and collect it through the customs authorities. For other harmful emissions, individual instruments for measurement and control have to be developed, which can lower public acceptance of the instruments.

For political acceptance of instruments that produce revenue, the way the revenue is used will be definitive. As a rule, redistributing the revenue to the public or using it in a targeted fashion to support environmental goals will increase acceptance (OECD, 1997a).

2.3. *Service and infrastructure instruments*

2.3.1. *General description and rationale*

Service and infrastructure instruments are goal-directed transformations of services or infrastructure for the promotion of certain behaviours. The term services refers to actions of individuals or organisations that enable or support other actors to achieve their goals. For example, a man who cares for an ailing father-in-law enables him to have his basic needs fulfilled and to stay alive, a hairdresser enables her clients to look nice, a fair-trade organisation enables consumers to support local farmers in developing countries: they all provide services. The term infrastructure denotes man-made, mobile or immobile physical objects that shape the actor's scope of opportunities for action. The bicycle that parents give to their daughter for her trip to school, a railway line, or a parking space are examples of infrastructure.

The existence of services and infrastructure determines to a large extent what actions are objectively possible or not for certain actors. If there is no bridge across a wide river, most people cannot cross it; if no travel guides offered their services, only very few Europeans would consider travelling to Africa or Tibet; if there was no health care system, people could not have their blood pressure checked regularly. In addition, the design and quality of services and infrastructures influence the salience of action possibilities and their potential to provide experiences of pleasure, excitement, healthiness, or security. For example, taking local trains late at night, where there are no train attendants apart from the driver in his cab and only a few passengers to inspire mutual confidence, can be unpleasant, frightening, and even dangerous.

Therefore, the rationale of this type of instrument is that by modifying services and infrastructure, desired actions can be made possible and undesired ones can be excluded, and that desired action possibilities can be made more salient and pleasant and undesired ones less salient and less pleasant. Environmentally relevant modifications of services or infrastructure can be *attraction based* (offering or improving environmentally sound products, services, and infrastructure that allow or facilitate environmentally sound action), or *repulsion based* (withdrawing products, reducing services, or degrading infrastructure that allow or facilitate ecologically desirable action).

Table 1A: A typology of policy instruments for sustainable development.

Command and Control Instruments	
Environmental quality standards (impact thresholds and standards) **Emission standards** • best available technology • prescriptive technology standard **Product standards and regulations for the use of pollutant substances** • restriction, rationing, or prohibition • product standards	**Licensing** • licence to construct • licence to operate • licence to sell **Liability regulations**[1] • strict liability • reversal of the burden of proof • compulsory third party liability insurance **Zoning** • land use regulations • water protection areas • nature conservation zones
Economic Instruments	
Subsidies • grants • tax allowances • soft loans • guarantees • compensation for foregoing use of the resource **Incentive taxes** • taxes on energy/resources • taxes on emissions • taxes on products/processes **Charges** • one-time charge for connection to services • recurrent charges for use • charges on advantages (value-added contribution) • prepaid disposal fees	**Deposit-refund systems** **Market creation** • tradable allowances or permits • joint implementation **Incentives as parts of action campaigns**[2] • rewards • lotteries • contests/benchmarking • discounts

1 Liability regulations are often classified as economic instruments.

2 These instruments – although not usually described as economic instruments – are placed here, because from the target group's perspective their rationale is the same as in the other economic instruments (see text for further explanations).

(Table 1A, cont.)

SERVICE AND INFRASTRUCTURE INSTRUMENTS	
Service instruments • offering or improving ecologically sound products • withdrawing environmentally undesirable products • offering or improving services that allow or facilitate ecologically sound action • reducing services that allow or facilitate environmentally undesirable action	**Infrastructure instruments** • offering or improving infrastructure that allows or facilitates ecologically sound action • dismantling or degrading infrastructure that hinders or inhibits ecologically sound action
COLLABORATIVE AGREEMENTS	
Public-private agreements • agreements on prepaid disposal fees on specific product groups • agreements on consumption goals or standards • formal agreements with individual companies	**Certifications and labels** • with legal compliance • without legal compliance
COMMUNICATION AND DIFFUSION INSTRUMENTS	
Communication instruments without a direct request • presenting facts • presenting options • presenting appraisals, goals, and norms • providing experience of reality • presenting model behaviour • giving feedback and enabling self-feedback **Communication instruments with a direct request** • persuading about facts • persuading about options • persuading about goals, appraisals, and norms • sending appeals • presenting prompts and reminders • stimulating self-commitment	**Diffusion instruments** • establishing direct personal contact • establishing contact via person-to-person media • establishing contact via mass media

Table 1B: Typologie der Instrumente für eine nachhaltige Entwicklung.

Gebote und Verbote	
Umwelt-Qualitäts-Standards (Immissionsbegrenzungen) **Emissionsbegrenzungen** • nach bestem Stand der Technik • gemäss zwingender technischer Vorschrift **Vorschriften für den Umgang mit umweltgefährdenden Stoffen und Produktestandards** • Beschränkungen, Rationierungen oder Verbote • Produktestandards	**Bewilligungspflichten** • Baubewilligung • Betriebsbewilligung • Verkaufsbewilligung **Haftungsrechtliche Vorschriften**[1] • Gefährdungshaftung • Umkehr der Beweislast • Obligatorische Haftpflichtversicherung **Raumwirksame Vorschriften** • Baurechtliche Ordnungen (Nutzungspläne) • Gewässerschutzzonen • Naturschutzzonen
Marktwirtschaftliche Instrumente	
Subventionen • à-fonds-perdu Beiträge • Steuererleichterungen • Darlehen/Zinsvergünstigungen • Bürgschaften • Abgeltung für Nutzungsverzicht **Lenkungsabgaben** • Abgaben auf Energie/Ressourcen • Abgaben auf Emissionen • Abgaben auf Produkte/Prozesse **Gebühren und verwandte Abgaben** • Einmalige Anschlussgebühren • Wiederkehrende Benützungsgebühren • Vorzugslast (Mehrwertbeitrag) • Vorgezogene Entsorgungsgebühren	**Pfandsysteme** **Einrichtung von Märkten** • handelbare Zertifikate/ Lizenzen/Konzessionen • Joint Implementation **Punktuelle Anreize im Rahmen von Aktionen und Kampagnen**[2] • Belohnungen • Lotterien • Wettbewerbe/Benchmarking • Rabatte

1 Haftungsrechtliche Vorschriften werden oft den marktwirtschaftlichen Instrumenten zugeordnet.

2 Wir ordnen diese Instrumente – obwohl nicht marktwirtschaftliche Instrumente im üblichen Sinne – hier zu, weil ihr grundlegender Wirkmechanismus aus der Perpektive der Zielgruppe derselbe ist wie bei den andern ökonomischen Instrumenten.

(**Table 1B,** cont.)

SERVICE- UND INFRASTRUKTURINSTRUMENTE	
Serviceinstrumente • Bereitstellung oder Verbesserung umweltschonender Produkte • Rückzug umweltbelastender Produkte • Bereitstellung oder Verbesserung von Dienstleistungen, die umweltschonendes Handeln ermöglichen oder erleichtern • Abbau von Dienstleistungen, die umweltbelastendes Handeln ermöglichen oder erleichtern	**Infrastrukturinstrumente** • Bereitstellung oder Aufwertung von Infrastrukturen, die umweltschonendes Handeln ermöglichen oder erleichtern • Rückbau oder Abwertung von Infrastrukturen, die umweltschonendes Handeln verhindern oder erschweren
VEREINBARUNGEN	
Vereinbarungen zwischen Staat und Wirtschaft • Vereinbarung über die Belastung bestimmter Produktegruppen mit einer vorgezogenen Entsorgungsgebühr • Vereinbarung von Verbrauchs-Zielwerten oder Normen • Verträge mit einzelnen Unternehmen	**Zertifizierungen und Labels** • mit legaler Übereinstimmung • ohne legale Übereinstimmung
KOMMUNIKATIONS- UND DIFFUSIONSINSTRUMENTE	
Kommunikationsinstrumente ohne direkte Aufforderung • Sachverhalte darstellen • Möglichkeiten darstellen • Bewertungen, Ziele und Normen darstellen • Realität erfahrbar machen • Modellverhalten zeigen • Feedback geben und Selbstfeedback ermöglichen **Kommunikationsinstrumente mit direkter Aufforderung** • von Sachverhalten überzeugen • von Möglichkeiten überzeugen • von Bewertungen, Zielen und Normen überzeugen • Appelle aussenden • Hinweise und Gedächtnishilfen anbieten • Selbstverpflichtung anregen	**Diffusionsinstrumente** • Direkten persönlichen Kontakt herstellen • Kontakt über Person-zu-Person-Medien herstellen • Kontakt über Massenmedien herstellen

Table 1C: Typologie des instruments pour un développement durable.

INSTRUMENTS RÉGULATIFS	
Normes de qualité environnementale (seuils d'impact et standards) **Normes d'émission** • meilleure technologie disponible • réglementation technique obligatoire **Règlements d'utilisation de substances toxiques et normes de produits** • limitations, rationnement ou interdictions • normes de produits	**Autorisations obligatoires** • autorisation de construction • autorisation d'exploitation • autorisation de mise en vente **Réglementation de la responsabilité civile**[1] • responsabilité à raison du risque • renversement de la charge de preuve • assurance responsabilité civile obligatoire **Règlements de planification** • plans d'utilisation et d'affectation des sols • zones de protection des eaux • zones naturelles protégées
INSTRUMENTS ÉCONOMIQUES	
Subventions • contributions à fond perdu • allégements fiscaux • prêts et taux d'intérêts avantageux • cautions • dédommagement pour renoncement d'utilisation de ressources **Taxes incitatives** • taxes sur l'énergie et les ressources • taxes sur les émissions • taxes sur les produits et processus **Redevances** • redevances uniques de raccordement • redevances périodiques d'utilisation • redevances sur des avantages (contribution à la valeur ajoutée) • redevances (avancées) d'élimination	**Système de dépôt** **Création de marchés** • certificats/licences/concessions échangeables • joint implementation **Incitations ponctuelles dans le cadre de campagnes d'action**[$] • récompenses • loteries • concours/benchmarking • rabais

[1] Les réglementations de la responsabilité civile sont souvent classées parmi les instruments économiques.

[2] Ces instruments – bien qu'ils ne soient pas des instruments économiques au sens usuel – sont classés ici parce que, dans la perspective des acteurs visés, leur mécanisme d'action fondamental est le même que celui des autres instruments économiques (cf. explications dans le texte).

(**Table 1C,** cont.)

INSTRUMENTS DE SERVICE ET D'INFRASTRUCTURE	
Instruments de service • offre ou amélioration de produits écologiques • suppression de produits nocifs • offre ou amélioration de services permettant ou facilitant un comportement écologique • réduction de services permettant ou facilitant un comportement néfaste à l'environnement	**Instruments d'infrastructure** • offre ou revalorisation d'infrastructures permettant ou facilitant un comportement écologique • réduction ou dévalorisation d'infrastructures empêchant un comportement écologique
ACCORDS LIBREMENT CONSENTIS	
Accords entre l'État et l'économie • accords sur des redevances d'élimination avancées pour des groupes de produits • accords sur des standards et des valeurs cibles de consommation • accords formels avec certaines entreprises	**Établissement de certificats et de labels** • avec conformité légale • sans conformité légale
INSTRUMENTS DE COMMUNICATION ET DE DIFFUSION	
Instruments de communication sans sollicitation directe • présenter des faits • présenter des options • présenter des évaluations, des buts et des normes • fournir l'expérience de la réalité • présenter des modèles de comportement • donner du feedback et permettre l'auto feedback **Instruments de communication avec sollicitation directe** • persuader sur des faits • persuader sur des options • persuader sur des évaluations, des buts et des normes • envoyer des appels • offrir des indications et des rappels • inciter à l'auto engagement	**Instruments de diffusions** • établir un contact personnel direct • établir un contact au travers d'un média de personne à personne • établir un contact au travers des mass-médias

2.3.2. *Actors*

Service and infrastructure instruments can be applied by private companies (public transportation firms, retail business companies, and so on), public authorities, or non-governmental organisations (NGOs), but also by individuals and informal groups. Public authorities are usually in control of the quantity and quality of most transport infrastructures (roads, railways, airports), while private companies decide what products to offer and how to present and market them. In private life, voluntary services provided to others and decisions on the purchase of household appliances and vehicles can be regarded as service and infrastructure instruments.

2.3.3. *Target groups*

Service and infrastructure instruments are often targeted to the general public. However, they may also aim at specific professional or lifestyle groups, or at groups defined by some common behaviour, e.g., cyclists or wine drinkers.

2.3.4. *Implementation and enforcement*

The setting up and modification of large infrastructures and public services, e.g., in the transportation or energy sector, usually require political decisions and often also corresponding legal bases. Therefore, service and infrastructure decisions may constitute hot political issues. On the other hand, more subtle forms of service and infrastructure instruments may be implemented very easily (the setting up of an adequate infrastructure for waste separation in households and communes, or the offering of home delivery services by a grocery store, for example). The *attraction* based forms of service and infrastructure instruments have the advantage of being easily accepted by the target persons. Travelling by train because it is comfortable and fast is perceived much more positively than travelling by train because there is no alternative. The *repulsion* based forms (such as the reduction of parking space) often meet as much resistance as command and control instruments and taxes or charges, because they are felt to restrict personal freedom and autonomy.

2.4. *Collaborative agreements*

2.4.1. *General description and rationale*

Collaborative agreements are legally binding or non-binding commitments by the private sector or parts of it (industrial sectors, their associations, or individual companies) made towards the government to enhance energy and resource efficiency or greenhouse gas abatement in sectors that are traditionally not regulated by public policy. Frequently, the environmental targets are defined through lengthy negotiations between the private sector and the government. The government may in turn forego the definition of environmental targets through statutes or detailed regulations. Formal, i.e., legally binding, agreements often have a regulatory component, and they are usually situated in a context of national or even international conventions (e.g., the UN Framework Convention on Climate Change, or the EU/UN Air Pollution Convention). However, in the international context no sanctions have been introduced so far. These conventions thus constitute only a moral obligation.

The rationale of legally binding collaborative agreements may come close to that of regulatory instruments, and sometimes the price mechanism is used in a way very similar to that in economic instruments. A clear distinction between the categories is not always possible. For example, agreements that are made under the pressure of subsidiary state interventions and use the price mechanism for promoting the desired behaviour may be regarded as economic instruments. But in comparison with authoritative state interventions, collaborative agreements contain – because of the importance of negotiations in reaching the agreements – an element of self-regulation of the actors involved. According to Dente (1995: 17), it is exactly this potential for self-regulation that needs to be explored further in the search for new environmental policy instruments (see also Glasbergen, 1998, for an in-depth analysis and evaluation of public-private agreements).

Collaborative agreements can also be distinguished from voluntary actions, such as those elicited by the Swiss government's "Energy 2000" Action Programme, which we treat as a set of communication and diffusion instruments.[3] In those cases there are no formal mutual obligations or mandates; they can consist in unilateral declarations about certain goals or programmes and concrete actions to be performed.

2.4.2. *Actors and target groups*

The actors involved are usually the government and parts of the private sector (the "polluters"); the target group itself participates as a partner in developing and implementing the instrument. An agreement can take the form of a mandate to follow certain objectives and programmes or to take certain measures in specific domains. Agreements can also be worked out between private organisations, such as between consumer and producer organisations, or between umbrella organisations and business and trade associations (BFE, 1998a; 1998b). In such cases, one party of the "iron triangle" of environmental policy (government, polluters or target group, and environmental organisations) has been left out. Such "private-private agreements" usually lead to the development of a label or a certification.

2.4.3. *Implementation and enforcement*

Although collaborative agreements have a touch of "voluntarism", the industries taking part in such agreements are generally put under slight pressure to participate. The threat of more restrictive statutes and regulations, if the environmental goal cannot be achieved by collaborative agreements, forces the industries to define the goals together with the government.

In Switzerland, there has been only limited experience with legally binding agreements in the environment sector. There have been several successfully implemented agreements with single in-

[3] For a description of this programme, see the chapter by BÄTTIG AND BALTHASAR.

dustries regarding the introduction of prepaid disposal fees (glass, aluminium cans, refrigerator or electronic scrap recycling). Other agreements, such as negotiations with the printing and graphic arts industry regarding targets for volatile organic compounds (VOC) emissions, have failed. In the energy sector the government was successful in signing an agreement with the electrical appliances industries that set targets for energy efficiency standards. The negotiated targets have also been integrated into the Ordinance on Energy Use. Negotiations with automobile importers on fuel consumption failed. The government thereafter issued regulations on fuel consumption targets (Bättig, 2001).

2.5. *Communication and diffusion instruments*

2.5.1. *General description and rationale*

Communication instruments aim at influencing the actors' internal conditions of action, especially their goals (their "preferences" in terms of economic theory), their knowledge, and their behaviour programmes. By influencing "mental skills" as well as "motor skills" of individuals, they also affect the social relations and communicative processes among different actors. The rationale of these instruments is that behaviour changes require a modification of the motivational, cognitive, and social preconditions of action, and that this can be achieved by stimulating the thinking of individuals and by shaping societal discourses on goals and options for action. [4]

Because the aim is not to affect single individuals, but rather to achieve blanket effects penetrating large groups, the interventions must be "transported" to affect the target groups through the aid of diffusion instruments. Diffusion is a problem that all instruments entail, but with communication instruments it is a special problem, because implementation of a communication instrument does not make any sense at all unless combined with a diffusion instrument (see the study by GUTSCHER ET AL.).

Typically, two phases of diffusion can be distinguished. In a first phase, *initial diffusion* is reached through the aid of targeted diffusion instruments. The second phase is one of *self-diffusion*, in which the contents spread either directly from person-to-person within the social surround of the primary target group or by means of the traditional mass media. An example of person-to-person dissemination is the passing along of particularly illuminating arguments – supporting participation in a boycott, for example – within the direct circles of family or workplace. Diffusion through the social surround can also be promoted specifically through concrete instructions and

---

4 In the literature, the term "persuasive instruments" is sometimes used (e.g. Dahme and Grunow, 1983). It has a more restricted meaning than our definition of communication and diffusion instruments, because it usually refers only to a few selected instruments to influence knowledge. Another frequently used term is "social marketing". Social marketing refers to the activities of non-commercial organisations in planning, organising, realising, and controlling. These are activities that aim – directly or indirectly – to solve a social task and/or to disseminate social ideas or behaviours within a community (Kotler, 1978; Kotler and Roberto, 1991; Geller, 1989; Prose et al., 1994).

activities. An important example is the *building and management of networks*. In the area of sustainability strategies, for instance, networks represent an important intermediate goal that can be achieved through communication and diffusion instruments. Self-diffusion through the effects of the mass media on its own, finally, also plays a role when the successful initial diffusion is itself a media event. This occurred in September 2000 in France, when the media gave coverage to the blockades of oil refineries and service stations that were organised all over the country as a reaction to increases in the price of petrol.

2.5.2. Actors

Communication and diffusion instruments are available to public authorities, to the entire educational system from kindergarten to the university level, to providers of continuing education as well as to all other actor groups, especially to the media, NGOs, informal social movements, private companies, and to individuals, especially to parents and teachers. The use of communication and diffusion instruments in private life is up to individuals or informal social groups. For instance, it is up to parents to decide what they want their children to know about environmental issues and how they will try to motivate them to behave in certain ways. However, public actors can try to influence the private use of communication and diffusion instruments by using communication and diffusion measures themselves. So, in a first step, private actors may be the target group of communication instruments applied by public actors, and in a second step, they may themselves be the actors. The use of communication instruments by teachers or the media can be influenced to some extent by public authorities, by using command and control or service instruments. On the other hand, private actors can use communication and diffusion instruments to put pressure on public authorities or on private companies to take certain actions.

2.5.3. Target groups

Communication instruments can be targeted at the general public, at specific professional or life-style groups, or at groups defined by some common behaviour, such as car drivers or tourists. Children and young people are often regarded as highly promising target groups. Target groups can also be companies, associations, or public authorities.

2.5.4. Implementation and enforcement

The use of communication instruments is indicated mainly in early phases of the implementation of strategies. One of the strengths of communication instruments lies in the fact that initial effects can be achieved relatively quickly, in contrast to creating legal bases for the implementation of classic instruments, which can entail years of political process. For this reason, it makes sense to use communication instruments to *prepare the way* for further intervention instruments.

Using communication instruments to *complement* other instruments is also generally indicated. The success of traditional types of instruments depends to a large part on the implementation of these instruments and their effects. Of course, not knowing about command and control instruments does not protect people from sanctions, and it is hardly possible to escape the economic effects of price changes. But still, it is communications about the instruments that cause a large extent of their effects, because people will take care not to perform certain actions or will plan and execute alternative actions ("announcement effect"). In this way, the effects of classical instruments can be reinforced and optimised. These reinforcing effects also base upon the fact that in addition to the mechanisms of classical instruments, communication instruments promote insight into, understanding about, and the legitimacy of external measures.

In principle, communication instruments – like the other instrument types – can be used for any purpose. Therefore, within a society an ethical discourse about tolerable communicated contents is necessary. This discourse is itself a communication instrument on a meta-level. As a result of such a discourse, command and control instruments can be applied in order to prevent harmful forms of communication. Every legal system is likely to contain regulations pertaining to contents that may not be legally expressed publicly in words or images. For example, the Swiss penal code prohibits the distribution of ideologies that are based upon systematic discrimination of racial or ethnical groups. Similar regulations exist with regard to pornography. Restrictions of advertising for cigarettes or alcohol are additional examples of controlling communicated contents. For promoting sustainable development as well, society may have to set careful norms to control the communication of certain ecologically dangerous contents.

Public discourse is also necessary on tolerable behaviour and – especially with regard to the young – on the topic of the various ways that behaviour is trained. Not always do training opportunities have socially or ecologically desirable effects. For example, allowing people to learn to drive contributes to a growth of motor transport. Grossman and DeGaetano (1999) claim that realistic video games with the imitation guns sometimes used allow players to practice shooting in a realistic way. They develop shooting skills that can be applied when holding a real weapon in hand for the very first time. This explains the horribly high success rate of children who are given a single opportunity to use a fire weapon after chronic video game training. Of course, not all trained behaviours will be performed in everyday situations. However, the likelihood that they *will* be performed rises with training. As with the control of communicated contents, every society and legal system has to limit and control training opportunities that are likely to lead to hazardous consequences with regard to social sustainability. To achieve this, communication instruments have to be combined with command and control instruments.

3. Detailed description of the instruments

Within each of the five types of policy instruments, a number of sub-types can be distinguished. Tables 1A-C (pages 40-45) provide a trilingual survey of the entire typology and the distinctions we make in each of the five classes. In sections 3.1 through 3.5 below, we will explain and describe these types and sub-types in more detail and give illustrative examples for some of them.

3.1. Command and control instruments

COMMAND AND CONTROL INSTRUMENTS	
Environmental quality standards (impact thresholds and standards) **Emission standards** • best available technology • prescriptive technology standard **Product standards and regulations for the use of pollutant substances** • restriction, rationing, or prohibition • product standards	**Licensing** • licence to construct • licence to operate • licence to sell **Liability regulations** • strict liability • reversal of the burden of proof • compulsory third party liability insurance **Zoning** • land use regulations • water protection areas • nature conservation zones

The distinctions we make within the command and control type of instruments are based on a classification developed by the International Energy Agency (IEA) of the Organisation for Economic Co-operation and Development (OECD) (IEA, 1989), which focuses on energy policy, but is highly relevant to other environmental policy domains. The IEA typology has been slightly adapted and illustrated by examples from Swiss environmental policies.

Example 1:
Regulatory measures to reduce polluting emissions of heating systems

Since the 1980s, the Swiss federal government has actively pursued a policy to reduce the emission of pollutants from heating systems. This policy was tightened in the 1992, revised Ordinance on Air Pollution Control. The programme bases upon three measures, which are regulatory in nature.

Emission thresholds: In an international comparison, the regulation sets very strict threshold values for the emission of nitrogen oxide (NO_X: 120 mg/m^3) and carbon monoxide (CO: 80 g/m^3) for heating systems up to 350 kW.

Systems inspections: An inspection of the system type is required for combustion installations, which guarantees adherence to emission thresholds. Through compulsory licensing for construction, permits are granted only for buildings with heating systems that have passed inspection.

Heating controls: The cantons responsible for compliance with the Ordinance on Air Pollution Control appoint heating system controllers, who check the emission levels of new and existing systems.

There is general agreement that this combination of various regulatory mechanisms has definitely contributed to a reduction of polluting emissions from heating systems. An important point is that the regulations do not stop at declaring acceptable emission levels, but rather that there are follow-up measures for implementing the regulations (for greater detail, compare Balthasar and Knöpfel, 1994). Balthasar and Knöpfel (1994) show, moreover, that the regulatory programme has also had positive side effects the economy and on policies that foster innovation.

3.1.1. Environmental quality standards (Impact thresholds and standards)

Environmental quality standards define "the allowable average concentrations over a specific time period for a given pollutant in a particular region" (IEA, 1989: 38). In Switzerland, this type of regulation is widely used in air policies. The Ordinance on Air Pollution Control defines impact thresholds (maximum concentrations) for nitrogen dioxide, sulfur dioxide, volatile organic compounds, ozone, and other pollutants. The Swiss experience shows, however, that such environmental quality standards are repeatedly infringed. While the standards may indicate that a level of concentration of pollutants has been exceeded, no immediate action is demanded unless a responsible party can clearly be identified. In the case of air pollution, this is often impossible because of the multitude of actors contributing to the violation of the emission impact threshold (e.g., all drivers of motor vehicles and all home-owners operating fossil fuel heating systems). If a responsible party can be identified (such as in a case of water pollution that is caused by the release of pollutant substances by an industrial company), it faces the sanctions as defined by law.

3.1.2. Emission standards

Emission standards "set a maximum allowable rate of pollution output for each generic type of source (transport, power plants, industry) by type of pollutant" (IEA, 1989:39). Emission standards are often closely related to technological procedures and are, for this reason, frequently referred to as technology standards, even though they do not actually prescribe the use of a particular technology. There are more flexible variants of the emission and technology standards called **"best available technologies"** or "best practicable means". These may help to increase the pressure to apply new technologies under an unchanged regulatory regime. The emission standards for heating systems imposed by the revised Ordinance on Air Pollution Control from 1992 in Switzerland is an example of such a regulation (see Example 1). The IEA lists another form of environmental regulation, namely the **prescriptive technology standard**. It defines the type of technology or method to be applied in a particular instance. Prescriptive technology standards are not widespread because of their inherent rigidity.

3.1.3. Product standards and regulations for the use of pollutant substances

Product standards define the quality, that is, the admissible concentration of certain substances that has to be met by any product produced and/or sold on the market. In Switzerland – as in many other OECD countries – **product standards** are used widely to regulate fuel quality, as in the case of the proscription of leaded gasoline. The Swiss Ordinance on Air Pollution Control sets detailed quality standards for transport and heating fuel, and the Water Protection Decree prohibits the addition of phosphates to the contents of laundry detergents.

Specific requirements can be set for the use of harmful or toxic substances; they can be **restricted to specific conditions, rationed, or prohibited entirely**. Restriction or rationing can also be based on more general criteria. During the oil crisis in the 1970s, for instance, the Swiss authorities prohibited the use of fuel driven cars on 12 Sundays per year.

Example 2:
Environmental impact assessment

In Switzerland, facilities that could have a considerable impact on the environment may be constructed and operated only if environmental impact has been assessed and approved through an official procedure. The assessment is required for an extensive range of facilities: transportation facilities, such as highways; harbour facilities; railway lines and airports; energy production plants; waste disposal facilities; military facilities, such as weapons ranges; tourist and recreational facilities; but also mainly industrial facilities that have a potentially negative environmental impact.

For assessment of environmental impact, a report must first be submitted. In particular, the report must describe the measures that are planned in order to protect the environment and describe environmental impacts that will occur despite these measures. The reports must fulfil very high standards with respect to content. For this reason, they are prepared as a rule by specialised experts. For large projects, this requires a great deal of effort and expense. This is notable, because it is the person or organisation that seeks to realise the project that must submit the report.

Reports are then examined by the authorities and serve – together with other information, such as petitions by environmental protection agencies or other experts – as a basis for the decision on whether the project satisfies environmental provisions and may thus be approved. Frequently, environmental impact assessments must be co-ordinated with other licensing procedures (for example, water conservation authorisations or land clearing grants).

Mandatory environmental impact assessments have been in effect in Switzerland since 1985. Following their introduction, there were initial complaints that the instrument placed too lengthy delays on the projects or halted them all together. Since then, criticism has largely ceased, not least because the authorities undertook various efforts to speed up the procedure. Today, as a necessary step in the planning of large projects, environmental impact assessments have become self-understood.

3.1.4. Licensing

In general, licensing is not used as an independent instrument, but rather to support the implementation of regulatory standards. Three types of licenses can be distinguished: **licences to construct, licences to operate, and licences to sell.**

Licences are of importance in the siting of new facilities. In most Swiss cantons, the license to build new housing facilities is bound to compliance with various impact thresholds and, less often, emission standards. The builder has to demonstrate that the building meets the terms of the SIA-380-1-norm[5]. As regards insulation, this norm is the benchmark corresponding to the prescription of the "best available technology". Furthermore, the builder has to prove that the heating system has no important negative impact on the environment (regarding fixed environmental standards). If the requirements are not fulfilled, an environmental impact assessment is mandatory. This proce-

[5] The SIA-380-1-norm is a recommendation edited by the Swiss Association of Engineers and Architects (SIA). It is used as a planning tool for energy efficient construction with respect to the maximal heating need per surface and the minimal heat rate. The norm by itself is not legally binding, but various cantons implement it as a binding guideline in their cantonal laws, mostly at the ordinance level.

dure is designed to guarantee that environmental quality standards are respected or, at least, not violated disproportionately during the construction and operating of new facilities (see Example 2).

Technical licensing procedures must be followed to obtain a licence to sell. This licensing is important to the implementing of product and/or emission standards. As described above in Example 1, Swiss environmental law prescribes a technical licensing procedure for heating systems up to 350 kW. This procedure serves primarily to check whether the system on the market is in accord with the legal emission thresholds. Technical licensing procedures are also prescribed for cars.

3.1.5. Liability regulations[6]

Whoever suffers damages must carry them. This old, general principle remains valid in Swiss law as well as in most European legal systems. Increasingly, however, the principle is being overturned through the creation of legal bases that hold the person who causes the damage liable to payment of compensation (the polluter-pays principle). Even in areas where there is a liability basis, the person causing the damage is usually liable only if blame can be established. If and only if it can be proved that the damaging party can be charged with fault, or in other words, if the damage occurred through negligence or intent, can there be a liability claim. The fact that the person suffering the damages as a rule must prove that the accused party is guilty often hinders the enforcement of rights to compensation.

Strict liability: Liability is tightened in areas where a particular facility or particular behaviour involves special risks. The legal system makes the provision in such areas that the originator of damages – for example, the operating authority of a factory that pollutes water – will be held liable even if it has undertaken all possible measures to avoid damage. This type of liability is independent of misconduct. The fact alone that people have created risk makes them responsible for any damage that arises. For this reason, such instruments are called strict liability or liability regardless of fault. Swiss environmental law includes strict liability (for example, special strict liability for nuclear plants or pipelines). If a special danger to the environment – ground, water, air – results from the operations of a company or plant, the owners are liable for damages if the risk becomes reality. Liability, however, is restricted to damage incurred to persons or property. In contrast, there is no liability provision that applies in the case of so-called damage to the environment alone – damage to the air, water, or wild plants and animals (with the exception of fish). Other countries extend liability for environmental damage much further than Switzerland.

Reversal of the burden of proof: In the case of normal liability for fault as well, the law can make provisions that facilitate the enforcement of claims by the injured party. Here it is not, as in the usual case, the injured party who must prove that the damaging party is guilty of fault, but rather the wrongdoer must prove that he is in fact blameless. However, reversal of the burden of

---

6 Liability regulations are often classified as economic instruments, especially by economists. In our interdisciplinary Synthesis Group, arguments were exchanged between lawyers and economists, and the Synthesis Group finally decided to list them with the command and control instruments.

proof is not significant in environmental law, because here liability regardless of fault usually applies. Legal systems in other countries sometimes make provisions for reversal of the burden of proof, or causation presumption, with regard to causation issues (a connection between operations of a facility and the onset of environmental damage).

Compulsory third party liability insurance: Environmental damage can in some cases be very expensive. In such situations the most stringent laws concerning liability will be of no use, if the damaging party cannot pay damages because he is insolvent. For this reason, for various, particularly risk-laden areas, liability laws make the provision that the owners of facilities and installations must be insured. In theory, compulsory insurance against liability (or comparable liability provisions) may be required for any area. In practice, however, certain risks are hardly insurable (for example, the risks incurred by refuse collection centres, or landfill, often show up after long periods of time have elapsed and are extremely difficult to appraise).

3.1.6. Zoning

Zoning laws regulate the admissibility of conduct in certain geographical areas. The content of the laws may be positive or negative, in that they may admit, or even prescribe, certain activities for one area that are prohibited in other areas or rule out activities that are approved elsewhere. In the area of zoning laws, a number of instruments can have a direct or indirect effect on sustainable development:

Land use regulations represent the most important category of zoning laws. The regulations aim to reduce environmental damage particularly in sensitive areas by, for example, separating industrial and business zones from residential zones or recreational areas. In this sense, the regulations are nothing more than a spatial differentiation, or specification, of emission and environmental quality standards. In combination with licensing procedures, zoning should help to ensure that certain levels of pollution control are achieved. At the same time, zoning can achieve spatial concentration of environmentally damaging activities, and thus shortened transport distances, which reduces emissions.

Water protection areas are designed to prevent pollution of ground water that is used to gain drinking water. They serve in the full sense of the word to protect the environment at the source.

Nature conservation zones serve special environmental concerns and efforts, such as protection of rare or endangered animal or plant species.

3.2. *Economic instruments*

ECONOMIC INSTRUMENTS	
Subsidies • grants • tax allowances • soft loans • guarantees • compensation for foregoing use of the resource **Incentive taxes** • taxes on energy/resources • taxes on emissions • taxes on products/processes **Charges** • one-time charge for connection to services • recurrent charges for use • charges on advantages (value-added contribution) • prepaid disposal fees	**Deposit-refund system** **Market creation** • tradable allowances or permits • joint implementation **Incentives as parts of action campaigns** • rewards • lotteries • contests/benchmarking • discounts

As mentioned above in the general description of economic instruments in section 2.2., there are three principal ways to intervene: (1) reduce the costs of environmentally sound behaviour, (2) raise the costs of pollutant behaviour, and (3) establish markets for polluting rights. We distinguish two groups of instruments by which the costs of environmentally sound behaviour are reduced (subsidies and incentives as parts of action campaigns), two groups of instruments that raise the costs of pollutant behaviour (incentive taxes and charges), and a group of instruments for the creation of markets for polluting rights. Deposit refund systems are treated as a group of its own.

Raising the costs of pollutant behaviour can be achieved by public taxation. Taxes comprise all payments imposed one-sidedly by sovereign authority of the government upon taxable persons. Taxation has to be legislation-based. Because of this requirement, in some cases taxes can be introduced only after a lengthy political decision making process. There are several categories of taxes. In terms of purpose, they can be divided into **incentive taxes** and **charges**. Incentive taxes aim to achieve efficient environmental protection by making detrimental practices expensive enough that there is an adaptation reaction and a change in practices. "Pure" incentive taxes are not levied in order to create government revenue (fiscal function), but rather to steer environmental practices via price adjustments. **Charges** represent recompense for special public services or special advantages.

3.2.1. *Subsidies*

Companies and households receive financial support from the government for environmentally sound practices. Financial subsidies can be granted by the federal government (FG), the canton

(C), or the municipality (M) and, depending upon amount and application area, must be approved by parliament or even by public vote. Increased or changing demand for these investment goods can also have an indirect effect on other economic sectors. Such effects may be one of the explicit goals of investment through subsidies. For instance, the Decree on Energy Investment within the Energy 2000 Action Programme has the aim of promoting energy-efficient building and renovation *and* giving new impetus to the economy.

Example 3:
Government subsidisation of wastewater disposal

In the last 25 years, the Confederation has massively subsidised investments in wastewater disposal by municipalities and associations (approximately 17% of the total investment). In addition to the federal government, the cantons have also granted subsidies. In sum, subsidies by the cantons totalled more than the government grants. Subsidisation has achieved the following results:

(a) **Increased tempo effect:** Expert opinion rates the incentive effect of subsidies on the willingness of municipalities and associations to build wastewater treatment facilities as strong. Without the subsidies, the current status of water protection could not have been achieved.

(b) **Compensation effect:** Subsidy levels were set according to the financial capacity of the cantons or municipalities. This resulted in indirect financial compensation.

(c) **Decrease in wastewater treatment costs** and subsidisation problems: With subsidies, wastewater treatment costs went down. This meant that polluters were not paying the full cost of disposal of wastewater. As the polluter-pays principle for a number of environment areas became anchored and gained widespread acceptance, government and canton subsidies were cut drastically. This presented a new challenge particularly to the municipalities, which now had to remediate or rebuild expensive wastewater infrastructure without the financial aid of government or canton. In some municipalities, over 90% of wastewater investments had been subsidised by government and canton. This led, in exceptional cases, to a situation where wastewater infrastructure had been created that the municipalities could not, in the long term, finance themselves (see the chapters by MÜLLER AND KRAMER and the comment by FRIEDERICH; ECOPLAN, 1993).

Grants are non-reimbursable contributions, that is, outright financial aid from the government. Grants may also be called subsidies in the narrowest sense of the word.[7] Examples include the Energy Investment Programme in the framework of the Energy 2000 Action Programme (C), investment grants to freight traffic facilities (C), to public transport services in municipalities (C), to wastewater treatment facilities (FG, C), and subsidisation of the remediation of public buildings (C, M) (see also Example 3).

Tax allowances: Environmentally acceptable practice is rewarded through tax advantages. This can be achieved by special tax deductions on general income tax or by special tax rates. Some ex-

7 In Germany and France, only contributions of this type are called subsidisation. Tax allowances and loans then fall into the general category of "financial assistance by the government". The Swiss use of the term corresponds more closely to English usage, where "subsidies" comprises both non-reimbursable financial assistance by the government (grants) and tax allowances and loans.

amples are tax deductions for the use of renewable energy sources and for investments in improving the energy efficiency of buildings (C), or reduced tax rate for solar power installations (see Example 4).

Example 4:
Tax allowances for energy measures in Switzerland

Since the late 1970s, more and more cantons have created the opportunity to deduct energy investments from income tax. Since 1995, the federal taxation authorities have granted a generous tax deduction for energy investments for privately owned properties. Tax deductions are very popular politically, but the environmental gains have been extremely modest:

(a) The tax deductions for energy measures are little known – only 27% of property owners are familiar with the allowable deductions.

(b) In 70% to 80% of cases, property owners would have implemented the energy measures even without tax deductions. The so-called "cash-in" effect is therefore considerable. A shortfall in tax revenue of 50 to 100 million Swiss francs annually does not constitute positive energy effects.

(c) The cost reduction effect of the tax deduction is strongly dependent upon income level. Households with high incomes profit most from the tax deduction. This differential price reduction effect is justified neither from the perspective of the polluter nor from the perspective of efficient energy policy (ECONCEPT and IPSO, 1997).

Soft Loans: This form of financial assistance is a repayable loan at much more favourable conditions, at very low or nil interest, than on the capital market. An example is a loan provided by the Swiss cantons to companies for purposes of instituting measures to prevent wastewater production.

Guarantees: If the government guarantees credits extended to an actor for the financing of environmentally sound measures, the actor can arrange for credit at more favourable conditions. Capital costs can be kept low. For example, the cantons may extend credit guarantees to companies to finance measures to prevent wastewater.

Compensation for foregoing use of a resource: Rather than penalising environmentally damaging practices (through environmental taxes) or prohibiting them (through laws and statutes), environmentally sound practices can also be promoted by paying the polluter to desist. Usually the environmental polluter is ordered to desist from the practices at least in part. Polluters are thus not completely free in their decision to take compensation rather than to continue to pollute. In this case, financial compensation takes on the character of a reimbursement for lost profits or ensuing costs, and the political motivation is social distribution. The lines are fuzzy in this area, and it is also not always possible to make a clear distinction between compensation and subsidies. As a rule, in practice the instrument mutates to reimbursement or subsidy. No true market arises. Some examples in agriculture include reimbursement for establishing ecological buffer land, compensation for improvement measures in the area of water inflow, water adits, and cachments.

3.2.2. Incentive taxes

Incentive taxes aim to achieve efficient environmental protection by offering price-based financial incentives, causing those who have to pay for the impact they have on the environment to adapt and change to more environmentally sound practices.

Taxes on energy and resources: The justification for taxing resources is that shortages of the resources (such as soil or drinking water) are not reflected adequately in market prices. The assumption is that market prices do not mirror the needs of future generations or of the third world. With regard to energy taxes, the problem of environmental pollution provides an additional argument. Examples: drinking water tax (C), tax on gravel (C), energy tax, electricity tax (C).

Taxes on emissions: Emissions taxes proper have to be linked directly to the polluting emissions (for example, through measuring wastewater volume). For enforcement to be realisable, however, taxes are frequently levied on certain products that are known to make a clear contribution to emissions. The line drawn between taxes on emissions and taxes on products is fluid. Examples: wastewater tax (C), VOC tax, CO_2 tax, NO_X tax on combustion installations, noise-related landing tax at airports (see also Example 5).

Taxes on products and processes: With the product taxes, tax is levied on products that cause environmental degradation through their production, consumption, or disposal. Process taxes are levied on certain types of processing or production and, in a broader sense, on certain branches of industry. The distinction between process and product taxes is not always clear. Product and process taxes are often levied when a direct tax on emissions would be too difficult or impossible to enforce for technical reasons. Examples: distance-dependent charge on heavy goods vehicles; consumption-dependent motor vehicle tax (C); tax on non-recyclable beverage containers; tax on fertilisers; energy efficiency tax on buildings (see the chapter by GARBELY AND MCFARLANE).

3.2.3. Charges

Charges *[Kausalabgaben]* are recompense for certain services or advantages provided publicly to private persons. These charges are based on the concept of *quid pro quo*, a thing given as compensation (by private persons) in return for services (public).

One-time charge for connection to services: This is a charge for connection to public services, such as public utility systems. The idea is that persons served by the service should buy into the existing infrastructure system. Examples: service connection charge for electricity, gas, long-distance heating from a district heating system; service connection charge for sewer services.

Recurrent charges for use: These are charges for continual use of public infrastructure facilities. They are often made up of a fixed amount (annual basic charge) and a variable, consumption-dependent amount (sometimes called the "working price" or "quantity charge"). Examples: waste disposal charges (M); charge for wastewater treatment; public utility charge for water (M). (see the chapter by MÜLLER AND KRAMER)

Charges on advantages (value-added tax): These are charged for the (potential) economic advantage enjoyed by a private person that stems from public institutions. In contrast to the charges for connection to services, VAT is not dependent upon actual use of the advantage. Example: property-owner charge for the street that provides access to the property.

Pre-paid disposal fees: These fees are not paid when the public service (disposal) takes place, but rather in advance at the time of purchase. For example, there are pre-paid disposal fees for refrigerators and batteries.

Example 5:
Incentive tax on volatile organic compounds in Switzerland

With the year 2000, Switzerland has become the first country to fight ground level ozone with an incentive tax on volatile organic compounds (VOC, also known as hydrocarbons). The tax level is substantial, as is the expected contribution to reduction in the use of substances containing VOC. Furthermore, redistribution of the revenue on a per-capita basis makes the tax highly innovative.

Together with nitrogen oxides, VOC contribute to the excessive formation of ozone near ground level. The Swiss Air Pollution Control Strategy's reduction targets for ground-level ozone were exceeded in past years. The excessive impact of VOC damages public health and the environment and must be reduced through a corresponding reduction of emissions. To keep within the impact thresholds, VOC emissions have to be reduced by 70-80%, i.e., by at least 70,000 tons annually.

To fight ground-level ozone emissions, the Swiss Federal Assembly adopted the Amendment of the Environmental Protection Law (EPL) and brought it into force on 1 July 1997. The Ordinance on incentive taxes on VOC – based on Articles 35a and 35c of the EPL – dates from 12 November 1997 and can be regarded as the first implementation of an effective incentive tax in Switzerland. Even though emission reduction targets are similar in the EU, the tax on VOC in Switzerland is the first of its kind in the world.

Starting on 1 January 2000, a tax on VOC is levied upon volatile organic substances such as solvents, paints and varnishes, glues, and cleaning materials. To meet a claim submitted by industrial representatives, a stepwise introduction of the tax is outlined. The Ordinance sets a rate of 2 CHF per kg VOC until 2002 and an increased tax of 3 CHF per kg VOC from 2003 on.

Only VOC emitted in Switzerland should be taxed. Levying the tax on import and on production in Switzerland means that VOC that are not emitted into the environment, or those that are exported, are exempted. Furthermore, the Federal Council may grant exemption from the tax to companies that have already reduced VOC emissions well beyond the legal requirements, to the extent of the additional costs incurred.

The incentive tax is expected to generate net revenues of 160 million CHF per year between 2000 and 2002 and 210 million CHF per year from 2003 on. The law requires that the revenue be distributed equally among all members of the population through the mandatory health insurance programme. Health insurance is simply being used as the administrative channel to distribute the monies; this model does not subsidise health insurance premiums. To maintain transparency, premiums and the VOC deduction will be shown clearly on insurance policies or accounts for premiums.

3.2.4. Deposit-refund systems

A deposit is added to the price of potentially polluting or resource consuming products. When pollution or resource consumption is avoided by returning these (often reusable) products, a refund of the deposit follows. In Switzerland, most existing deposit-refund systems are the result of negotiations.

3.2.5. *Market creation*

Environmental polluting can be traded on markets. For this, there has to be an at least implicit definition of ownership rights to environmental resources. A polluter has to buy, for example, a license from the owner of the environmental goods (the government). If ownership rights to environmental goods are extended to the polluter, the polluter will demand recompense for discontinuing the polluting practices.

Tradable allowances or permits: Here the total acceptable extent of environmental impact (pollution standard) is established. The polluter has to acquire the necessary certificates. Often certificates are implemented in combination with emission standards: provided that polluters do not reach the prescribed emission allowances, they may "bank" their "credits" or sell the allowance in excess of their own needs to other facilities or industries. In this way, certificates can be traded, and the price is the result of supply and demand. Examples: Tradable emission allowances or certificates for water pollution or air pollution; individually transferable fishing quotas. The Kyoto Protocol to the United Nations Framework Convention on Climate Change (UNFCCC) provides for emissions trading, i.e., the trading of CO_2 emissions between two countries that have committed to reduce emissions.

Flexible financial mechanisms under the Kyoto Protocol to the United Nations Framework Convention on Climate Change (UNFCCC): With flexible mechanisms, such as Joint Implementation, industrialised countries can finance measures towards greenhouse gas (GHG) abatement in developing countries and countries in transition. The reduced greenhouse gas emissions abroad are banked into the account of the industrialised country towards its commitment under the Kyoto Protocol to reduce GHG emissions. The pilot phase of Joint Implementation – Activities Implemented Jointly (AIJ) – was finalised in 2000; JI will probably start only in 2008. Clean Development Mechanisms (CDM) was begun in 2000.

3.2.6. *Incentives as parts of action campaigns*

The main target group of such incentives is consumers. These instruments are usually not included among economic policy instruments. But from the point of view of the target group, they rely on the same mechanism as subsidies. They reduce – temporarily – the price of environmentally sound goods relative to conventional ones in an attempt to influence the consumers' purchase decisions. The incentives can be distinguished from subsidies as follows: Subsidies are classical instruments of financial support of companies and/or households by the state, and a formal political decision is required for their implementation. Usually such support is provided with a medium or long-term perspective (months or years). In contrast, the incentives as described below are provided as parts of campaigns or other well defined programmes, they are usually rather small in amount, and they are often provided by actors other than public authorities, especially by private enterprises or organisations. Their main goal is to stimulate people to rethink their usual routines and to encourage them to try new products or behaviours. The assumption is that after the incentive campaign has ended, people will maintain their new behaviours. The immediate goal of such incentives may also be to successfully introduce new, resource efficient technologies or products on the market and to accelerate their diffusion.

Rewards: People who act in desired ways or achieve a particular behavioural goal receive rewards. Rewards may be financial, remuneration in kind, or symbolic, such as cash prizes, free services, parking permits, and so on. Quite common are coupons for certain actions. The coupons may be exchanged for products in certain stores, for instance. Long-term and lasting implementation of a reinforcement plan is a problem (usually unresolved). All studies show that the desired behaviour disappears as soon as rewards come to an end. This can be explained by the fact that people justify the new behaviour on the basis of the expected reward. This external justification for an action makes a change in attitudes unnecessary. The motivation for a change in behaviour comes from the outside, and there is no inner acceptance. For this reason, it is better to offer carefully apportioned, rather small rewards that by themselves alone do not cause a massive change in behaviour. Too strong external justification hinders the emergence of additional motivational and self-justification processes that are founded in a person's own thinking (Diamond and Loewy, 1991; Everett et al., 1974; Katzev and Pardini, 1987-1988). Examples: Free bus tickets to a sports stadium to people who have arrived by train; certificates, medals, or other recognition awards.

Lotteries: People who act in the desired way receive a lottery ticket. The principle here is related to the principle of rewards. The difference is that reward is not certain and will be received only with a certain probability. Even though long-term success after a lottery is no more assured than with other reward models, this method is at least an affordable alternative to widely distributed rewards, especially if the intervention is planned for a longer time period. It is important to set an ideal probability level for winning. If chances of winning are too high, an external justification has been created for the behaviour (the prize). If chances are too low, people can become frustrated due to lack of success. Ideal chances of winning would result in a prize now and then, but not often enough that winning becomes the only justification for the behaviour. Comparative studies show that the effect of expected possible rewards (lottery) is greater than that of regularly received rewards (Jacobs and Bailey, 1982-1983; Witmer and Geller, 1976). Example: a lottery ticket in every carton of eggs from free-range chickens.

Contests/benchmarking: Contests are held among persons, households, or other groups. The winner of the prize is the person or group who reaches the set goal first. All contestants should have the same chance of winning. If there seems to be no chance of winning, participation and the willingness to make an effort will drop off rapidly. A contest where the winner is established only near the end of the contest is recommended. This sustains the excitement of the contest and keeps the effort level high. As a form of intervention, contests show the same disadvantages as other rewards: The more attractive the prize, the more it will become the main cause of the behaviour change. This in turn hinders the development of a real change in underlying attitudes. If the prize is not attractive enough, only extremely achievement-oriented and competitive persons will take part. Again, it is important to find the right balance, for the main goal is the *long-term* establishment of a new behaviour (Witmer and Geller, 1976). Examples: The first person to reach an energy-saving goal, or the person who saves the most energy in a certain period of time, wins the prize; solar power prizes for municipalities.

Discounts: Temporary price reductions for environmentally sound products may be offered by retailers or producers in order to draw consumers' attention to the products and to make them more attractive relative to conventional products. In special cases, sellers themselves may benefit from the increased diffusion of such products. For instance, it may be more profitable for a public electricity utility to sell energy-saving household appliances at a discount price than to buy and sell additional amounts of electricity.

3.3. *Service and infrastructure instruments*

SERVICE AND INFRASTRUCTURE INSTRUMENTS	
Service instruments • offering or improving ecologically sound products • withdrawing environmentally undesirable products • offering or improving services that allow or facilitate ecologically sound action • reducing services that allow or facilitate environmentally undesirable action	**Infrastructure instruments** • offering or improving infrastructure that allows or facilitates ecologically sound action • dismantling or degrading infrastructure that hinders or inhibits ecologically sound action

Service and infrastructure instruments are goal-directed transformations of services or infrastructure for the promotion of certain behaviours. They can be used in order to make desired actions possible, more salient, or more pleasant, or to exclude undesired actions or to make them less salient and less pleasant. Environmentally relevant modifications of services or infrastructure can be *attraction based* (offering or improving environmentally sound products or services and infrastructure that allows or facilitates environmentally sound action), or *repulsion based* (withdrawing products, reducing services, or degrading infrastructure that allows or facilitates environmentally undesirable actions).

3.3.1. *Service instruments*

Services are actions of individuals or organisations that enable or support other actors to achieve their goals. A service may consist in offering a physical product for sale, in providing information or guidance, in offering special skills, and so on.

Offering or improving ecologically sound products: A common barrier to environmentally responsible purchase behaviour is the unavailability of respective products, their lower quality in terms of aesthetics or convenience, and the lack of reliable information about their environmental quality. Therefore, improving the quality of such products, improving their accessibility, and reducing the uncertainty about their environmental qualities will increase their attractiveness. Example: Offering organically grown food at or very near the places where people do their daily shopping, displaying it in an attractive way, and taking care that taste and visual appearance are at least as good as in conventionally produced food.

Withdrawing environmentally undesirable products: A retailer may discontinue to offer certain environmentally undesirable products, thus making them impossible to buy. Example: A drugstore discontinues the sale of body care products that have been tested on animals.

Offering or improving services that allow or facilitate ecologically sound action: Using public transportation is in general less damaging to the environment than travelling by car. However, it is often less comfortable, and it can even be distressing. Overcrowded trams are neither pleasant nor healthy. Travelling in a nearly empty suburban railway car with a few drunken youngsters who are smoking in the non-smoking compartment, without any attendants who could provide assistance, can even be intimidating. For economic reasons, the number of attendants at railway stations and in trains has been cut drastically in recent years. This has led to a deterioration of the service quality of public transportation. On the other hand, improving public transportation by attractive rolling stock, ample seating, and a train staff that cares about passengers would make travel by train more pleasant and healthy and would thus facilitate the consumer's choice of the environmentally better alternative. A further example: Car sharing organisations offer an attractive service that makes it easy for people to get rid of their own cars without suffering a loss of personal mobility (see also Example 6).

Example 6:
Encouraging car-pooling: the CARLOS project

With the CARLOS project, an innovative transportation option is being designed and tested in the region of Burgdorf in the German-speaking part of Switzerland. Car-los is a play on words, for in German it means "carless", i.e., to have no car. The project aims to build a pick-up system on the basis of private automobiles in thinly settled regions that have limited access to public transportation. The system is a kind of highly developed hitchhiking system. Special car stops form the core of the new services and infrastructures. The car stops are equipped with an electronic display board that shows approaching cars where a potential passenger would like to go. Women may signal on the board that they wish to be picked up by women only. The users activate the display board by paying a small fee into a meter. They receive a ticket that they hand to drivers who stop to provide a lift. Drivers can use these tickets as gift certificates, lottery tickets, or similar things. Video cameras record the automobile, license number, time of departure, and destination.The new services and infrastructures offered are meant to be taken over by public transportation providers in future in order to provide a denser network of transportation routes in areas with limited public transportation. CARLOS allows people to be more mobile without owning cars, while it at the same improves the poor occupation ratio of vehicles. In regions where regulations give priority to highly occupied vehicles (HOV) – i.e., on certain roads and in parking areas – CARLOS represents a useful complement to such measures. (Hasler and Wälti, 1999)

Reducing services that allow or facilitate environmentally undesirable action: Christmas shopping flights to far away destinations or heli-skiing in the Alps are ecologically quite harmful behaviours; by taking them off their programmes, tourism or travel agencies are applying a repulsion based service instrument.

3.3.2. Infrastructure instruments

Infrastructure means man-made, mobile or immobile physical objects that shape the actor's scope of action possibilities. Physical aspects of the environment influence human behaviour very directly by encouraging certain behaviours (Kaufmann-Hayoz et al., 1996).

Offering or improving infrastructure that allows or facilitates ecologically sound action: Whether a certain service, e.g., an additional train in the evening rush hour, can be offered, also depends on the infrastructure. It may be that additional trains to supply more seats can only be introduced if more tracks are built or if the signalling installations allow shorter distances between the trains. Other examples: In many older toilet models, the entire contents of the toilet tank empty with every flush. Water-saving behaviour would be possible if there were an interruption button or if toilets were so constructed that one flush would empty only part of the water in the tank (in most cases sufficient). Only with repeated pushing of the flushing handle would the toilet empty the entire tank.

Example 7:
New design of main roads in town centres: The "Bern Model"

In Switzerland, many national highways go directly through town centres that grew up historically. In the centres, there are restaurants, shops, and the like on these roads. In past decades, some of these roads were enlarged to accommodate multiple-lane traffic. In many towns, 20,000 cars a day pass through, and during rush hours traffic jams hold up public buses and trams. This produces not only air and noise pollution detrimental to public health, but has other negative effects on the quality of life and the well being of residents and visitors to the centres. The roads cut the town centres in two, and pedestrian crossings are dangerous and involve long waits. Pavements in front of restaurants and shops are small and uninviting to pedestrians. Bicyclists are endangered by the heavy traffic that takes over the whole of the road.

The "Bern Model" is a modern traffic-planning concept that seeks to correct the negative consequences of previous design decisions and, at the same time, to create options for future developments.

On the basis of participatory planning processes, structural measures appropriate to the traffic situations are realised to reach the desired goals (Kobi et a., 1995). The "Bern Model" was applied to redesigning the Seftigenstrasse in Wabern near Bern. Structural measures (roundabout, centre lane, bicycle lane, enlargement of pavements and pavements in front of shops) succeeded in enlarging the space for pedestrians and bicyclists and making it more attractive (the number of pedestrians and bicyclists increased). At the same time, traffic flowed more steadily, so that emissions were reduced (Haefeli et al., 2000, Tiefbauamt des Kantons Bern, 2000).

Dismantling or degrading infrastructure that hinders or inhibits ecologically sound action: As long as there are plenty of free parking spaces in a city and traffic runs smoothly, a large proportion of people will not use public transportation, even if it maintains high standards. By delimiting the capacity or quality of the road infrastructure, the relative attractiveness of public transportation increases. Other examples: The availability of a clothes dryer in a household invites residents to use it; removing it would constitute a repulsion based infrastructure measure (see also Example 7).

3.4. *Collaborative agreements*

COLLABORATIVE AGREEMENTS	
Public-private agreements • agreements on prepaid disposal fees on specific product groups • agreements on consumption goals or standards • formal agreements with individual companies	**Certifications and labels** • with legal compliance • without legal compliance

Collaborative agreements are legally binding or non-binding commitments of the private sector or parts of it (industrial sectors, their associations, or individual companies) to the government to enhance energy and resource efficiency or greenhouse gas abatement in sectors that are not regulated by public policy. An important part of the agreements are the negotiations about the targets to be achieved.[8]

3.4.1. *Public-private agreements*

These are legally binding commitments by single enterprises, industry associations, or full branches to achieve jointly defined environment and climate targets at a certain point in time so that the government does not have to implement the threat of regulations and law. This does not necessarily require an agreement between the government and all the market partners. An agreement with the market leader(s) may suffice, as the announcement effect of such agreements is considerable. Additionally, it is necessary in any case to carry out further (flanking) measures by the government or the industries (such as informing customers) in order to guarantee the agreement's successful implementation.

---

[8] More detailed overviews of different types of collaborative agreements in OECD countries are given in IEA (1997) and OECD (1999).

Agreements on prepaid disposal fees for specific product groups: Technically seen, this is a charge in the sense of economic instruments. If the charge is finalised through negotiations between the concerned industries or importers and the government, this charge can be called a collaborative agreement. In turn, the government foregoes the introduction of more compelling instruments. Examples of this policy process have been the recycling of glass, aluminium cans, PET bottles, batteries, and refrigerators.

Example 8:
The large-scale consumer model in the canton of Zurich
(legal compliance model for large-scale consumers)

The new energy law of the canton of Zurich allows large-scale consumers, singly or groupwise, to make a commitment to the canton to increase their energy efficiency. In return, the canton exempts large-scale consumers from detailed regulations. With this new execution regulation, large-scale consumers may choose the most economic way to achieve the energy targets, and by doing so they gain more flexibility. For reasons of efficiency, agreements with groups of large-scale consumers are desirable. By the end of 1999, four groups had signed agreements with the canton of Zurich:

- In mid-1997 eleven four- and five-star hotels together with the Kongresshaus (the so-called "Convention Pool") committed to increasing their energy efficiency within the next 20 years by 27%.
- The Airport of Zurich decided in 1998 to enter into a 10-year agreement.
- By the end of 1998, 12 enterprises working together in the framework of the Zurich Energy Model committed to increasing their energy efficiency by 2007 by 15%.
- The "Group 4" constitutes a holding with headquarters in the United States. The holding is not allowed to join any group, and therefore a separate agreement with the canton had to be signed. The holding is also not allowed to publish any sensitive data.

The canton of Zurich foresees to expand the agreements to include a possible thousand large-scale consumers in the canton.

The quantitative impact on energy efficiency gains of this large-scale consumer model can not yet been assessed. Not enough time has elapsed, and enterprises are participating only voluntarily. Initial experiences, though, indicate that the agreement allows an additional, operationally still profitable energy efficiency potential to be realised. Success factors for the large-scale consumer model include a cost-effective impetus to participation from outside the enterprise, clear target negotiation, intensive communication between the parties to the agreement, stringent control and sanction mechanisms, and the implementation of an internal energy management system.

Agreements on consumption goals or standards: The Swiss Decree on Energy Use and its implementing provisions in the Ordinance on Energy Use provide for joint definition of consumption goals regarding plants and equipment by government and concerned organisations. Negotiations between the government and branches of trades not only define target consumption, but also the time period in which these goals are to be achieved. For such agreements to succeed, it is essential that the agreed upon goals are not softened with time, but are continuously intensified. Examples of successful negotiations are the agreements on consumption goals for electrical household and office appliances. A negative example is the failure of the negotiations between the Swiss government and importers of cars and other actors (Swiss Touring Club, automotive trade association) on consumption goals for fuel. Following the failure of these negotiations, the Swiss government enacted an ordinance[9] in 1995, which prescribes a reduction of the fuel consumption of new cars by 15 % within five years.

Formal agreements with individual companies: In the framework of the execution of energy regulations, a few cantons foresee the possibility of agreements with individual large-scale consumers or groups of large-scale consumers in order to oblige them to fulfil specific targets. In return, the canton foregoes the introduction of detailed energy regulations.

3.4.2. Certifications and labels

Labels and certifications can guarantee compliance with legal environment standards (inclusive of control) or can even go beyond these legal standards. If the standards are not fulfilled, however, the government cannot impose sanctions. **Labels** are product oriented environmental standards. The processes behind the product do not necessarily have to be environmentally friendly. **Certifications** are enterprise or organisation related labels that show that the enterprise's premises and its production processes aim towards environmental quality. Certifications do not stand for the quality of single products, but rather certify that the environmental impacts of the whole production and life cycle process are documented and controlled.

Example 9:
ISO 14001 certification

The ISO 14001 norm certifies an environment management system that aims to reduce energy consumption and environmental impacts by constantly upgrading the processes of production, product design, provision, disposal, and services. The environment targets have to be clearly defined. Continuous and systematic monitoring and documentation of the processes have to provide evidence of the efforts. Not the government, but independent, privately accredited certification institutions periodically check compliance with the targets (in Switzerland about six institutions exist, such as SQS, SGS Group, TÜV Thun).[10] The certificate ISO 14001 is not put on the certified enterprise's products, but it can be used for purposes of marketing and advertisement on company stationery, for instance. In 1999 approximately 340 enterprises in Switzerland had obtained certification under the 14001 norm.

With legal compliance: Certifications and labels with legal compliance have to guarantee compliance with legal environment standards (inclusive of control by the government). Such prod-

9 *Verordnung über die Absenkung des spezifischen Treibstoffverbrauchs von Personenwagen (VAT).*

10 The ISO 14001 certification is not classified in the category "with legal compliance" unlike the EMAS certification, which is based on European Law and which is executed by institutions accredited and certified by the government. In addition, enterprises with EMAS certification have to sign a publicly accessible environment declaration that is not required for ISO 14001 certification.

ucts and processes are promoted, supported, and subsidised by the government. Examples: a label for organically grown food (*Bio-Knospe* label), for food grown by integrated production (IP label), Eco-Management and Audit Scheme (EMAS) certifications, green electricity label (see the contribution by GEHRIG AND NORTH).

Without legal compliance: Labels and certifications without legal compliance go beyond legal environment standards by setting higher environmental goals. The government does not control compliance. Conformity with law is thus a necessary but not sufficient condition. Examples: *KAG* and *agri-natura* (meat), *M-Sano* (food), *Öko-Tex* (textiles), *Max Havelaar* (fair trade), *Minergie* (low energy houses) (see also Example 9).

3.5. *Communication and diffusion instruments*

COMMUNICATION AND DIFFUSION INSTRUMENTS	
Communication instruments without a direct request • presenting facts • presenting options • presenting appraisals, goals, and norms • providing experience of reality • presenting model behaviour • giving feedback and enabling self-feedback **Communication instruments with a direct request** • persuading about facts • persuading about options • persuading about goals, appraisals, and norms • sending appeals • presenting prompts and reminders • stimulating self-commitment	**Diffusion instruments** • establishing direct personal contact • establishing contact via person-to-person media • establishing contact via mass media

Communication instruments aim at influencing the actors' internal structure, especially their knowledge, goals, and behaviour programmes. We distinguish between communication instruments that contain a direct request to adopt certain views, goals, or behaviours, and instruments that contain no such direct request. Diffusion instruments are used to "transport" an instrument to as many actors in the target groups as possible.

In ordering and describing communication and diffusion instruments, we have taken into account that here – in contrast to the other types of instruments – there are far fewer established policy instruments to be found and categorised. Moreover, some of the usual names given to communication instruments, such as "information campaign" or "advisory service", actually refer to a variety of instrument functions. Depending upon their purpose and design, for example, information campaigns and advisory services may simply distribute facts or also contain options, appraisals, and requests. The effects produced will vary accordingly. The decisive factor used for the typology was not the concrete design of the instruments as measures (e.g., courses and other educational offerings, advice, network formation, pilot and demonstration facilities or projects, information campaigns, and so on) and not the details of their physical and formal realisation. It is, after all, not crucial whether paper or e-mail serves as the medium. What is decisive are the *functions* that are thus put into effect. For these reasons, we settled on a representation that is systematic, functional, and grounded in social psychology (cf. Mosler and Gutscher, 1998). The advantage is that not only can tools be positioned that are already familiar in the world of practice, but also that functional gaps in an existing combination of instruments can be revealed. The form chosen for the typology thus also serves the needed goal of stimulating social creativity in policy strategy planning and in the development of new instruments.

3.5.1. Communication instruments without a direct request

Presenting facts: Facts are presented by making verbal statements about reality without presenting any appraisals and by presenting visual or auditory images (such as pictures of deforested areas in Brazil, sound recordings of the traffic noise along the Brenner motorway) and pieces of evidence (for example, an illegally shot lynx, a species protected by law in Switzerland). Since statements and images of reality are not the reality itself, they can be true or false. To deliberately make false statements about people (to slander) or non-social aspects of the world (to claim that CO_2 emissions have no effect on the global climate) or to show fake images is a powerful means of social influence. For promoting sustainable development, of course, the only ethically justified statements are statements of specific, true facts.

Presenting facts influences the target persons' knowledge. This serves not only the purpose of mediating new knowledge, e.g., about ecological phenomena, such as the destruction of the ozone layer, but also of destroying false knowledge, e.g., about the seeming harmlessness of certain everyday behaviours, such as throwing batteries into household trash. Knowledge can provide target persons with new action options. Knowledge also has an effect upon individual mental skills. Facts can be presented by using a wide variety of diffusion instruments (see section 3.5.3). Examples: Newspaper reports about the melting of the polar ice caps and its causes; school lessons about the principles of climatology; publication of violations of air pollution impact thresholds.

Although the presentation of facts is a communication instrument that can be used on its own, its psychological effects (see the chapter by FLURY-KLEUBLER AND GUTSCHER) are not limited to modifications of knowledge, but also include the experience of appraisals, especially emotions. That is, the mere presentation of facts can evoke emotions, since the facts presented are evaluated by the target persons in relation to their pre-existing goals or – in other words – their values. Contents that are already associated in the target person to a certain emotion can trigger that emotion. If this happens simultaneously with the presentation of new contents, these contents will in the future be associated with the emotions triggered. Thus, appraisal of the facts presented can be influenced.

Presenting options: Not only factual reality can be presented, but also options, or action possibilities and possible future states. Presenting how reality could be, if we so desired, and what we could do to realise it, is one of the most important ways of influencing the thought and action of others. Examples: Describing scenarios of the development of road traffic depending on different economic, political, social, or technological decisions; pointing out how household waste can be reduced or energy can be saved in the private home.

Presenting appraisals, goals, and norms: In general, the presentation of facts or options is accompanied by appraisals. Every appraisal refers to a certain mental content and implies **goals**, since it is always at least implicitly related to a goal. Besides knowledge, goals are a second key target of communication instruments. Like influencing knowledge, influencing goals also has an effect upon behaviour, because goals have both an *appraisal* function and a *motivation* function. Therefore, they influence how we interpret stimuli and how we act in a situation. If goals are very general and (relatively) stable in time, we can call them *values*. If they refer to human mental activity or behaviour, it is common to call them norms. We have to distinguish *prescriptive* and *descriptive* norms. The first refer to what thought or behaviour is supposed to be, the latter to how performance really is. Norms shared by several individuals are social norms.

Appraisals can be expressed and experienced in two forms. First, they can be experienced as verbal concepts. Second, they can have the phenomenal quality of *emotions*. Within a target person, the presentation of appraisals can trigger both verbal concepts and emotions. Communication instruments can allow people to categorise certain facts using verbal labels. But they can also trigger emotions; they can evoke fear or joy. Fear appeals are among the most efficient communication instruments, if properly designed. They should not only trigger fear, but also provide concrete, detailed, available, and efficient coping actions (Leventhal, 1970: 157f.). In addition, overly intense negative or positive emotions disturb the ability to process information systematically (Jepson and Chaiken, 1990; Smith and Mackie, 2000: 277ff.). Appraisals are communicated not only verbally, but also by associating music or other non-verbal signs with the facts or options presented. This provides immediate clues about the relevance and positive or negative valence of an event. The potential in music is widely used, such as for propagating religious ideas, but also in advertising (cf. Zimbardo and Leippe, 1991: 250-251). Examples: A television commercial can accompany the scene of a person travelling by train with a musical sound track that is expected to trigger pleasant emotions in the target groups. In a leaflet, an appeal to avoid wearing fur products can be accompanied by photographs of suffering animals.

Communication instruments can activate pre-existing **norms** to strengthen their action-guiding effect. They can also transform norms or even establish new ones (Cialdini et al., 1990; Reno et al., 1993; Hopper and Nielsen, 1991). By presenting appraisals, everybody can try to set and communicate prescriptive social norms. One way of expressing them is to claim that many, most, or at least relevant people behave in a certain manner, that is, to claim the existence of a certain descriptive social norm. This means is quite popular in advertising, but also in media reports. Examples: Setting thermostats to 20 degrees C in sitting rooms and to 18 degrees in bedrooms; an advertisement with the message that "brains take trains"; ethical codices of specific professional groups or economic branches.

In the context of setting and communicating norms, it is also important to make sure that certain norm contents are *not* set or communicated. Television stations, web sites, or political communicators should not set or communicate norms that are hazardous in content. To prevent this, communication instruments can be used to a certain extent. However, if the content of a certain norm is extremely hazardous, this is also a field of command and control instruments. For instance, a topic worth discussing is whether the glorification of dangerous driving in movies should be prohibited.

Providing experience of reality: The presentation of facts can be designed so as to involve all the senses and give target persons a comprehensive experience of a certain piece of reality. One means of making the experience of realities that are to be communicated more intensive consists in moving into the situation that is the object of communication. One can go with people to the environments and objects one would like to show them. The providing of experiences of reality is also accompanied by presenting appraisals. It is often a combination of a communication instrument and a service instrument (see section 3.3), since, e.g., the moving into a situation and the guiding of people are services. Examples: A wilderness camping tour; a guided excursion to a slum, a waste dump, or a polluted beach; an open house in a low-energy house or in a wind power plant; enabling "experimental" realities in all sorts of pilot and demonstration projects (see Example 10).

Example 10:
Light electromobiles pilot and demonstration programme

New technologies for energy utilisation aim to reduce the damage to the environment that arises from motor vehicles. Light electromobiles (LEM) and vehicles with low emission rates, such as hybrid cars and cars running on natural gas, can become important elements in a future traffic concept that accents choosing the most energy-efficient vehicle for the particular need. The "light electromobile" pilot and demonstration programme was started in Switzerland in 1992. Within the programme, a large-scale trial was started in 1994. This large-scale trial with LEM in Mendrisio and partner communities was designed to test the effects of light electromobiles on traffic and on the environment and to examine any issues pertinent to their introduction on the market. At the official opening of the trial on 23 June 1995, Federal Councillor Adolf Ogi outlined its three main goals:

- Demonstrate and evaluate the appropriate implementation of LEM
- Test individual and combined measures to promote the introduction of LEM on the market
- Demonstrate the role of LEM as elements of an environmentally friendly and forward-looking mobility concept.

Results of an accompanying study show that the large-scale trial was highly successful with regard to raising people's awareness and knowledge about this new technology. The findings reveal clearly that with increasing proximity to LEM projects, people's knowledge and evaluation of the new technology increase positively (Ulli-Beer and Haefeli, 1999).

Presenting model behaviour: Whenever human individuals are perceivable by others or when people report about human behaviour, model learning can take place, independently of the model's intention to influence others.[11] When seeing, hearing, or reading about the behaviour of another person, we can learn about the person's situation, the goals he or she pursues, what means he or she uses, and what consequences the use of those means in that situation has. Model behaviour influences our perceived descriptive norms and thus indirectly our prescriptive norms. Presenting model behaviour can be used deliberately as a means of planned social influence. To influence norms, a model person can show desired behaviour directly in everyday life, on television, posters, or in other media. Both real and fictitious persons or even animals (in comics and cartoons) can serve as models. The higher the status and charisma a person has – in the subjective experience of the model learner – the more likely it is that the model behaviour will be performed by the learner. A large number of models has a stronger effect than a small one. What models have the greatest impact depends on the target group. The target group determines who has high status and charisma, who is trustworthy and thus norm-defining (Wagstaff and Wilson, 1988; Winett et al., 1982; Winett et al., 1985). Example: Famous athletes who travel by train or bus instead of by car or plane whenever possible.

Giving feedback and enabling self-feedback: Giving target persons, groups, or organisations feedback about their individual or collective behaviour and/or its environmental impact can change their future behaviour. The more individual the feedback is and the more immediately it follows the behaviour, the more effective it is. This instrument can only work by referring to norms that are already internalised by the target groups. If the members of a household are informed about their consumption of electricity or water during a certain period of time and about the cost, their consumption may change. The success of this instrument depends on a fast, regular, specific, and trustworthy feedback (Midden et al., 1983; Pallak and Cummings, 1976; Pallak et al., 1980; Rothstein, 1980; Mosler and Gutscher, 1998).

Feedback is not just a way of confronting a person with contradictions between goals and behaviour (compare section 3.5.2.), but it helps the actor to keep the goal in sight and to be able to react quickly to discrepancies between goals and reality. Feedback information is also efficacy information about individual or collective behaviour. Self-feedback is based on a person's own behaviour records, such as diaries about daily travel behaviour (times and places of start, destinations, distances, and times of arrival). Examples: Fuel consumption display in the car; energy consumption account for a certain time period; driver's logbook; "CO_2 calculator", a model for determining personal CO_2 emission and energy consumption.[12]

[11] Note that model learning is a broader term than Bandura's (1986) notion of observational learning, which does not include language-bound learning about behaviour experiences of others.

[12] An example can be found online in the WWW at <http://CLEAR.eawag.ch/models/index.html>.

3.5.2. Communication instruments with direct request

Persuasion is the attempt to influence another person's beliefs about factual reality, possible reality, and options for action, and about appraisals, goals, and norms. In addition to presenting certain contents, the communicator accompanies the presentation with the more or less overt demand to adopt the presented contents.

In the case of **persuading about facts**, the communicator attempts to persuade somebody of the *truth* of certain facts. Persuasion is used when target persons are not automatically willing to believe the presented facts. For instance, a minority of scientists tries to persuade people that human release of carbon dioxide is not a relevant cause of global warming.

When persuading someone about **options**, the communicator tries to present a possibility not only as a theoretically imaginable reality, but also as a realistic option. The objective is to either show that a certain desirable possibility can be achieved realistically, or to persuade the target person that we risk a specific undesirable development or outcome becoming reality. This implies that persuasion about desired or undesired possibilities has to be combined with the pointing out of behaviour options that can promote or prevent these possibilities.

To persuade people that something is the case or might be the case is one thing. To persuade people about what ought to be the case – i.e., **persuading about goals, appraisals and norms** – is another thing. If people have learned that releasing greenhouse gases leads to global warming, they also have to be persuaded that global warming is hazardous, and that one has to take measures to reduce greenhouse gas emission. Since goals affect appraisals, persuading people successfully about goals will have an effect upon their appraisals. On the other hand, communicators can also try to persuade people about appraisals. They can try to persuade people that something has to be appraised in a certain way, that something is good or bad, right or wrong, important or irrelevant.

In persuading about facts, possibilities, goals, or appraisals, the success chances of a communicator depend – among other things – on his or her trustworthiness. The communicator can, but need not, be a famous person. Neighbours and other familiar persons can also be persuaders. Success depends also on direct personal contact, which is much more effective than written persuasion attempts. Communicators can be trained to use vivid language, to use information that is relevant for the addressee, and to involve him into the conversation (Burn, 1991; Burn and Oskamp, 1986; Gonzales et al., 1988; Hopper and Nielsen, 1991). Persuasion is only possible if the arguments used are compatible with at least some of the values that the target person already has. Examples: Information exchange within GAP groups (see the chapters by BRUPPACHER AND ULLI-BEER and by VATTER ET AL.); television commercials; "ambassadors" in the 30-kilometers-per-hour campaign in Münsingen (see the chapter by GUTSCHER ET AL.).

In order to persuade people of the need to change goals it may be helpful to confront them with contradictions between a goal they pursue and their behaviour. Such discrepancies are experienced as unpleasant, so that people are motivated to resolve them through either a change in attitude or a change in behaviour (Kantola et. al., 1984). However, it is important not to build up too much pressure and to leave enough face-saving opportunities. Otherwise target persons are likely to react by rationalising and playing down the conflict, so that the desired behaviour change does not occur. Examples: Persuasion about the fact that road traffic is still increasing; persuasion about the option to replace a considerable share of road traffic by rail transport or to avoid it by new modes of organising work (e.g., telecommuting); persuasion about the appraisal that a further increase of road traffic is undesirable and about the goal that it should be avoided by specific means, such as higher fuel prices and better rail infrastructure.

Sending appeals: An appeal is a general request to behave in a certain way. In contrast to "prompts", these messages are not delivered at the very place of action. Appeals can be expressed with or without arguments. Sending an appeal to do or not to do something is actually the same as persuading somebody about a behaviour related goal. Examples: Summer smog poster campaign in Bern in 1996: "Drive at low engine revolutions or leave your car at home".

Example 11:
Stimulating self-commitment as a means against traffic jams

In this study the task was to reduce, during a special action week, morning and evening traffic congestion that regularly formed at the Baregg highway tunnel near Zurich. Traffic back-ups result from local and temporal overuse of the common pool resource "traffic space". By means of a publicity campaign and public collection of signed statements of self-commitment from a core group of about 18,000 regular commuters it was attempted to establish a kind of rotation system in avoiding travel at peak traffic times. The main intervention instrument consisted of public self-commitments, in writing, that were available through newspaper advertisements and the Internet. To support diffusion and as a reminder during the experimental week, drivers were also encouraged to display their participation in the form of bumper stickers on their vehicles.Compared to the baseline week, the total reduction of traffic congestion amounted to 10% or two hours respectively. Per peak hour in the morning, 100 cars travelled at other, less busy times. The study demonstrated that the contributions in the form of signed self-commitments were in fact effective and that the statistical levelling – regular distribution of the uncoordinated contributions to reduce traffic during the periods of heavy traffic – functioned reliably. The publicity and public acceptance of the principles of this campaign based on voluntary contributions were very large in scale (Gutscher et al., 2000).

Presenting prompts and reminders: Prompts and reminders tell target persons what behaviour they are expected to show. The medium for these messages can be signs, stickers, posters, flyers, brochures, oral speech, and so on. It is important that the messages are placed where the behaviour takes place or is supposed to take place (Aronson and O'Leary, 1983; Baltes and Haywood, 1976; Hopper and Nielsen, 1991). These instruments activate norms that already exist in the target individuals. They are reminders about commitments or earlier intentions. Examples: "Turn the motor off at red lights", "Turn the computer screen off before your coffee break".

Stimulating self-commitment: Self-commitment means that people oblige themselves to behave in a certain way. It is expected to make an individual's environmental goals conscious and thus behaviour-guiding. Self-commitment is probably one of the most efficient instruments with regard to fostering a desired behaviour (Bachmann and Katzev, 1982; Pardini and Katzev, 1983-1984; Wang and Katzev, 1990). Commitments can be practised publicly or privately. Public self-commitment involves communicating the commitment made to the public, such as by means of an advertisement or a sticker. Example: The director of a company commits himself to travel to work by public transportation for a month and publishes this in the company magazine (at the same time, this is an example of model behaviour); public commitment of a large retail trade company that 75% of the transports from distribution centres to the shops will take place by train (see Example 11).

3.5.3. Diffusion instruments

The diffusion instruments described here are used primarily for the first step of initial diffusion (see section 2.5. above and the chapter by FLURY-KLEUBLER AND GUTSCHER). We order diffusion instruments according to whether diffusion takes place in direct personal encounters, via person-to-person media, or via the mass media.

Diffusion instruments are in fact the "shipping containers" to transport communication instruments, or their mechanisms. As a rule, they require a "transmitter", or *communicator*. These are usually real persons, or where diffusion occurs in written form, organisations or institutions. The actors responsible for the intervention are not always the transmitters or communicators. The responsible parties may employ the help of transmitting communicators or so-called "multipliers".

Multipliers can be recruited to use their social contacts or social status to support the goals of a planned intervention. No matter how contact is established – direct personal contact through person-to-person media or contact through mass media – multipliers bring communication instruments into effect with single individuals or several persons at a time by means of monologues or dialogues or through offering directly observable model behaviour. (Multipliers may, for example, talk directly with acquaintances, others at work, fellow club members, or they may give interviews in a local newspaper.)

The effectiveness of multipliers can be increased if they receive training or some kind of information support. If there is a need to systematically contact large numbers of strangers, multipliers generally receive training and are paid for their work. For multipliers working in this manner, the term *activators* has also been used in the literature (Gonzales et al., 1988). The efforts of multipliers can also take on the form of low-threshold advice, whereby the driving force of the diffusion of new behaviours takes place through active consulting (examples: energy use advice, advice on composting).

Direct personal diffusion and diffusion based upon person-to-person media make use of the personal relationship of the transmitter, or communicator, to the receiver, or recipient. It is important to take advantage of existing contacts within the multiplier's social network, so that recipients are contacted by familiar communicators. Familiarity also plays a role when communicators appear in the mass media; people are receptive to messages from well-known and well-respected communicators, and in this way, target groups are reached more easily and efficiently.

Establishing direct personal contact: In the case of non-media communication, diffusion typically occurs face to face. In this setting *monologues* play a part in the form of announcements, lectures, speeches, or theatre pieces. Forms of *dialogue* may include anything from a conversation between two persons to group discussions or discussions with moderators – among other things, for example, in the framework of *participatory planning processes* or various forms of initiating *public discourse* (for example, Round Tables, Public Discussion Fora, or Local Agenda 21 Processes.) Social changes can not simply be "put through". Social change has to go down the tedious and risk-laden path of democratic, participatory, and argumentative procedures. "Solutions" to environmental problems require adaptations and modifications that people understand, accept, and carry out. This is the reason why it is absolutely essential that people from all walks of life, through extensive dialogue, reach a common determination of the goals and contents (see Example 12).

Besides face to face settings in a narrow sense, there is another type of non-media diffusion, whereby organised *models* carry out certain model actions publicly (highly visibly to others).

A further special form that is based for the most part on non-media diffusion are so-called *diffusion tasks* (Prose and Wortmann 1992; Prose et al., 1994; Mosler and Gutscher, 1998 and Mosler and Gutscher, in prep.). Here a number of target persons, usually simultaneously, are asked or instructed to pass along a certain object to friends and acquaintances. The object passed along may be, for example, a brochure that presents facts, interpretations, and an appeal. Diffusion as a rule then takes place as a chain reaction; if the instructions include photocopying the object, the spreading may take the shape of an avalanche.

The *initiation, building, and management of networks* are also based mainly on non-media contacts, or sometimes on contacts via person-to-person media. Here there is the additional goal that individuals or corporate actors in a chain or network introduce themselves and get to know each other. In this way, a virtual collective is formed, which facilitates the exchange of information and experience (see the chapter by BÄTTIG AND BALTHASAR on network management within the Swiss federal government's Energy 2000 Action Programme). Networks can also support the building of special consulting and continuing education offerings that individual members could not afford to pay for on their own. Networks can have further synergy effects, such as in shared financing of demonstration projects, the development of employee training programmes, public campaigns, and so on. Similarly, networks can combine the political efforts of their members and thus increase their effectiveness.

Establishing contact via person-to-person media: Diffusion can also be designed using person-to-person media; most important are letters, telephone conversations, faxes, or e-mail. We are already seeing newer forms of person-to-person media today, such as the popular Short Message Service on mobile phones. Many other new possibilities for audio-visual media communication will be available in the near future. Common to them all is the fact that they are used generally by individuals to communicate with other individuals. The transition to mass media is fluid, however; fax, letter, and e-mail can also be used to contact a large number of people. In principle, person-to-person media are suited to facilitating a range of forms of communication, from both monologues and dialogues to moderated or non-moderated discussion forums or mail groups.

Establishing contact via mass media: The World Wide Web offers innumerable variants of contacting target groups. Its technologies also allow for relative spontaneous, more personal forms of dialogue, through e-mail, for example. The possibility to implement communication instruments via the Web are virtually unlimited: NGOs, environmental agencies, universities, research groups, and private persons present a large body of facts, options, appraisals, model behaviours, and action possibilities as well as various forms of persuasion and appeals. However, as the Internet does not as a rule represent a push technology, as a first step diffusion of the offerings is needed, either through Web portals or via conventional mass media.

Example 12:
Paragliders and wildlife

Paragliding has become very popular since the late 1980s (in Switzerland, 10,000 pilots were trained in a five-year period). Ethological research has shown that in some areas, wild animals, especially ibex, do not adapt to the paragliders: they flee every time a paraglider flies overhead, and they end up spending their days in wooded areas at lower elevations. Here they do not find appropriate fodder and instead start to eat young trees. Also, observations show that paragliders disturb some rare eagle species during breeding periods (see Ingold, 1999).

Alarmed by these findings, the national wildlife protection agency set up a national task force for in-depth investigation of the problem. At the local level, the authorities started a participatory problem-solving procedure involving all relevant local actor groups.

This group made a careful analysis of the local problem situation, bringing in not only the point of view of the wildlife specialists, but also the perspectives of the paragliders, farmers, tourist managers, and others. This led to solutions that were adapted to the specific local context and acceptable to both wildlife protectionists and paragliders. Agreements were made on certain rules that paragliders will observe on a voluntary basis (e.g., to avoid flying over specific areas during certain periods). Control is exercised by the local authorities, and as long as the rules are observed, they will forego decreeing legal regulations (for details on the procedure and its effects see Ingold, 1999).

Television or radio shows, newspapers, and books are conventional mass media that serve well as vehicles of communication instruments. They include editorial reports that base on research or on media conferences, but also paid advertising in the form of commercials or other ads. Printed products comprise brochures, flyers, and the like; they continue to carry the heaviest load in traditional information communications. Examples: Leaflet for people who have moved to a new town and who do not yet know what to do with different kinds of trash, such as recyclable glass, paper, or used batteries.

There are various technical possibilities for distributing printed products: they can be distributed by people (common for flyers and newspapers) or by mail, or be available in dispensers. Dispensers refer to all possible physical arrangements that serve to offer, or distribute, communication instruments that base mainly upon printed products.

Diffusion by means of films at cinemas or other image and sound recordings represent another type of mass media diffusion. The poster is a prototype of an information carrier that uses pictures and language in all sorts of designs. Dispensers, in addition to facilitating distribution, can also take on mass media functions. Through their design, they can attract attention and present appeals. This makes dispensers themselves vehicles of communication instruments, and they fulfil the same tasks as the poster.

Various goods produced by people, whether through mass or unique production, are also possible vehicles of communication instruments. At present, this potential is still being utilised mainly for image advertising and for firmly establishing brand names.

There is another aspect of contacting that plays a role in diffusion via the mass media: the *degree of specificity of addressing*. Contact can be established *personally* (through personally addressed direct mailings, for example) or *impersonally* (anonymous mailings, television and radio shows, printed materials, and so on). Here the issue is the extent to which people feel that the communication is personally addressed to them and that they are obligated to give the matter attention. Certainly, personal address is preferable, such as when all residents of a town receive a letter. If contacting is impersonal, a certain degree of control in reaching a target group is possible through positioning the message at sites and places assigned to particular persons (mailboxes, for example) or through the use of target-group specific media channels. If this is not possible, it is up to individuals in the target group themselves to attend to the message (self-selection). In such cases, efforts are made to tailor-design the communication instruments to appeal to a specific target group. Gaining the attention of a target group is thus an issue for diffusion instruments (certain diffusion instruments attract more attention than others) and also a question of the design of communication instruments.

3.5.4. General remarks on the application of communication and diffusion instruments

The various diffusion strategies have advantages and disadvantages: Diffusion via direct personal contact or person-to-person media requires enormous efforts, but it guarantees targeted access to difficult to reach persons or corporate actors. Diffusion via the mass media suffers the most from the problem of loss of spread.

In the work of diffusion, therefore, it seems generally advantageous to use the most familiar persons as possible as communicators and to address the target group in the most specific and personal way possible. The scope of the present discussion does not allow us to go into the details of further dimensions of communicators, such as expert status, competency, charisma, credibility, trustworthiness, engagement, similarity, physical attractiveness, and sociometric status (O'Keefe, 1990).

Table 2 shows the basics that are required of diffusion and communication instruments (compare the chapter by FLURY-KLEUBLER AND GUTSCHER; Kok and Siero, 1985; Zimbardo and Leippe, 1991: 137). In a first step, diffusion instruments guarantee efficient and rapid "transmission" of the intervention to the individuals in the target group and secure their initial attention. Widespread norms of interaction, or politeness, guarantee that diffusion through personal contact, or through person-to-person media, will gain at least a minimum of initial attention. The continued success of diffusion depends upon the design of the communication instruments. Design involves formulating the topic and/or "packaging" the topic in a way that attention is sustained. The content of the communications must be understandable; that is, it must be adapted to knowledge and language of the target group. In order to reach the necessary level of acceptance of the content and conclusions, communications have to base upon the target group's existing structure of values and norms. Content and consequences also have to be able to become anchored in memory as well as influence the forming of intentions. Finally, and most important, intentions have to be expressed in actual new and changed behaviours, whereby stabilisation of a new behaviour will place additional requirements on communication instruments.

Table 2: Requirements of diffusion and communication instruments.

Type of instrument	Required
Diffusion instruments	**Exposure and initial attention:** The instrument must gain the attention of all individuals in the target group and thus expose them to the effects of the communication instrument.
Communication instruments	**Attention:** The instrument must direct target persons' attention to its content and goals.
	Comprehension: The content and goals communicated by the instrument have to be understandable.
	Acceptance: The content and goals communicated have to be able to be accepted, that is, receive a positive evaluation.
	Retention: The new content and goals must become anchored in memory.
	Intention: The new content and goals should lead to the forming of intentions.
	Behaviour: The new goals should be expressed in behaviour in a successful and lasting way.

The requirements can not always be fulfilled through a single combination of a communication instrument and a diffusion instrument. As a rule, a mixture of different, complementary combinations – in temporal succession – is implemented. In designing the communication instruments it is very important to have knowledge of the values, norms, knowledge, and acceptance conditions of particular contents in various segments of the population (compare the chapter by VATTER ET AL.). Moreover, it is necessary to be aware of the contents and main thrust of communications that are currently being communicated by other sources. When using communication instruments, we have to reckon with the fact that a large number of other actors are using such instruments as well. The effects of all these attempts at social influence interact with each other. So it is not unlikely that a sole caller in the desert remains unheard.

In order to maximise the probability that particular segments of the population will be exposed to the effects of a certain intervention, costly combinations or repeated strategies may be required. In order to select a diffusion instrument and design it in concrete form, it is therefore important to have information about the best way to reach particular target groups. Where this information is lacking, it should be generated by preparatory studies, such as surveys or other means of collecting data. Only then can decisions be made on whether differing, target group specific variants of the communication instruments should be implemented; whether in the framework of certain strategy goals it is sufficient to address only a particular segment of the population; or whether blanket, non-selective diffusion of a single communication instrument will suffice.

4. Comparison and preliminary evaluation of the instrument types

As stated above, the typology of instruments that we present should help us to look at classic policy instruments as well as new instruments from an *actor-oriented perspective,* and it should allow instruments to be evaluated with respect to issues of *practical interest* (effectiveness, costs, and acceptance). We will now try to make a preliminary comparison and evaluation of the instrument types with these goals in mind.

Looking at policy instruments from an actor-oriented perspective means that one has to refer to a general conceptual framework or model of human action. Such a conceptual framework was outlined in the Introduction chapter by KAUFMANN-HAYOZ AND GUTSCHER (see Figure 1, page 23). The primary targets of the different types of instruments can be explained according to this conceptual framework. In Figure 1 below we have visualised them by adding the instrument types to the model of human action. Of course, only the *primary* points of attack are shown in the figure. Obviously, if an instrument achieves primary changes in internal or external structures, this may have a number of secondary effects, since the actors respond and adapt to these changes.

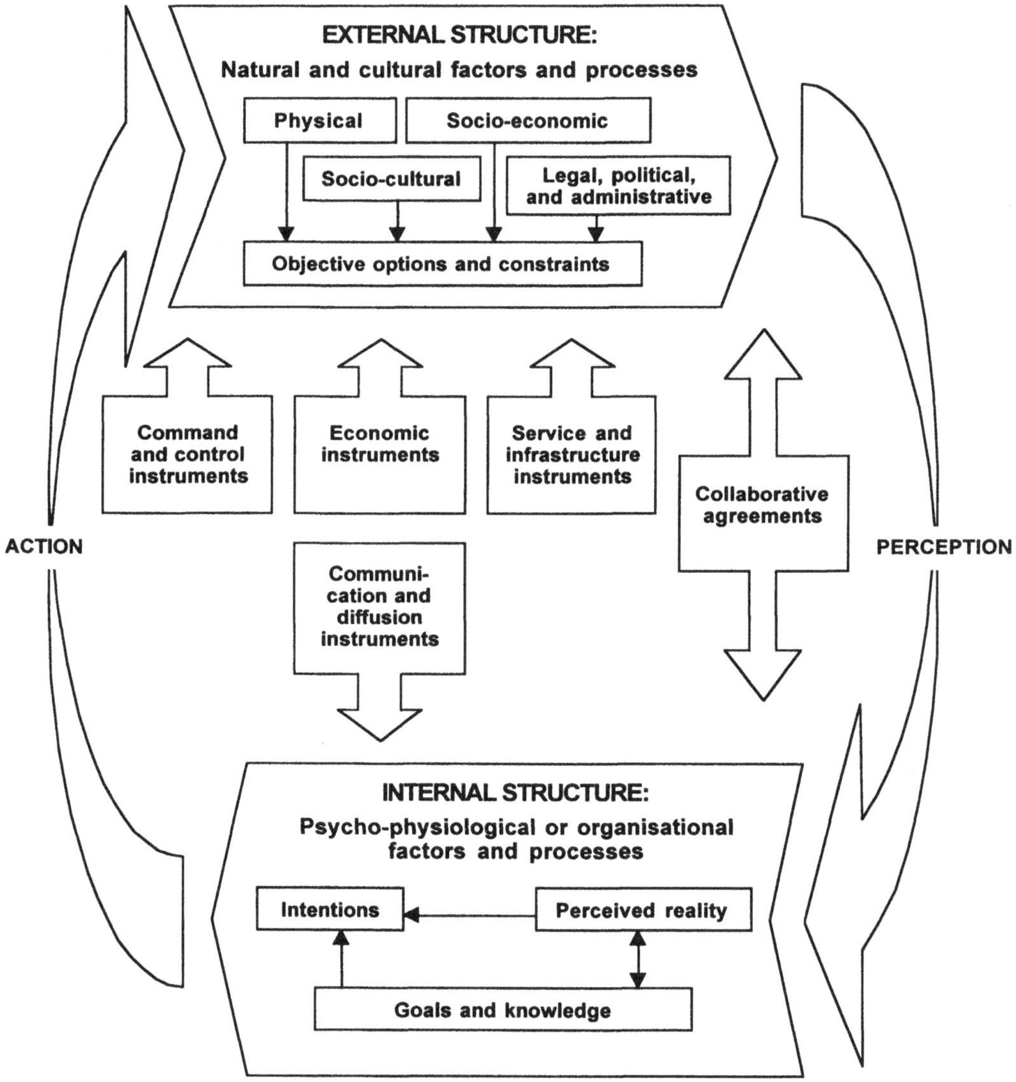

Figure 1: Primary targets of the different types of instruments.

Does an instrument aim primarily at modifying the individual or organisational actors' internal structures (their goals and their decisions about how to achieve these goals), or does it target the external structure (the physical, cultural, socio-economic, and political, legal, and administrative framework of action)? The first three types – command and control instruments, economic instruments, and service and infrastructure instruments – clearly aim primarily at modifying different aspects of the external structure: the political, legal, and administrative, the socio-economic, and the physical environment. Most command-and-control instruments as well as many service and infrastructure instruments directly restrict agents' options, while economic instruments and some service and infrastructure instruments make environmentally desirable actions more attractive or rewarding and undesirable actions less attractive. In contrast, most communication and diffusion instruments clearly aim at influencing the actors' internal structures. They attempt to influence the knowledge and goals of individuals and organisations and their respective ways of achieving them as well as to affect their perceptions and appraisals of the social and physical reality. The *design* or *re-design of institutions* is initially based on the interaction of individual and organisational actors, and communication and diffusion instruments predominate. However, every one of the instruments can in principle support the concrete, blanket administration of institutional arrangements. Some types of collaborative agreements aim at influencing the goals and norms that companies set for themselves and can therefore be regarded as influencing the internal structure of companies. Others influence primarily the economic and legal-administrative framework of action and the relations between companies and other actors (especially public authorities and NGOs).

We will now try to discover further differences and similarities among the five types of instruments by answering the following seven questions. Table 3 (page 90f.) summarises the questions and the main elements of the answers.

- Why is the instrument effective?
- Who can apply it in the main?
- What are the most important target groups?
- What can be said about its effectiveness?
- What innovations does it promote?
- How well is it accepted?
- What effort and costs does it involve?

4.1. *Why is the instrument effective?*

The assumptions and hypotheses about the actors' responses to the introduction of an instrument constitute its *rationale*. This explains the mechanisms according to which an instrument is supposed to work, i.e., to have an effect on the target group's goals and their ways of achieving them. The basic rationales of the different types of instruments were described above (sections 2.1 through 2.5), and they are summarised in the first line of Table 3. It can be seen that the different types of instruments are supposed to influence the actors' behaviour for different reasons. While command and control instruments are believed to have an effect because people want to avoid the penalties associated with the violation of the prescriptions, economic instruments are supposed to steer behaviour because the actors make decisions according to economic rationality. Service and infrastructure instruments have an effect by making certain behaviours possible, salient, or attractive, and they exclude others or make them unattractive. Collaborative agreements work by generating – through negotiations – a more or less binding commitment of (collective) actors to behave in a certain way. Finally, communication instruments influence people's values and knowledge, which are assumed to be important determinants of intentions and the choice of goals.

4.2. Who can apply it in the main?

Public authorities can implement all instruments, whereas private actors can apply only some of them. In Western democracies, only legislative authorities are legitimised to use command and control instruments, and only they can create the legal basis necessary for most economic instruments and for many important infrastructure instruments.

4.3. What are the main target groups?

With respect to the target groups, there are no clear distinctions among the five types of instruments. All of them – except collaborative agreements – may be directed to collective actors, especially companies, as well as to individuals or unorganised target groups. In collaborative agreements, the companies or associations involved are at the same time actors and target groups. The distinction between the actors who apply the instrument and the target group disappears. The same applies to some communication and diffusion instruments: In a participatory planning process, for example, the public authority may be the actor who initiates the process, but at the same time it is also part of the target group.

4.4. What can be said about its effectiveness?

In the literature on policy evaluation, a distinction is often made between *impact* and *outcome*. When evaluating the impact, we ask whether or not the application of an instrument has the intended effects on the behaviour of the target group. When evaluating the outcome, we examine all direct and indirect, intended and unintended effects as a result of or in addition to the consequences of the target group's behaviour change. In the case of environmental policy, the effects on the environment are of course among the most important outcomes. Two essential aspects of evaluation are, therefore, (a) the scale according to which the effects are measured, and (b) how large the target group is that has been impacted by the measure. An electricity saving campaign in an office building may, for instance, have a strong impact, because all employees change their habits and switch off computers and light bulbs whenever they are not in use. As an outcome, the electricity consumed by these appliances in this particular building may be reduced by 30%. However, this may be only a fraction of the building's total electricity consumption, and only one out of 1,000 companies may run such a campaign, hence the ecological outcome of the instrument on a national scale, for example, will be judged quite small. In the following, we make some rough, preliminary comments on what is known about the effects of the different types of instruments.

4.4.1. Command and control instruments

Environmental policy in Switzerland and elsewhere is still essentially based on command and control instruments, and the positive effects are acknowledged (see, e.g., Jänicke and Weidner, 1995). A main advantage is that their effectiveness is highly foreseeable. Economists like Vallender (1990) or Kirchgässner (1997) emphasise that these instruments are successful under certain conditions: "Successes are achieved where standards or prohibitions can easily be implemented and where they are effective for regionally limited problems" (Kirchgässner, 1997: 185). Nonetheless, there is a general consensus that traditional environmental policy has somewhat reached its limits (Kirchgässner, 1997: 187). The main drawback of command and control instruments from an economic standpoint is their rigidity, which may hinder the efficiency and effectiveness of environmental policies.

Basically, command and control instruments make the same prescription for every member of the target group. Taking the example of emission standards for industries, every company has to limit the emission to the same level, regardless of the individual cost. It would be more efficient, however, if companies with lower expenses for reducing emissions would observe lower emission levels, whereas firms with higher costs would be allowed to pollute more. Moreover, under a regulatory regime, companies that already comply with the emission standards do not have any incentive to invest in further improvements. Theoretically, command and control instruments could be tailored according to the conditions of different target groups. However, this approach is exposed to the lobbying of powerful interest groups and demands a lot of resources and technical expertise within the public administration (Kirchgässner, 1997). In domains with rapid technological change, such instruments are not recommended, whereas prescriptions of a technologically up-to-date standard can stimulate the diffusion of an innovation and have significant secondary economic benefits (Balthasar and Knöpfel, 1994).

Command and control instruments apply – in theory – uniformly to everybody, independently of individual economic strength. But this does not imply that everyone is affected alike; command and control instruments are not a priori compatible with *social* objectives. They can, however, support or encourage the application of other types of instruments. Existing legal norms can accelerate the development of economic instruments, of collaborative agreements, of service and infrastructure, and of communication instruments. For example, the decision for a ten-year nuclear moratorium in Switzerland has fostered an expansion of the federal energy action programme Energy 2000.

4.4.2. Economic instruments

The impacts of economic instruments on the behaviour of target groups can be assessed for various time ranges. If, for example, a tax on non-renewable energy is introduced, a *short-term* effect may be that the existing infrastructure is used less extensively and that energy waste is being avoided. In the *medium term,* investments in more energy efficient infrastructure or in appliances powered by renewable energy will be made. In the *long term,* research and development of such alternative techniques and appliances is stimulated. When considering the possible introduction of economic instruments (such as a tax on carbon dioxide), many countries have used modelling and forecasting techniques to investigate their likely costs and benefits. But even if the number or practical applications of economic instruments is growing, there is still very little systematic evidence about the performance of such instruments in practice. This is due first of all to the fact that experiences are too recent, and some of the effects take quite a long time to become evident. Thus there is little empirical evidence about the impact on technological change. Second, suitable data on emission sources is often lacking. Finally, there is an important conceptual difficulty in evaluation: the need to define the alternative (What would have happened without this instrument?). Despite these uncertainties, economic instruments now enjoy the reputation of "proven successes in reducing pollution at low cost" (Stavins, 2000) and they have moved into the centre of political debate. This debate is quite different from the controversy in earlier times, when these concepts were considered to be "licenses to pollute". In 1997 the OECD concluded that "greater efforts to evaluate the efficiency and effectiveness of economic instruments actual in practice would contribute greatly to improve future policy" (OECD, 1997b: 132).

4.4.3. Service and infrastructure instruments

In general, if environmentally sound action is unpleasant or even risky due to infrastructure and service conditions, only idealists and people who can not afford the more pleasant alternatives choose the environmentally preferable alternative. In a country or area with poor public transportation, only persons who do not have access to a car will use it. If environmentally sound food is only available in a few small shops far away from where the people do their everyday shopping, only few people will buy it (and those who do may have to drive their private cars to get it...).

The connection between infrastructures and services that enable or encourage environmentally undesirable behaviours and the frequency of these behaviours is obvious from past developments. Consider the following examples: Water consumption per capita rose at least ten times after the installation of running water in houses and flats, particularly after the introduction of the flush toilet. Traffic volume usually grows with the increasing supply and quality of roads. Shopping centres and leisure facilities at the periphery of cities attract motor traffic, and the availability of parking spaces at the destination is one of the major determinants of the choice to use a private vehicle. On the other hand, an attractive and safe net of bike trails in cities encourages the use of bicycles as a means of transportation. It has also been shown that the design of roads influences the driving style of car drivers: If a road affords slow and steady driving instead of a stop-and-go pattern, traffic becomes more fluent, and emissions are significantly reduced (see, for example, Haefeli et al., 2000). Finally, the demand for organically grown food in Switzerland has increased considerably since the big grocery stores started to offer it along with their "traditional" vegetable supply. The effectiveness of many infrastructure and service instruments, especially of attraction based forms, depends on whether demand can be created. Also, infrastructure instruments that are physically built structures and facilities often have very long-term effects, and they are hardly reversible within a few years or decades.

4.4.4. Collaborative agreements

An advantage of collaborative agreements is that they allow environmental targets to be reached without unnecessarily restricting entrepreneurial freedom of action and without unnecessary additional cost. Generally, agreements can only have higher conformity with the goals than command and control instruments if government and industry have identical interests, if an already ongoing development and modernisation process is accelerated or supported, and if the industry can achieve the targets without economic disadvantages and without distortions in competitiveness (Rennings et al., 1997). However, the incentive to enter into agreements can be diminished once the target is achieved, if no further targets are set (Kohlhaas and Praetorius, 1994). As Kirchgässner puts it:

> "[Voluntary agreements] are actually a substitute for command and control policies and not for market oriented measures. Such agreements are rather popular with producers, as they are able to use their information advantage in the bargaining process with the public bureaucracy. They may be of some use as they can – compared with one-sided bureaucratic measures – lead to lower (economic) costs of environmental policy, possibly at the cost of lower standards and, therefore, lower environmental protection. More important is, however, that they share the basic weakness of bureaucratic measures: because prices below a certain emission limit are zero, there is no incentive to further reduce emissions below these limits and to develop new techniques with lower emissions. Moreover, as the literature about rent-seeking tells us, given such possibilities there are strong incentives to invest in political bargaining instead of emission reduction"(Kirchgässner, 1997: 187).

A considerable disadvantage of agreements is their legal non-commitment and their non-suability, when, for instance, the economic situation does not allow participants to continue to pursue the set targets. The reason for this is that the binding factor of agreements is only of a political, but not of a legal nature. Experience indicates that the binding effect is actually stronger if the number of actors involved is smaller and if communication among the parties is intensive. However, the realisation of environmental goals through collaborative agreements still lags behind realisation by command and control instruments (Rennings et al., 1997).

4.4.5. *Communication and diffusion instruments*

The impacts of communication and diffusion instruments used within action campaigns to influence the behaviour of specific target groups are well documented in many cases (see, for example, Dwyer et al., 1993; Gardner and Stern, 1996; Hopper and Nielsen, 1991). However, the effects are sometimes small or questionable in terms of environmental outcomes. This has to do with the fact that the potential influence of behaviour change on resource consumption and environmental pollution is often very limited. For instance, household energy consumption is determined to a large extent by the physical characteristics of a building and by the technology for regulating room temperatures. "Energy saving behaviour" of residents, even if exemplary, has only a small influence on the overall energy consumption of the household (Ligteringen, 1999). Also, interventions of this type often reach only a small segment of the population. For this reason, efforts to move the broader public to use, for example, infrastructures in a more conscious and energy-saving manner are frequently seen to be an inefficient alternative (compare here, for example, Stern 2000b: 410). But the fact should not be overlooked that such campaigns are useful in creating the motivational groundwork preparatory to promoting the rapid spread of new, more efficient technologies through other types of instruments (compare the chapter by MAUCH ET AL.).

However, communication and diffusion instruments are not confined to short term campaigns. In a larger view, communication is also a matter of building up a comprehensive and high quality educational system that is available to the entire population. And it is a matter of a responsible media policy that does not leave everything to market forces. Last but not least, it is a matter of promoting problem-oriented research activities. In addition, communication and diffusion instruments may play an important role in mitigating some of the drawbacks of classic policy instruments. The neglect of psychosocial and behavioural aspects accounts for a number of the difficulties encountered with the policy instruments described above: (1) Command and control instruments are often perceived as constraining personal freedom and autonomy; they may therefore cause the behavioural phenomenon of reactance. In addition, the enforcement of legal prescriptions directed towards barely accessible target groups is often very difficult, because control is almost impossible. Therefore, the motive to avoid sanctions disappears. (2) Economic instruments – besides the problem of acceptance, especially among the industry – can lead individuals to act illegally in order to avoid paying fees (illegal waste disposal may have this motive), and they may destroy intrinsic motivations to act environmentally responsibly ("I pay for having my trash collected, therefore I can legitimately produce as much trash as I want"). (3) Collaborative agreements, such as ISO 14001 certification, require implementation within the enterprise down to the level of the daily routines and decisions of individual employees. Thus, communication and diffusion instruments may be very helpful in participatory policy development, in accompanying the implementation of collaborative agreements or economic instruments, and in enforcing command and control instruments.

4.5. *What innovations does it promote?*

A special and very important outcome aspect of policy instruments is their potential to promote innovations. We can examine this potential of the various types of instruments and find out what kind of innovations it promotes. Command and control instruments apparently have the potential to promote the diffusion of the best available technology at a given time, but their potential to promote continuous improvement seems to be limited. In contrast, a major advantage of economic instruments is seen in their very potential to stimulate continuous technical and organisational improvements in order to achieve environmental targets as efficiently as possible. They provide an incentive to reduce all units of pollution rather than to simply comply with the legal requirements. Service and infrastructure instruments are often themselves social and institutional innovations. Collaborative agreements may not only stimulate technical and organisational innovations, but they may also lead to new institutional arrangements (for instance, for the control of certifications and labels). Finally, communication and diffusion instruments promote the diffusion of available new technologies, and they play a decisive role in the initiation of new institutional arrangements. To the area of communication and diffusion instruments fall the transmission and evaluation of new knowledge and the consequently derived judgements and setting of goals that will shape the design of new resource regimes.

4.6. *How well accepted is it?*

In a democracy, the implementation of classic policy instruments depends upon public acceptance. Therefore, the question of the acceptance of an instrument type is of practical importance.

4.6.1. *Command and control instruments*

As the main elements of traditional environmental policy, command and control instruments are still quite accepted, although there is a growing resistance against the increasing density of regulations, because they restrict entrepreneurial freedom and entail complicated bureaucratic control mechanisms.

4.6.2. *Economic instruments*

Economic instruments have been propagated by economists for decades (Pigou, 1952). Although usually believed to be effective, they are often not well accepted. Opponents are afraid that they will weaken a nation's industry's position in international competition – especially if economic instruments are to be implemented in a single-handed effort of one nation alone.

In contrast, those in favour of the instruments argue that early implementation of economic instruments forces a nation to make the necessary investments and to stimulate technical progress.

Table 3: Summary of the main features of the different types of policy instruments.

	COMMAND AND CONTROL INSTRUMENTS	ECONOMIC INSTRUMENTS
Why is the instrument effective?	*...because people want to avoid the penalties for non-compliance.*	*...because people want to achieve maximal benefit at minimal cost.*
Who can apply it in the main?	• Public authorities	• Public authorities[1]
What are the most important target groups?	• Individuals • Companies • Public and private organisations	• Companies • Individuals
What can be said about its effectiveness?	• Effectiveness highly foreseeable • Effective if compliance easily controlled • Relatively inflexible – not very adaptable to specific situations • Can adapt to changes only slowly	• Reaches goals at minimal cost • In practice often limited effectiveness due to low taxes, undifferentiated charges, and approval of exemptions • Provides incentives for improvements beyond the prescriptions
What innovations does it promote?	• Spread of best available technologies	• Development and spread of new technologies, processes, and products • Innovations in company organisation
How well is it accepted?	• Accepted as the main element of current environmental policy • Resistance to increasing regulation	• Resistance to new costs • Disagreement over the utilisation of new revenue • Increasing political acceptance
What efforts and costs does it involve?	• Creating and instituting laws is time-consuming • Administrative costs	• Creation of legal bases is time-consuming • Administrative costs covered by revenues from taxes and charges • Costs for subsidies

[1] Incentives can also be implemented by private organisations in the framework of campaigns and deposit-refund systems.

(**Table 3,** cont.)

SERVICE AND INFRASTRUCTURE INSTRUMENTS	COLLABORATIVE AGREEMENTS	COMMUNICATION AND DIFFUSION INSTRUMENTS
...because people can carry out an action only if there is the possibility to do so, and because they prefer to make use of attractive as opposed to unattractive offers.	*...because people uphold contracts.*	*...because people pursue goals in accordance with their perceptions, categorisations, appraisals, and knowledge.*
1. Public authorities 2. Companies 3. Public and private organisations 4. Individuals	• Public authorities together with companies[2]	1. Public authorities 2. Public and private organisations 3. Individuals
1. Individuals 2. Companies 3. Public and private organisations	• Companies	1. Individuals 2. Public and private organisations
• Effective if demand can be created • Once created, infrastructures are as a rule not reversible	• Effective if legally binding and if there are sanctions for non-compliance • Reaches only those companies/branches involved • Relatively flexible – quite adaptable to specific situations	• Effectiveness difficult to measure • Effective quickly, but with limited reach if applied in the framework of action campaigns • Effective in the long term if there is a corresponding trend in society • Suited to supporting the acceptance and effectiveness of other instruments
• New companies • New institutions	• Optimisation of processes • Innovations in company organisation • New institutional arrangements	• New forms of social organisation • New institutional arrangements • Development and spread of new technologies, processes, and products
• Acceptance of attraction-based forms • Resistance to repulsion-based forms	• High acceptance by companies, as long as less expensive than other measures	• High acceptance of traditional forms • Hesitant acceptance of unfamiliar forms • Resistance to new forms of participation possible
• Investment and operating costs	• Negotiations can be tedious and lengthy	• Costs for planning and executing projects and action campaigns • Costs for education and research

[2] Or, for labels, public and private organisations together with companies.

When the measures and instruments are introduced internationally at a later date, a country may profit from its "first mover advantage".

Despite the objections, the implementation of economic instruments has increased rapidly world wide in recent decades. In 1994 the OECD determined that, in general, OECD countries resort to measures of the command and control type to achieve changes in polluting behaviour. (OECD, 1994). Since then, market-based policy instruments have been on a steady rise. In a detailed review of existing measures, Stavins (2000) cites examples in dozens of countries, particularly European countries, that are being applied in the most diverse areas. We have to keep on mind that this impressive list of concrete applications of economic instruments looks at market-based instruments in the widest sense, that is, it includes instruments whose primary goals are fiscal. The Swiss Agency for the Environment, Forests and Landscape (SAEFL) reports that economic instruments are rapidly replacing statutory command and control instruments, especially since the Amendment to the Federal Environmental Protection Law in December 1995 (SAEFL, 1997).

4.6.3. Service and infrastructure instruments

The attraction based forms of service and infrastructure instruments have the advantage of being easily accepted by the target persons. Travelling by train because it is comfortable and fast is much more positive than travelling by train because there is no other alternative. However, the acceptance of the financing of service and infrastructure instruments cannot be taken for granted. The repulsion based forms of service and infrastructure instruments are no more easily accepted than command and control instruments and repulsion based economic instruments.

4.6.4. Collaborative agreements

Collaborative agreements are generally well accepted by industry, their main target group. The motivation of companies or industrial sectors to enter into negotiations often lies in the government's commitment to forego implementing other, more restrictive instruments. In contrast to taxes or other economic instruments, the industry also perceives an advantage in the fact that with agreements, no transfer of finances between the industry and the government takes place. Environmental organisations, on the other hand, are often sceptical of agreements.

4.6.5. Communication and diffusion instruments

Most forms of communication and diffusion instruments are generally well accepted by target groups as well as by political decision-makers, although there may be hesitancy when the instruments are not yet well known. Participatory processes may be rejected due to their inherent potential to alter the power relations between different actor groups.

4.7. *What effort and costs does it involve?*

The application of policy instruments for achieving environmental goals always carries monetary and non-monetary costs. Of course, the amount of costs varies within each instrument type, but a few general observations can be made. Command and control, as well as economic and some infrastructure and service instruments, require legal bases, and the creation of these may be very time-consuming. In addition, command and control and economic instruments generate administrative and technical costs, although in the case of taxes and charges these can be covered by the revenue created. Service and infrastructure instruments generate investment and operation costs. For reaching collaborative agreements, prolonged negotiations are sometimes required. Expenditures for implementing communication and diffusion instruments vary enormously and depend upon both the type of instrument and the size of the target group. Generally, it seems advisable to procure consulting services by specialists once the size of the target group reaches several hundred persons.

4.8. *Conclusion*

Clearly, the different types of instruments rely on quite different rationales, they have different ranges, and different actors are involved in their design and implementation. Another look at Table 1A-C (pages 40-45) and at Table 4 (page 90f.) reveals that – going from command and control to communication and diffusion instruments – there is a decrease in the instruments' capacity to influence the sphere of action of the target groups and a decrease in formal obligation created by the instrument. On the other hand, there is an increase in context-specificity and in multi-actor participation. It is clear that each of these features may have its specific advantages and disadvantages depending on the political situation and the specific problems that are addressed. There can be no single "best instrument" or general "best policy-mix". The "appropriate mix" has to be selected according to policy objectives and under consideration of the possible interactions (facilitating or inhibiting) between the different types of instruments. In other words, the appropriate selection of instruments, their design, and their timing must be embedded into a *strategy*, which requires first of all that policy goals be defined (see also Jänicke et al., 1999).

5. Relationships to other typologies and classifications

Knowledge on the rationale of different instruments and knowledge on their modes of operation is of course founded in various scientific disciplines and their respective theories, methods, and empirical findings. It would go beyond the scope of this volume to review this multidisciplinary background. However, we would like to provide the interested reader with some guidance by pointing out the major sources and relevant approaches. They are selected mainly from jurisprudence, the political sciences, economics, and psychology (see also König and Dose, 1993, for a detailed review of various classifications used in jurisprudence, economy, political science, sociology, the sociology of law, and administrative science).

5.1. Jurisprudence

The legal perspective focuses first of all on *who* implements a certain policy instrument and on the legal consequences of the instrument. In other words, in view of the law, it is important to determine if the instrument creates rights and obligations of the target group, how the legal relationships arise, and what the content of the rights and obligations are. "Law" signifies a *binding* and – as a rule – *enforceable* social arrangement. There is thus an important differentiation between legal obligations and behavioural norms based upon other premises not enforceable by the state. These may involve, for example, ethical or moral considerations or obligations.

It is the duty of the state to implement consensus arrived at through the political process. This duty must be expressed in its regulatory activities, even if instruments other than statutory prohibitions and prescriptions are implemented. Self-regulation is politically popular, because it seems to be a way to turn over the regulation of a problem to the actors and all those involved themselves. In terms of enforceability and implementation, however, the concept of self-regulation has inherent limits: a person not willing to voluntarily adhere to a regulation can not be forced to comply as long as there is no legal, enforceable basis.

5.1.1. Forms of government action

In a legal perspective, government actions at the levels of federal government, canton, and municipality can take various forms, but there are no limits to the number of forms of action. Traditionally, the following categories are distinguished:

Statutes: Through statutes, or laws, the state regulates a certain area in a *general and abstract way for all involved.* This is the wider sense of the term "law" that comprises constitutions, formal laws and statutes, ordinances, and more. Law making is a classic function of the state that can be exercised at various democratic levels (all eligible voters, parliament, government).

Orders, directions, acts decreed: Through orders, the state defines the individual rights and obligations of private persons in individual cases. Orders may involve burden (assessing taxes, assessing of contested fees) or favour (license to construct, license to operate). Through orders the state again acts on the basis of the legitimate power monopoly and sovereign capacity to which it is entitled, independently of the will of the person involved. Orders therefore always require democratic legitimatisation and a legal basis. The administration is responsible for the issuing of orders.

Transitional forms between statutes and orders: Some forms show the features of laws and orders. Some examples are land use plans, or zoning regulations, which define the options for land use in particular areas, and general decrees (such as prohibition of driving in a neighbourhood street).

Contract: Similar to orders, through contracts the state determines the rights and obligations of private persons. In contrast to the order, however, contracts are not issued in a sovereign manner, but are drawn up with the agreement of the other party. Contractual regulations do not necessarily require an explicit legal basis, but they must not contradict the law.

De facto administrative action *[Realakt]*: Unlike the above forms, administrative action does not define rights or obligations, but rather supplies services. For example, the state supplies drinking water and energy, purifies wastewater, and provides public education.

It is relatively easy to assign classical instruments to the traditional categories of government activity. Command and control instruments in our typology belong primarily to the category of statute, or law. An exception is licensing (license to construct, license to sell), which in the concrete case may be issued in the form of an order. The economic instruments are implemented in the main in the form of orders (which, as mentioned, require a legal basis). With regard to content, economic instruments may be categorised as burden or favour and also according to the nature of the rights and obligations (financial, other). For instance, subsidies represent a financial favour, while taxes and charges entail a financial burden. Licenses are favourable orders with a different content, and burden may be attached to them. Market creation can be assigned in part to the legal form of a law, not an order, as it regulates a "competitive system" in a general and abstract manner.

The newer instruments, in particular communication and diffusion instruments and collaborative agreements, can be categorised as contracts insofar as binding rights and obligations are contracted. Otherwise, in the traditional categorisation they fall under "de facto administrative actions". It is often not possible to draw exact lines, because various instruments also contain contractual elements. In the science of jurisprudence, government activities in the area of information, motivation, and concrete recommendations have recently come to be known as "informal administrative actions". From the perspective of jurisprudence, these actions constitute a grey zone of administrative law (although not outside the realm of the law) that stands in a relationship of potential tension to the legality principle as defined in the constitution (Müller, 1995: 361).

5.1.2. *Legal base of government action*

For the state there is no sphere of action outside a legal framework. State actions are admissible only if they serve directly or indirectly to fulfil public duties or functions. Public duties and functions are determined by the law as a "concentrated" result of a political process. According to the legality principle, academic law and court decisions today demand that all administrative action, not only administrative interference (interference with civil liberties and other rights of citizens), but also administrative benefits and services (such as state services in education or public assistance), have a legal basis. This legal basis may in some cases be very general in nature. New instruments of informal administrative action as well as the collaborative agreements and communication and diffusion instruments described in this chapter also ultimately require a (more or less general) legal basis.

Besides this required legal basis, communication and diffusion instruments may be implemented only under consideration of certain legal principles. In all cases, public authorities are bound to observe the fundamental *rights of citizens, equal treatment* of all citizens, and the *prohibition against arbitrariness*. For example, tax reductions for environmentally sound practices may not be granted at will and from case to case, but only on the basis of special statutory power and in factually justified cases. If the state issues awards or certificates for a certain product or activity, it may not put competitors at a disadvantage for no evident reason. Depending upon relevance to fundamental civil rights, the legal demands will be formulated more or less strictly. In this respect, it is relatively unproblematic for a Federal Councillor to demonstrate how eggs can be cooked using less energy. But if a public authority warns the public about consumption of a particular sort of soft cheese in an effort to ward off a listeriosis epidemic, and this results in a massive drop-off in sales of the product, the situation is very different (see Schweizerische Bundeskanzlei, 1987: 97ff.).

It is also important to note that various instruments require a special legal basis. For implementation of instruments that have fiscal consequences, for instance, a decree by the fiscal authorities is always a prerequisite. Government information and motivating programmes, particularly government recommendations, raise some special issues. The body politic may represent certain "values", such as ecological concerns, of course, but it must observe the principle of neutrality (in worldview) and the factual nature of the information. In referendums, information provided by the government about the bill that goes beyond the official statement is admissible only as an exception, because public authorities may not interfere with the freedom of the public to form an opinion. The Federal Court has declared it inadmissible, for example, for cantonal authorities to publish quite objective information in the newspaper prior to the referendum vote. In this connection, recent voices in the field of law have suggested that more weight should be placed on public authorities' obligation to inform the public (Decurtins, 1992).

Many instruments of informal administrative action in the areas of communication and diffusion instruments and collaborative agreements are relatively new to the field of jurisprudence and have not yet undergone extensive legal examination (but see Müller, 1995; Nützi, 1995; Pfenninger, 1996; and their references to the literature). Up to now, it has been the exception and not the rule for these instruments to be regulated in an expressly legal and binding manner. There is, however, the "co-operation principle", which has been anchored in the Environmental Protection Law in 1997 (Art. 41a of this law). In line with this provision, the Confederation and the cantons work together with organisations in the economy. Through setting quantity targets and corresponding time limits, they can promote agreements with various branches of industry. Prior to issuing regulations for execution, they examine voluntary measures by the economic partners, and whenever possible and necessary the agreements are incorporated into the implementing law either in part or in their entirety.

5.2. *Political Sciences*

In the political sciences, especially in recent research on the policy implementation process, there has been lively discussion on the classification of policy instruments. The focus is on the state as actor who tries to influence the behaviour of actors outside of the political-administrative system. Usually the action possibilities of public authorities are looked at in detail, while the scope of action of those to be influenced is not the object of reflection. However, there is neither consensus on a specific typology nor dominance of one of them.

For analytical purposes, Varone (1998: 33 ff.) distinguishes two kinds of typologies: one focuses on *nominal categories*, the other on *fundamental attributes*. Nominal categories may be elaborated on the basis (1) of resources needed for the implementation of an instrument (Hood, 1983), (2) of the degree of legitimate constraint (Doern and Phidd, 1983), (3) of the expected effects of an instrument (McDonnell and Elmore, 1987), or (4) of the behavioural aptitudes of the target groups (Ingram and Schneider, 1990). Among the typologies concentrating on attributes, Varone mentions Sabatier and Pelkey's (1987) advocacy coalition approach, Salamon's (1989) public management approach, Linder and Peters' (1989) policy design, and Trebilcock's et al. (1994) public choice theory. Our typology shows some analogies to the constraint-based continuum proposed by Doern and Phidd (1983), but combines that with further elements from various other typologies. With a particular focus on the German literature and the traditional instruments, Kaufmann and Rosewiz (1983) give an overview of classifications that have been important in the discourse within the political sciences. One of their conclusions is that attempts to develop very generalised typologies that are independent of the domain, goals, and contents of policy-making might not be as fruitful as less generalised classifications that are closer to reality. Their own typology is on social policy, where they distinguish among legal, economic, ecological, and educational instruments (Kaufmann and Rosewitz, 1983: 45). Dahme and Grunow (1983) have presented a thorough analysis of the implementation of "persuasive programmes" that use the more or less conventional instruments for influencing knowledge and capacities (providing information, education, training, consulting, and so on) in the domain of health and consumer education.

For the domain of *environmental policy*, the usual distinction made is between regulatory and economic instruments as "classical" policy instruments, with mention of a heterogeneous class of "new" instruments (Dente, 1995; Jänicke and Weidner, 1995; Jänicke et al., 1999: 101) point out that in the political sciences, instruments are usually ordered according to the degree of governmental regulation of behaviour, ranging between the poles of coercion and voluntariness. Their proposal distinguishes the following main types of environmental policy instruments: regulatory law instruments (high degree of state control), planning instruments (state control medium to high), economic instruments (state control medium), co-operation (state control medium to low), and information (low degree of state control). It is obvious that this typology is quite similar to the one we propose, the main difference being that service and infrastructure instruments are not included, and that our "communication and diffusion instruments" are broader and go beyond "information".

Klok (1995) criticises existing typologies for classifying instruments according to one or a few differentiating characteristics and neglecting characteristics that they have in common, and for being more or less closed systems unable to incorporate new instruments that emerge in the political praxis. His understanding is that a classification "is not a set of sharply deliminated and exclusive cells, but a method for the characterisation of policy instruments (showing differences and similarities)" (p. 21). He then introduces a simple model of human action that assumes that all human activity is the result of (internal and external) motivations and resources (physical goods, skills, information, time, and social resources, such as money, legal rights, or trust). Using this model, political measures can be characterised as influencing the actors' resources and/or motivation by conditional or unconditional provision or deprivation of resources. This approach shares some similarities with our own, as it relates the description of instruments to what they should influence, namely human action. Klok's extremely broad concept of "resources" is interesting theoretically in view of economic as well as psychological action models. For practical purposes, however, it might be too one-dimensional, because the concrete measures that provide or withdraw various types of resources are probably very different: providing trust, for example, requires completely different measures than providing money.

Another analysis and classification of public policy instruments has been presented by Vedung (1998). As in our approach, the dimension underlying his division is the degree of constraint intended by the policymakers. Attempting to offer a parsimonious classification, he suggests that not more than three basic types of instruments be distinguished: regulations (the "sticks"), economic means (the "carrots"), and information (the "sermons"). It is obvious that these types correspond more or less to our command and control, economic, and communication and diffusion instruments. Our service and infrastructure instruments would be classified by Vedung as economic instruments, as they can be regarded as providing or withdrawing non-monetary resources: "A bump in the road to prevent motorists from speeding is an economic instrument just as a tax levied on gasoline is" (Vedung, 1998: 33). Negotiations leading to agreements or labels would be regarded as "governing through persuasion" (ibid: 37), i.e., as belonging to the information type of instruments. Vedung and van der Doelen (1998) analyse and discuss "information" as a public policy instrument that "covers government-directed attempts at influencing people through transfer of knowledge, communication of reasoned argument, and moral suasion in order to achieve a policy result" (Vedung and van der Doelen, 1998: 103). Concentrating on the forms of information-dispensing available to public authorities, they distinguish several sub-types quite similar to the ones that we propose for diffusion instruments.

5.3. *Economics*

The extensive literature on economics – and in particular on market-based instruments – presents a myriad of typologies too numerous to describe in detail. They often differ only on minor points. There are three basic approaches: (1) theory based approaches, (2) approaches based upon concrete criteria, and (3) experience based approaches.

5.3.1. *Theory based approaches*

This approach is often found in basic research, which describes the economic fundaments and mechanisms of these instruments. The objective is "to provide a systematic treatment of the theory of externalities and its implications" (Baumol and Oates, 1993: 2) in order to facilitate discussion of the theoretical advantages of economic instruments in regard to eco-efficiency and their effect upon income distribution.[13] The types of instruments are divided into broad categories. Tietenberg (1990) uses the categories of Emissions Trading and Emissions Charges; Baumol and Oates (1993) add subvention as a third category. Tietenberg also makes a distinction between "efficiency charges", designed to produce an efficient outcome to totally compensate for damage, and "cost-effective-charges" to achieve a predefined goal at the lowest possible cost. This distinction is theoretical, however, and does not correspond to instruments that are in fact applied. This is indeed the main difference between this approach and our own typology, which does not base upon theoretical aspects, but rather upon concrete applications and mechanisms.

5.3.2. *Approaches based upon concrete criteria*

A second approach comprises typologies that, while they do employ concrete criteria, do not differentiate among all possible applications. For example, Lohman (1994) distinguishes among "financial instruments", which are all introduced with the exclusive purpose of financing specific, environmental tasks, "economic incentives", designed to create an incentive effect, and a mixed category for financial instruments that produce incentives as a side-effect.

Turner and Oschor (1994) distinguish instruments according to the type of influence on the market. These influences are direct modification of prices and costs, indirect influencing through fiscal means, and finally, market creation and support. To the direct modifications belong charges on products and processes (emissions charges) and deposit-refund systems. Indirect changes result from subvention and also from enforcement incentives (non-compliance fees and performance bonds). Market creation includes emissions trading and quota auctioning. Stabilisation of prices, such as for recycling products, is described as market support. Jeanrenaud (1997) forms three categories of environmental policy instruments according to the role of the state: (1) direct government intervention, that is, expenditures by the state to create, operate, and maintain infrastruc-

---

[13] For the reader interested in economic theory, Bohm and Russel (1985) and Christiansen and Tietenberg (1985) provide extensive lists of references.

tures (public anti-pollution investment) and to finance research; (2) measures to change practices (behaviour modification), including information and moral suasion, command-and-control, incentive-based regulations, and privatisation, whereby the incentive-based regulations include all of the "classical" economic instruments (charges, tradable permits, deposit-refund systems, joint implementation); and (3) "laissez-faire", where the actors influenced participate voluntarily, and the role of the state is restricted to defining clear property rights and liability for damages.

Our typology, which distinguishes among the instruments according to the effecting mechanisms, makes use of these approaches in part. For instance, our distinction of taxes and charges corresponds to distinction of financial versus economic instruments made by Lohman (1994). The categorisations by Turner and Oschor (1994) and Jeanrenaud (1997) are also reflected in our typology, although in more detailed forms. With regard to Jeanrenaud's typology, we have to say that in our understanding, all environmental policy instruments aim for behavioural modification, whether they are coerced or voluntary.

5.3.3. Experience based approaches

The third approach distinguishes among types of instruments that find application in reality. Here we find the various OECD instrument evaluation studies (for example, OECD, 1997b), whose typology has often been adopted, expanded, and modified. The OECD names four categories of economic instruments: charges/taxes (emission charges/taxes, user charges/taxes, product charges/taxes), subsidies (grants, soft loans, tax allowances), tradable emission permits, and deposit-refund systems. It should be noted that the terms "charges" and "taxes" are used synonymously, whereby the sub-category "user charges" is defined as "charges that are payments for the costs of collective or public treatment of effluent or waste" (OECD, 1997b: 16), which corresponds to our sub-type of charges.[14]

One of the most detailed typologies with the most examples of applications is that by Stavins (2000). Stavins divides the instruments into the four categories of charge systems, tradable permit systems, reducing market barriers, and reducing government subsidies. Charge systems comprise taxes, charges, and subventions (with definitions similar to our own typology), and also insurance premium taxes, which are designed to encourage companies to internalise environmental risks in their own decision making. Tradable permit systems subsume credit programmes (an emissions

---

[14] Other OECD publications distinguish:

Taxes: The OECD classification (as used in the annual OECD publication, *Revenue Statistics*) defines taxes as "compulsory, unrequited payments to general government. Taxes are unrequited in the sense that benefits provided by government to taxpayers are not normally in proportion to their payments" (OECD, 1997a: 18). Note that a tax (unrequited) *can* be earmarked if it is decided that a certain percentage of the revenue will be used for a specific purpose (e.g., when part of a gasoline tax is earmarked for road construction).

Charges/fees: The OECD classification also uses the terms "fees" and "user charges" (as opposed to "taxes") and "levies" without giving a precise definition of these terms. In practice, the terms "charges" and "fees" are often used interchangeably: "Charges and fees are compulsory requited payments, i.e., a service is provided in proportion of the payment. Charges can also be paid into specific "funds" and earmarked for specific environmental purposes" (OECD, 1997a: 18).

source gets credits if it can reduce its polluting below a set level; credits can be used by other emission sources) and cap-and-trade systems (a general pollution threshold is set and assigned to companies in the form of tradable licenses). Reducing market barriers includes market creation (measures to facilitate voluntary trading of water rights), liability rules, and information programmes (information obligation, labelling). Reducing government subsidies applies mainly to third world countries, in cases where individual production branches were protected by subventions despite their environmental impact (coal production in China, for instance).

This latter typology includes all the economic instruments in our own typology with the exception of incentives as parts of action campaigns. This is because their influence can not be measured at the macroeconomic level, and because they are not normally implemented by the state. The instruments are described in more detail and finer distinctions are made, such as for tradable permits and reducing subventions. Our typology focuses upon applications and application possibilities at the local and national levels, and for this reason it places less weight upon instruments that at this time are implemented mainly in North America and in third world countries. Insurance premium taxes, liability rules, and information programmes are instruments that in our typology do not fall in the category of economic instruments.

5.4. *Psychology*

Psychology focuses on describing, understanding, predicting, and eventually influencing individual behaviour. Traditionally, it has not dealt with policy instruments, because these are regarded as part of the milieu or situation within which individual behaviour takes place, and as such are not taken as psychological factors. Within the relatively recent development of environmental psychology, however, psychologists have begun to think about applying their knowledge on how to induce behaviour change to the field of achieving environmental goals (Mosler and Gutscher, 1998). New forms of interventions based on principles of behavioural and social psychology have been developed in order to complement traditional environmental policy instruments or to enhance their effectiveness (see the chapter by FLURY-KLEUBLER AND GUTSCHER for an overview and discussion of such principles and Stern, 2000a, for a general discussion of the contribution of psychology to the resolution of environmental issues). Geller and his colleagues (Geller et al., 1982; Geller, 1987; Geller, 1989) were among the first to present an approach based on behavioural analysis, which makes use of well founded principles of behaviour modification. Combining such an approach with principles used in marketing leads to "social marketing" which is finding increasing application in environmental contexts (Kotler and Roberto, 1991; Prose et al., 1994; McKenzie-Mohr and Smith, 1999). Social marketing mainly applies instruments that we subsume under "communication instruments", "service and infrastructure instruments", or – in cases of monetary incentives – under "economic instruments".

Gardner and Stern (1996) have presented a very broad overview of psychological knowledge relevant to understanding and solving environmental problems. Without setting out an explicit typology of instruments, they strongly recommend a combination of "solution strategies" for promoting behavioural changes, whereby emphasis is placed on the importance of participatory methods and networking. Gardner and Stern outline four types of approaches to be used in combination:

By **religious and moral approaches** they understand all attempts to change values, beliefs, and worldviews, especially the influence of pro-environmental movements on public opinion about environmental issues.

Educational interventions aim at providing information effectively by using feedback, modelling, and framing messages,[15] and at tightening the links between attitudes and behaviour through reminders and prompts, public commitment, and highlighting attitudes and norms.

By **changing incentives,** Gardner and Stern (1996) refer not only to the classic economic instruments (like taxes and subsidies), but also to what we mean by service and infrastructure instruments: They stress the importance of making environmentally responsible behaviour as convenient – and thus rewarding – as possible. Their understanding of "incentive" has its roots in the concept of operant conditioning (Skinner, 1969), which states that behaviour is shaped by the consequences it has. Therefore, the authors also seem to view the command and control instruments of our typology as "incentives", although they do not treat them explicitly.

Community management of the commons is treated as a means to overcome the socio-ecological dilemmas leading to the overuse of common resources and the "tragedy of the commons" (Hardin, 1968). Gardner and Stern (1996) recommend the application of participatory methods and networking and an approach to the problems that is – wherever possible – at the local level rather than at the state or national level.

Based on their analysis of a number of model intervention programmes, Gardner and Stern (1996) list the principles or "success factors" for interventions that aim at changing environmentally undesirable behaviour as presented in Table 4. These principles call for the intelligent application and combination of different instruments. Principle C requires the use of communication and diffusion instruments (which – as we have seen – influence primarily the internal (psychological) structure of the actors), while for realising principle D, instruments targeted primarily at the external structure are needed.

---

15 Framing refers to how pro-environmental behaviours are described, such as referring to "energy efficiency", a desirable goal, instead of "conservation", which many people associate with sacrifice.

Table 4: Principles for intervening to change environmentally destructive behaviour (from Gardner and Stern, 1996: 159).

A.	Use multiple intervention types to address the factors limiting behaviour change • Limiting factors are numerous (e.g., technology, attitudes, knowledge, money, convenience, trust) • Limiting factors vary with actor and situation, and over time • Limiting factors affect each other (interactive principle)
B.	Understand the situation from the actor's perspective • Conduct surveys or experiments • Participatory approach to program design
C.	When limiting factors are psychological, apply understanding of human processes • Get the actors' attention; make limited cognitive demands • Apply principles of community management (credibility, commitment, face-to-face communication, etc.)
D.	Address conditions beyond the individual that constrain pro-environmental choice
E.	Set realistic expectations about outcomes
F.	Continually monitor responses and adjust programs accordingly
G.	Stay within the bounds of the actors' tolerance for intervention
H.	Use participatory methods of decision making

References

Aronson, E. and O'Leary, M. (1983) The relative effectiveness of models and prompts on energy conservation: A field experiment in a shower room. *Journal of Environmental Systems, 12*: 219-224.

Bachmann, W. and Katzev, R. (1982) The effects of non-contingent free bus tickets and personal commitment on urban bus ridership. *Transportation Research, 16A*: 103-108.

Baltes, M.M. and Haywood, S.C. (1976) Application and evaluation of strategies to reduce pollution: Behavioral control of littering in a football stadium. *Journal of Applied Psychology, 62*: 501-506.

Balthasar, A. and Knöpfel, C. (1994) Umweltpolitik und technische Entwicklung. Eine politikwissenschaftliche Evaluation am Beispiel der Heizungen. Helbing and Lichtenhahn, Basel.

Bandura, A. (1986) Social foundations of thought and action. Prentice-Hall, Englewoods Cliffs, NJ.

Bättig, Ch. (2001) Fallstudie Evaluation der Verbrauchszielwerte, Fallstudie in Netzwerkmanagment als neues Instrument der Umweltpolitik. Interface, Luzern; *in prep.*

Baumol, W.J. and Oates, W.E. (1993) The theory of environmental policy. Cambridge University Press, Cambridge, New York.

Bemelmans-Videc, M.-L., Rist, R.C. and Vedung E. (eds.) (1998) Carrots, sticks and sermons. Policy instruments and their evaluation. Transaction Publishers, New Brunswick, New Jersey.

BFE/Bundesamt für Energie (1998a) Freiwillige Massnahmen unter besonderer Berücksichtigung von Vereinbarungen. Internes Arbeitsdokument BFE, Januar 1998, Bern.

BFE/Bundesamt für Energie (1998b) Organisation und Vollzug des Energiegesetzes und Zusammenarbeit mit Externen (OVE). Internes Arbeitsdokument BFE, Juni 1998, Bern.

Bohm, P. and Russel, C.S. (1985) Comparative analysis of alternative policy instruments. *In*: Handbook in natural resource and energy economics, Kneese, A.V. and Sweeney, J.L. (eds), North Holland.

Burn, S.M. (1991) Social psychology and the stimulation of recycling behaviors: The block leader approach. *Journal of Applied Social Psychology, 21*: 611-629.

Burn, S.M. and Oskamp, S. (1986) Increasing community recycling with persuasive communication and public commitment. *Journal of Applied Social Psychology, 16*: 29-41.

Christainsen, G.B. and Tietenberg, T.H. (1985) Distributional and macroeconomic aspects of environmental policy. *In*: Handbook in natural resource and energy economics, Kneese, A.V. and Sweeney, J.L. (eds), North Holland.

Cialdini, R.B., Reno, R.R. and Kallgren, C.A. (1990) A focus theory of normative conduct: Recycling the concept of norms to reduce littering in public places. *Journal of Personality and Social Psychology, 58*: 1015-1026.

Dahme, J. and Grunow, D. (1983) Implementation persuasiver Programme. *In*: Implementation politischer Programme II, Mayntz, R. (ed.), Westdeutscher Verlag GmbH, Opladen.

De Clercq, M. (1996) The implementation of green taxes: The Belgian Experience. *In*: Environmental taxes and charges. National experiences and plans, The European Foundation for the Improvement of Living and Working conditions (eds.), Luxembourg.

Decurtins, G.-A. (1992) Die rechtliche Stellung der Behörde im Abstimmungskampf. Universitätsverlag, Freiburg (Schweiz).

Dente, B. (ed.) (1995) Environmental policy in search of new instruments. Kluwer Academic Publishers, Dordrecht, Boston, London.

Diamond, W.D. and Loewy, B.Z. (1991) Effects of probabilistic rewards on recycling attitudes and behavior. *Journal of Applied Social Psychology, 21*: 1590-1607.

Doern, B.C. and Phidd, R.W. (1983) Canadian public policy. Ideas, structure, process. Methuen, Toronto.

Dwyer, W.O., Leeming, F.C., Cobern, M.K., Porter, B.E. and Jackson, J.M. (1993) Critical review of behavioral interventions to preserve the environment. Research since 1980. *Environment and Behavior 25/3*: 275-321.

ECONCEPT and IPSO (1997) Evaluation energiepolitisch motivierter Steuererleichterungen. Studie für das Schweizer Bundesamt für Energie im Rahmen der Evaluation des Aktionsprogramms Energie 2000, Zürich und Dübendorf.

ECOPLAN (1993) Abwasserabgabe für die Schweiz – Lenkung, Finanzierung und Ausgleich in der Abwasserentsorgung. Schriftenreihe Umwelt, Nr. 203. Bundesamt für Umwelt, Wald und Landschaft, Bern.

Everett, P.B., Hayward, S.C. and Meyers, A.W. (1974) The effects of a token reinforcement procedure on bus ridership. *Journal of Applied Behavior Analysis, 7*: 1-9.

Gardner, G.T. and Stern, P.C. (1996) Environmental problems and human behavior. Allyn and Bacon, Boston.

Geller, E.S. (1987) Applied behavior analysis and environmental psychology – From strange bedfellows to a productive marriage. *In*: Handbook of environmental psychology, Stokols, D. and Altmann, I. (eds.), John Wiley and Sons, New York.

Geller, E.S. (1989) Applied behavior analysis and social marketing: An integration for environmental preservation. *Journal of Social Issues 45*: 17-36.

Geller, E.S., Winett, R.A. and Everett P.B. (1982) Preserving the environment. New Strategies for behavior change. Pergamon Press, New York.

Glasbergen, P. (ed.) (1998) Co-operative environmental governance. Public-private agreements as a policy strategy. Kluwer, Dordrecht.

Gonzales, M.H., Aronson, E. and Costanzo, M.A. (1988) Using social cognition and persuasion to promote energy conservation: A quasi-experiment. *Journal of Applied Social Psychology, 18*: 1049-1066.

Grossman, D. and DeGaetano, G. (1999) Stop teaching our kids to kill: A call to action against TV, movie and video game violence. Crown, New York.

Gutscher, H., Keller, C. and Mosler, H.-J. (2000): Roads as new common pool resources, speed reduction as a public good – 2 case studies in organizing large-scale collective action. Available: CD: 8th Biennial conference of the International Association for the Study of Common Property (IASCP), Indiana University, Bloomington, Indiana. Online in the WWW at: <http://129.79.82.27/IASCP/Papers/gutscherh041300.pdf>

Haefeli, U., Matti, D. and Seewer, U. (2000) Die Sanierung und Umgestaltung der Seftigenstrasse. Auswirkungen auf Lebensqualität und Einkaufsverhalten der NutzerInnen (mit besonderer Berücksichtigung des Langsamverkehrs und der Ertragssituation des Detailhandels). Schlussbericht der Wirkungsanalyse. IKAÖ, GIUB, Universität Bern.

Hardin, G. (1968) The tragedy of the commons. *Science 162*: 1243-1248.

Hasler, P.D. and Wälti, M. (1999) Machbarkeitsstudie Autohaltestellen. büro für utopien, Markwalder und Partner AG, Burgdorf.

Hood, C. (1983) The tools of government. MacMillan, London.

Hopper, J.R. and Nielsen, J.M. (1991) Recycling as altruistic behavior: Normative and behavioral strategies to expand participation in a community recycling program. *Environment and Behavior 23*: 195-220.

Ingold, P. (1999) Freizeitaktivitäten und Naturschutz. *Mitteilungen der Naturforschenden Gesellschaft in Bern, Neue Folge 56*: 77-158.

Ingram, A. and Schneider, H. (1990) Improving implementation through framing smarter statutes. *Journal of Public Policy 10/1*: 333-356.

IEA/International Energy Agency (1989) Energy and the environment: Policy overview. OECD, IEA, Paris.

IEA/International Energy Agency (1997) Voluntary actions for energy-related CO_2 abatement. OECD, Paris.

Jacobs, H.E. and Bailey, J.S. (1982-1983) Evaluating participation in a residential recycling program. *Journal of Environmental Systems, 12*: 141-152.

Jänicke, M. and Weidner, H. (eds.) (1995) Successful environmental policy. Rainer Bohn Verlag, Berlin.

Jänicke, M., Kunig, P. and Stitze, M. (1999) Umweltpolitik. Politik, Recht und Management des Umweltschutzes in Staat und Unternehmen. Dietz, Bonn.

Jeanrenaud, C. (1997) Economic instruments for environmental policy. *In*: Environmental policy between regulation and market, Jeanrenaud, C. (ed.), Birkhäuser, Basel.

Jepson, C. and Chaiken, S. (1990) Chronic issue-specific fear inhibits systematic processing of persuasive communications. *Journal of Social Behavior and Personality, 5*: 61-84.

Kantola, S.J., Syme, G.J. and Campbell, N.A. (1984) Cognitive dissonance and energy conservation. *Journal of Applied Psychology, 69/3*: 416-421.

Katzev, R.D. and Pardini, A.U. (1987-1988) The comparative effectiveness of reward and commitment approaches in motivating community recycling. *Journal of Environmental Systems, 17*: 93-113.

Kaufmann, F.-X. and Rosewitz, B. (1983) Typisierung und Klassifikation politischer Massnahmen. *In*: Implementation politischer Programme II. Ansätze zur Theoriebildung, Mayntz, R. (ed.), Westdeutscher Verlag, Opladen.

Kaufmann-Hayoz, R., Häuselmann, C. and Gessner, W. (1996) "Eco-Design" – die wahrnehmungspsychologische Erweiterung eines technischen Konzepts. *In*: Naturbilder. Ökologische Kommunikation zwischen Ästhetik und Moral. Themenhefte zum Schwerpunktprogramm Umwelt, Lesch, W. (ed.), Birkhäuser, Basel.

Kirchgässner, G. (1997) Environmental policy in Switzerland: Methods, results, problems and challenges. *In*: Economic policy in Switzerland, Bacchetta, P. and Wasserfallen, W. (eds.), Macmillan Press, Houndmills.

Klok, P.J. (1995) A classification of instruments for environmental economic policy. *In*: Environmental policy in search of new instruments, Dente, B. (ed.), Kluwer Academic Publishers, Dordrecht

Kobi, F., Künzler, P. and Dietiker, J. (1995) Das Berner Modell – Die angebotsorientierte Verkehrsplanung als Instrument zur Reduktion der Umweltbelastungen durch den Verkehr. *VDI Berichte, 1228*: 195-215.

Kohlhaas, M. and Praetorius, B. (1994) Selbstverpflichtungen der Industrie zur CO_2-Reduktion. Sonderheft 152. Deutsches Institut zur Wirtschaftsforschung, Berlin.

Kok, G. and Siero, S. (1985) Tin recycling: Awareness, comprehension, attitude, intention and behavior. *Journal of Economic Psychology, 6*: 157-173.

König, K. and Dose, N. (1993) Instrumente und Formen staatlichen Handelns. Carl Heymanns Verlag, Köln.

Kotler, P. (1978) Marketing für Nonprofitorganisationen. Poeschel, Stuttgart.

Kotler, P. and Roberto, E. (1991) Social marketing. Econ, Düsseldorf.

Leventhal, H. (1970) Findings and theory in the study of fear communications. *In*: Advances in experimental social psychology, Vol. 5, Berkowitz, L. (ed.), Academic Press, New York.

Ligteringen, J.J. (1999) The feasibility of Dutch environmental policy instruments. Decreasing the environmental impact of households. Twente University Press, Enschede.

Linder, S.H. and Peters, G.B. (1989) Instruments of government. Perceptions and contexts. *Journal of Public Policy, 9/1*: 35-58

Lohman, L.d.S. (1994) Economic incentives in environmental policy: Why are they white ravens? *In*: Economic incentives and environmental policies: Principles and practice, Opschoor, J.B. and Turner, R.K. (eds), Kluwer Academic Publisher, Dordrecht, Boston, London.

Mayntz, R. (1980) Implementation politischer Programme: Empirische Forschungsberichte. Verlagsgruppe Athenäum Hain Scriptor Hanstein, Königstein/Ts.

Mayntz, R. (1983) Implementation politischer Programme II. Westdeutscher Verlag, Opladen.

McDonnell, L.M. and Elmore, R.F. (1987) Alternative policy instruments. Center for Policy Research in Education, Santa Monica.

McKenzie-Mohr, D. and Smith, W. (1999) Fostering sustainable behavior. An introduction to community-based social marketing. New Society Publishers, Gabriola Island B.C.

Midden, C.J., Meter, J.E., Weening, M.H. and Zieverink, H.J. (1983) Using feedback reinforcement and information to reduce energy consumption in households: A field experiment. *Journal of Economic Psychology, 3*: 65-86.

Mosler, H.-J. and Gutscher, H. (1998) Umweltpsychologische Interventionsformen für die Praxis. *Umweltpsychologie, 2*: 64-79.

Mosler, H.-J. and Gutscher, H. Der Einsatz von Selbstverbreitungs- und Interventionsinstrumenten bei einer Energiesparaktion; *in prep.*

Müller, M. (1995) Staatliche Empfehlungen. Gedanken zu Rechtscharakter und Grundrechtsrelevanz. *Zeitschrift für Schweizerisches Recht, I*: 357-405

Nützi, P. (1995) Rechtsfragen verhaltenslenkender staatlicher Information: Strukturen, Zulässigkeit, Haftung. Verlag Stämpfli, Bern.

OECD/Organisation for Economic Co-operation and Development (1994) Managing the environment. The role of economic instruments. OECD, Paris.

OECD/Organisation for Economic Co-operation and Development (1997a) Environmental taxes and green tax reform. OECD, Paris.

OECD/Organisation for Economic Co-operation and Development (1997b) Evaluating economic instruments for environmental policy. OECD, Paris.

OECD/Organisation for Economic Co-operation and Development (1999) Voluntary approaches for environmental policy: An assessment. OECD, Paris

O'Keefe, D.J. (1990) Persuasion. Theory and research. Sage, Newbury Park.

Pallak, M.S. and Cummings, W. (1976) Commitment and voluntary energy conservation. *Personality and Social Psychology Bulletin, 2*: 27-30.

Pallak, M.S., Cook, D.A. and Sullivan, J.J. (1980) Commitment and energy conservation. *Applied social psychology annual, 1*: 235-253.

Pardini, A.U. and Katzev, R.D. (1983-1984) The effect of strength of commitment on newspaper recycling. *Journal of Environmental Systems, 13*: 245-254.

Pfenninger, H. (1996) Rechtliche Aspekte des informellen Verwaltungshandelns: Verwaltungshandeln durch informell-konsensuale Kooperation unter besonderer Berücksichtigung des Umweltschutzrechts. Universitätsverlag, Freiburg (Schweiz).

Pigou, A.C. (1952) The economics of welfare, fourth edition. MacMillan, London

Prose, F., Kupfer, D. and Hübner, G. (1994) Social Marketing und Klimaschutz. *In*: Gesellschaftliche Aspekte von Klimaveränderungen, Fischer, W. and Schütz, H. (eds.), Forschungszentrum Jülich, Jülich.

Prose, F. and Wortmann, K. (1992) "Negawatt statt Megawatt": Eine Energiesparlampen-Aktion. *In*: Jahrbuch Ökologie, Altner, G., Mettler-Meibom, B., Simonis, U.E. and v. Weizäcker, E.U. (eds.), Beck'sche Reihe, München.

Rennings, K., Brockmann, K.L., Koschel, H., Bergmann, H. and Kühn, I. 1997) Nachhaltigkeit, Ordnungspolitik und freiwillige Selbstverpflichtung, Umwelt- und Ressourcenökonomie. Schriftenreihe für Europäische Wirtschaftsforschung. Physica Verlag, Heidelberg.

Reno, R.R., Cialdini, R.B. and Kallgren, C.A. (1993) The transsituational influence of social norms. *Journal of Personality and Social Psychology, 64*: 104-112.

Rothstein, R.N. (1980) Television feedback used to modify gasoline consumption. *Behavior Therapy, 11*: 683-688.

Sabatier, P.A. and Pelkey, N. (1987) Incorporating multiple actors and guidance instruments into models of regulatory policymaking. An advocacy coalition framework. *Administration and Society 19/2*: 236-292.

SAEFL/Swiss Agency for the Environment, Forests, and Landscape (1997) Sustainable development in Switzerland: Strategy of the Federal Council. SAEFL, Bern.

Salamon, L.M. (1989) Beyond privatization. The tools of government action. The Urban Institute Press, Washington.

Schweizerische Bundeskanzlei (ed.) (1987) Geschäftsbericht des Bundesrates. Eidgenössische Drucksachen- und Materialzentrale EDMZ, Bern.

SFSO/Swiss Federal Statistics Office and SAEFL/Swiss Agency for the Environment, Forests and Landscape (1997) The environment in Switzerland 1997 – Facts, Figures, Perspectives. SFSO and SAEFL, Bern. Online in the WWW: <http://www.buwal.ch/e/themen/umwelt>.

Skinner, B.F. (1969) Contingency and reinforcement. A theoretical analysis. Appleton-Century-Crofts, New York.

Smith, E.R. and Mackie, D.M. (2000) Social psychology, 2nd edition. Psychology Press, Philadelphia.

Stavins, R.B. (2000) Experience with market-based environmental policy Instruments. *In*: Handbook of environmental economics, Mäler, K.-G. and Vincent, J. (eds.), North-Holland; *forthcoming*.

Stern, P.C. (2000a) Psychology and the science of human-environment interaction. *American Psychologist, 55/5*: 523-530.

Stern, P.C. (2000b) Toward a coherent theory of environmentally significant behavior. *Journal of Social Issues, 56/3*: 407-424.

Tiefbauamt des Kantons Bern, Oberingenieurkreis II (2000) Erfolgskontrolle bei Strassenbauprojekten am Beispiel der Seftigenstrasse (Köniz). Verkehrs- und umwelttechnische Auswirkungen und Überprüfung der Instrumente des Berner Modells. Arbeitspapier. Arbeitsgemeinschaft Planungsbüro Jürg Dietiker, Windisch, Künzler Bossert und Partner GmbH, Bern.

Tietenberg, T.H. (1990) Economic instruments for environement regulation. *Oxford Review of Eoconomic Policy 6/1*: 17-33.

Trebilcock, M.J., Hartle, D.G., Prichard, J.R.S., and Dewees, D.N. (1994) The choice of governing instruments. A study prepared for the Economic Council of Canada. Canadian Government Publishing Center, Ottawa.

Turner, K. and Oschor, H. (1994) Environmental economics and environmental policy instruments: Introduction and overview. *In*: Economic incentives and environmental policies: Principles and practice, Opschoor, J.B. and Turner, R.K. (eds.), Kluwer Academic Publisher, Dordrecht, Boston, London.

Ulli-Beer, S. and Haefeli, U. (1999) Evaluation der LEM-Partnerschaftsprojekte des Grossversuchs mit Leicht- und Elektromobilen in Mendrisio. Bericht über die Bevölkerungsbefragung 1997/1998 und 1999. Teil B: Auswertung und Analyse. IKAÖ, Universität Bern.

Vallender, K.A. (1990) Ziele und Instrumente des Schweizerischen Umweltrechts. *In*: Festschrift zum 65. Geburtstag von Mario M. Pedrazzini. Stämpfli and Cie, Bern.

Varone, F. (1998) Le choix des instruments des politiques publiques. Haupt, Bern.

Vedung, E. (1998) Policy instruments: Typologies and theories. *In*: Carrots, sticks and sermons. Policy instruments and their evaluation, Bemelmans-Videc, M.-L., Rist, R.C. and Vedung, E. (eds.), Transaction Publishers, New Brunswick, N.J.

Vedung, E. and van der Doelen, F.C.J. (1998) The sermon: Information programs in the public policy prosess – choice, effects, and evaluation. *In*: Carrots, sticks and sermons. Policy instruments and their evaluation, Bemelmans-Videc, M.-L., Rist, R.C. and Vedung, E. (eds.), Transaction Publishers, New Brunswick, N.J.

Wagstaff, M.C. and Wilson, B.E. (1988) The evaluation of litter behavior in a river environment. *Journal of Environmental Education, 20*: 39-44.

Wang, T.H. and Katzev, R.D. (1990) Group commitment and resource conservation: Two field experiments on promoting recycling. *Journal of Applied Social Psychology, 20*: 265-275.

Weber, M. (1904) Die Objektivität sozialwissenschaftlicher und sozialpolitischer Erkenntnis. *Zitiert nach*: Gesammelte Aufsätze zur Wissenschaftslehre, Tübingen, 3. Edition, 1968: 146-214.

Winett, R.A., Hatcher, J.W., Fort, T.R., Leckliter, I.N., Love, S.Q., Riley, A.W. and Fishback, J.F. (1982) The effects of videotape modeling and daily feedback on residential electricity conservation, home temperature and humidity, perceived comfort, and clothing worn: Winter and summer. *Journal of Applied Behavior Analysis, 15*: 381-402.

Winett, R.A., Leckliter, I.N., Chinn, D.E., Stahl, B. and Love, S.Q. (1985) Effects of television modeling on residential energy conservation. *Journal of Applied Behavior Analysis, 18*: 33-44.

Witmer, J.F. and Geller, E.S. (1976) Facilitating paper recycling: Effects of prompts, raffles, and contests. *Journal of Applied Behavior Analysis, 9*: 315-322.

Zimbardo, P.G. and Leippe, M.R. (1991) The psychology of attitude change and social influence. McGraw-Hill, New York.

Changing Things – Moving People.
Strategies for Promoting Sustainable
Development at the Local Level
R. Kaufmann-Hayoz/H. Gutscher (eds.)

Psychological Principles of Inducing Behaviour Change

Peter Flury-Kleubler and Heinz Gutscher

In the comprehensive typology of instruments for promoting sustainable development presented above, the instruments described are attempts to induce behaviour change. How the behaviour of others can be changed has been a central issue in the field of psychology since its beginnings. Psychologists have dealt with inducing behaviour change in various contexts – from psychotherapy, education, and marketing to military training and political manipulation. In this chapter, we discuss the basic possibilities of changing the behaviour of others and develop a comprehensive and internally consistent psychological view of environmentally relevant behaviour change. We focus on the analysis of those principles upon which the instruments presented in the typology are based. We do not intend to present a synopsis of all psychological theories and empirical findings relevant to the issue, nor do all our thoughts represent the consensus view of the discipline. Following an introduction, we discuss a number of external interventions and analyse the internal psychological effects of these interventions. This is followed by a look at the conditions of efficacy and acceptance of the different attempts to induce behaviour change. The chapter concludes with a summary.

1. Introduction

1.1. Goals as motives and values

The existence, health, and well being of human beings are always at risk and depend on a number of preconditions within the environment, the organism, and the person's subjective state. Human beings have **needs** (such as water, nutrition, shelter, social contact, self-esteem, and others.). To satisfy needs, people must pursue **goals** (drink, eat, find good company). Needs and goals are not the same. Rather, the pursuit of goals is a necessary instrument within individuals that enables them to satisfy their needs (cf. Ryan et al., 1996; Tesser et al., 1996). Not all goals are equally well adapted to fundamental needs, and thus not all goals are equally legitimate in an ethical sense. The goal to drive at a speed of 200 kilometres per hour is not as legitimate with regard to basic needs as the goal to avoid being hit by a car.

Modern advertising sometimes gives the impression that human needs and goals can be formed at will. However, the basis of the situation of human existence is timeless: In the 21st century, a person who does not eat or drink will still die, and without social recognition and affection, people still perish (cf. Bronfenbrenner, 1979, about Spitz's work on hospitalism).

Two classes of **means** – mental processes and behaviour – are available to achieve goals. In this context, the relevant mental processes are imagination and thinking. Behaviour is muscle activity, such as speaking, typing computer keys, gesticulating, manipulating objects, or locomotion. Of course, the production of behaviour depends on prior mental processes. However, some goals can be achieved by mental processes alone, without subsequent behaviour. If we want to feel happy, it is possible to achieve this goal by imagining a situation that makes us feel good. Here there is no need to produce behaviour.

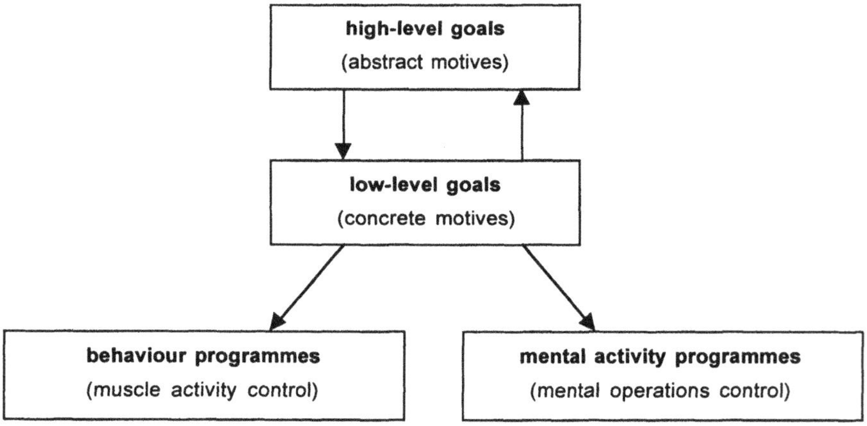

Figure 1: Goal hierarchy and means.

Goals are related to each other within a hierarchical structure. This means that high-level goals are superior to a number of low-level goals. Sometimes a high-level goal can be achieved by one of its inferior, or subordinate, goals. Sometimes a combination of inferior goals is necessary. Since the goal hierarchy consists of many levels, a goal can be both the super-goal of a number of sub-ordinate goals and a means to reach its super-ordinate goal. At higher levels, goals are more final, while at the lower levels goals tend to be instrumental. Low-level goals, or concrete motives, refer to specific mental processes or behaviours. They trigger specific mental activity programmes or behaviour programmes (Figure 1). High-level goals, or abstract motives, refer to the effects that are to be achieved through mental or behavioural means (compare, for example, Baars, 1988, for a thorough theoretical discussion of goal hierarchies). What means are suitable to achieve a certain high level goal depends on individual appraisal patterns. For some people, lying on a crowded beach is a good means to achieve the high-level goal of feeling comfortable, while others find the same behaviour very unpleasant, so that to them it is an unsuitable means for the same high-level goal.

As mentioned above, human needs and goals cannot be manipulated at will. The higher a goal is within an individual's goal hierarchy (e.g., the goal to eat, to survive, to protect one's autonomy), the more it is determined by universal biological necessities. The lower it is (e.g., the goal to go eat at a specific restaurant, to stay at a specific hotel), the more it is idiosyncratic and the result of an individual adaptation to a specific, temporary situation and the more it can be influenced by others. The higher a goal is, the more stable it is – the lower, the more volatile. If pro-environmental attempts of inducing behaviour change are to achieve persistent effects, they have to target medium level goals somewhere between unchangeable universals and labile ad hoc goals (e.g., the goal to minimise energy consumption).

1.1.1. The motivation function of goals

From a functional perspective, the goals of individual persons are – on the one hand – **motives** (Figure 2). That is, they drive the individuals to organise – consciously or unconsciously – mental activity and behaviour towards the goal content of the motive, that is the desired object, process, or state. A large proportion of everyday behaviour is routine behaviour that does not involve reflection upon goals and means (compare Bargh and Barndollar, 1996). It is based upon highly automatic processes (see, for example, Schank and Abelson, 1977; Baars, 1988). Nevertheless, these automatic processes are goal-directed and intended. In general, only if our routine patterns of goal-achievement fail do we start to think about our mental activity and behaviour.

What motives are active depends strongly on perception and appraisal of the current objective situation and on individual coping strategies. The motivational process consists of setting goals. In general, there are a number of goals to pursue. However, they cannot all be followed simultaneously. Therefore, setting and organising goals entail resolving goal conflicts. Goal setting can be either a conscious, voluntary process or an unconscious, automated one. When encountering a particular situation for the first time, people do not have well-established strategies for coping with it. Or if a previously familiar situation has changed, these strategies may fail. A new strategy must be developed and evaluated. In this case, the production of behaviour follows principles described by the theories of rational choice and planned behaviour (e.g., Ajzen and Fishbein, 1980; Luce and Raiffa, 1990). When consciously seeking the means to achieve goals, a limited number of behaviour options come to mind. People evaluate the imagined effects each option is likely to have and choose the option they believe to be the best. There are internal and external resources and restrictions that determine what mental and behavioural means are available to a person and which of them promise to be successful (cf. Gutscher et al., 1998). Resources as well as restrictions shape an individual's scope of action. This aspect is the subject of the "ipsative theory of action" proposed by Foppa (Frey and Foppa, 1986; Foppa, 1988; Frey, 1990; Ulli-Beer and Kaufmann-Hayoz, 1998; Tanner, 1998; 1999).

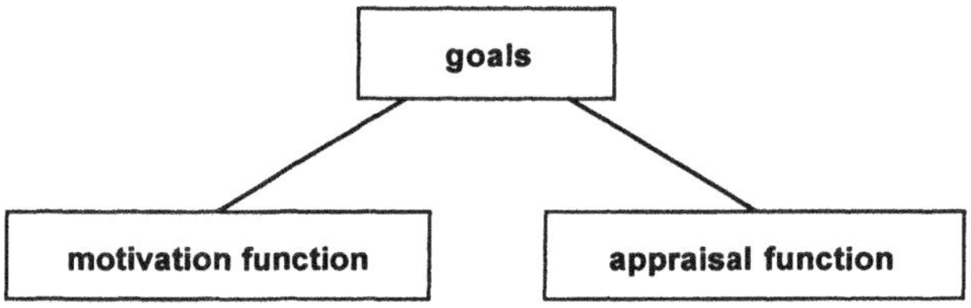

Figure 2: Psychological functions of goals.

1.1.2. The appraisal function of goals

Besides the motivational aspect, goals also have an **appraisal** aspect. The goals people want to achieve affect the way they perceive and appraise reality. What they view as good or bad, right or wrong, useful or useless, depends on their goals. According to the definition introduced in the description of communication and diffusion instruments, relatively stable goals are **values**. And goals that refer to mental activity or to behaviour are **norms**. This implies that values and norms are points of reference for our appraisals.

1.2. Principle ways of influencing the behaviour of others

As human individuals we have no direct access to what other individuals experience, nor to what they intend to do. To know what others may be experiencing, we depend on what they express through their behaviour, mainly verbal behaviour, i.e., what they say about their subjective reality. We never really know how true this behavioural expression is and whether we understand what others mean. Just as we cannot look directly into the minds of others, we cannot directly control another person's will and, thus, his or her behaviour.

When trying to change the behaviour of others, we can only influence the stimuli that the other person is exposed to. For an organism, stimuli are those aspects of the physical reality that it can register. The stimuli an individual is exposed to are the properties of his or her **physically objective situation**. This includes substances, inanimate objects, living organisms and their behaviour, most particularly the presence and behaviour of other human beings. Some stimuli are relevant to a person's existence and thus have **meaning**. This implicates that they shape the individual's action possibilities or have an influence on the effects of any actions performed.

Both in everyday life and when designing interventions for inducing behaviour change psychologically, we have theories about the functioning of human individuals. We have an idea of how others perceive and interpret specific stimuli. And we have assumptions on the effect that this will have on their goals and behaviour. We can design changes of stimuli so as to cause, indirectly, specific changes within individuals, i.e., to cause specific psychological effects. By influencing stimuli, we can try to influence the perceptions and appraisals of others, the goals they pursue, and the mental and behavioural means they use to achieve them. Thus these psychological effects, finally, have an effect upon their behaviour. According to learning theories, we also suppose that some changes of stimuli have long-lasting effects that persist even when our intervention on the stimuli has ended.

Like the behaviourist psychologists (e.g., Watson, 1970), we could limit the analysis of behaviour change strategies to the relation between external interventions upon stimuli and their effects upon behaviour. But by doing so, we would never understand the complexity of the laws of this relation and would, in many situations, not be able to design powerful strategies for achieving desired behaviour change. It is for this reason that we will discuss the internal psychological effects of external interventions.

Changing stimulus conditions, i.e., people's objective physical situations, often also implies a change in the relation between possible behaviours and their effects. Building a bridge over a river changes the effect of walking from one bank to the other. Before the bridge, the effect of this behaviour consists in getting wet. After the bridge has been built, a person can stay dry by walking above the surface of the water. This change of effects implies in particular that there is a change of the **success or failure conditions** of behaviours with regard to personal goals. The situational reality influences the effects that a person can produce by means of a certain behaviour.

Even if our main intention is to influence the goals of individuals and the means they intend to use in order to achieve their goals, we have to modify the objective stimulus situation of people, since the transfer of information depends on a physical substrate. By talking to people, we produce auditory stimuli within their environment. By writing messages, we produce visual stimuli on paper or computer screens. Our immediate point of attack lies in the environment of people, because we do not have direct access to their brains. Of course, we do not want to reduce communication to a physical perspective. Communication originates from a communicator's goals and has psychological effects within the target persons. But it is helpful to recognise that all interventions are built upon specific physical tools. Figure 3 presents a synopsis of all the interventions and psychological effects discussed below.

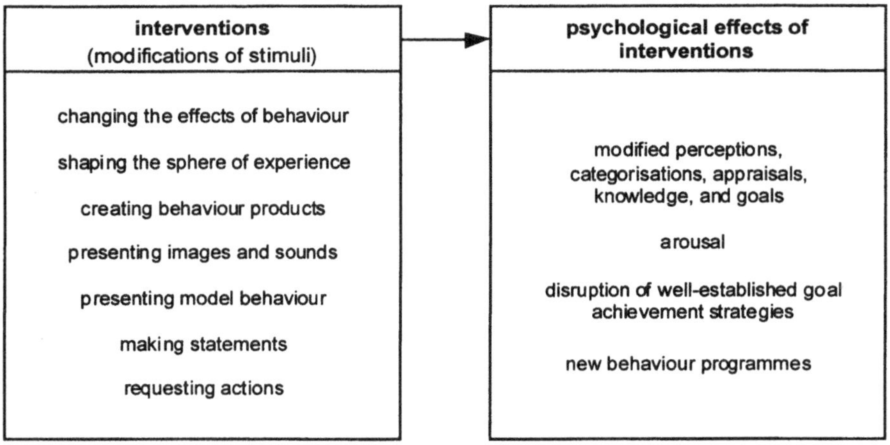

Figure 3: Prototypes of interventions and psychological effects.

1.3. Environmentally significant interventions

The psychological principles discussed in this chapter do not only apply to environmentally significant behaviour – they hold for any type of behaviour. However, in order to design effective interventions for the promotion of environmentally sound behaviour, not only their **efficacy** in changing goals and behaviour must be taken into account, but also the **environmental significance** of those changes. This is true both in the sense of an impact-oriented definition of environmentally significant behaviour – i.e., behaviour that has environmentally significant effects – and in the sense of an intent-oriented definition (Stern, 2000) – i.e., behaviour that is intended to have such effects. Not only behaviours that are primarily motivated by pro-environmental motives (e.g., within Stern's typology of environmentally significant behaviour, "environmental activism" and "nonactivist behaviours in the public sphere") are environmentally significant and therefore a possible target of environmentally significant interventions. Equally important are all those everyday behaviours that are primarily motivated by non-environmental motives, such as the goals to travel to work, to buy food, or to spend free time in a pleasant way. Although these behaviours are not intended to have any negative effect upon the environment, many of them – because many people behave similarly – are the most important causes of today's ecological problems. This can change, however, if people are motivated to incorporate pro-environmental motives into their personal sets of everyday motives. They can commute to work in an environmentally friendlier way and buy environmentally friendlier food than before, when they had not included pro-environmental goals into the set of conflicting goals that have to be co-ordinated with each other in everyday life.

2. Interventions: Modifications of stimuli

We can distinguish two different types of interventions to change stimulus conditions. First, stimuli can be influenced in a way that changes, even before an action is performed, a person's perception of the current situation or imagined anticipation of the situational outcome if the action is performed. Geller et al. (1982) call this type of intervention "antecedent strategies for modifying behaviours". Second, stimuli can be influenced in a way that becomes effective while or after an action is performed. Geller et al. call this type "consequence strategies for modifying behaviours". However, as soon as the consequences of an action are known in advance, they change the antecedent action conditions. The **announcement** of rewards or punishments for a certain behaviour is an antecedent strategy. However, the **reward or punishment itself** is a consequence strategy. Most interventions are based upon a combination of antecedent and consequence strategies.

Although all intervention takes place within the external structure, the focus of the instruments varies. While command and control instruments, economic instruments, and service and infrastructure instruments focus on the external structure (see Figure 1 on page 82), communication and diffusion instruments target the internal structure. The focus of collaborative agreements is on both.

Note that the interventions in the following list are not mutually exclusive. Some interventions belong to more than one of the types listed below.

2.1. *Changing the effects of behaviour*

Human behaviour is guided to a large extent by the expected effects of behaviour options. If an intervention upon the physical situation ("external structure" in Figure 1, page 82) changes the expected effects of a specific behaviour option, this will change the probability that it will be performed. When designing changes of behaviour effects, it is important not to limit design considerations to one, single behaviour option. Rather, the relative attractiveness of a possible behaviour has to be considered in comparison with the repertoire of behaviour alternatives the target individuals have at their disposal. For instance, the absolute price of motorised individual mobility is not a sufficient factor to determine the modal split between the different means of transportation. Price, comfort, and the availability of relevant alternatives have to be considered as well. Selection among several behaviour alternatives depends on the subjective costs and benefits of each option (which include benefits and costs not only in a monetary sense).

The effects of behaviour can be changed in various ways. We can transform a person's environment so that certain behaviour possibilities become more suitable means for achieving a certain personal goal. These interventions can be described metaphorically as **"bridge building"**. Interventions that make certain behaviours less suitable for achieving specific goals can be called **"barrier building"** interventions. The suitability of action options is experienced as subjective cost and benefit. Whether the result of an external intervention is perceived as a "bridge" or a "barrier" depends on individual goals. The same intervention can operate as a bridge for some people, as a barrier for others, and even be irrelevant to a third group of people. For instance, a reduction of train fares on weekends is a bridge for people who want to travel economically to visit friends on a Sunday. The same discounts can be barriers for people who want to work on their laptops while travelling, because the trains may be more crowded than usual. For people who plan to stay at home, the fare reduction may be totally irrelevant.

Changes in a person's environment may not only make behaviours more or less suitable for achieving certain goals, but they may even constitute absolute prerequisites or insurmountable obstacles. There is a distinction between **relative** bridges and barriers and **absolute** ones. An example of an absolute barrier would be if there were no means of public transportation from point A to B within a given interval of time. On the other hand, relative bridges and barriers only affect the relative suitability of a certain behaviour to achieve a certain goal. An increase in train fares does not make it impossible for most people to travel by train. But the number of train trips that can be taken with finances remaining constant becomes smaller. Or, paying the higher fare for the trips can prevent successful achievement of other goals, such as going out to dinner.

The instruments discussed in this book offer possible ways to build bridges and barriers. Command and control instruments define behaviour rules and building barriers (fines and imprisonment) where those rules are disobeyed. Economic instruments can be used both for building bridges (such as subsidies) and for building barriers (taxes and charges), thus changing the relations between the prices of different behaviour alternatives. Service and infrastructure instruments can also be used in the two ways. Better services and improved infrastructure can serve as bridges. The reduction or worsening of services or the dismantling of infrastructure can have barrier effects. Collaborative agreements are based, as stated above, on external or internal change. The external change consists in the fact that other individuals, groups, or organisations respect certain mutually valid rules, which transforms the conditions for the success of specific actions. If others respect certain rules, such as, for example, when other restaurant managers use only organic products, a particular restaurant manager no longer has the disadvantage of higher production costs when he respects those rules himself. This co-ordinated change of the behaviour of others can serve as a bridge for achieving personal goals by certain behaviours. The internal change consists in the fact that the person himself respects mutually agreed upon rules.

The building of bridges and barriers can have a number of effects for people. Bridges and barriers can affect **freedom of movement**. Imprisonment as a sanction associated with command and control instruments, of course, limits individual freedom of movement dramatically. But also the more subtle barriers imposed by economic barriers limit a person's freedom of movement, wherever that freedom is dependent on the use of means of transportation or on paying entrance or membership fees. On the other hand, financial bridges enlarge freedom of movement. While some measures affect individual freedom of movement directly, others do so indirectly by affecting financial resources. The functioning of command and control instruments is based on the deterrent effect of barriers that only become effective in the case where a person performs an action that is known to produce (with a certain probability) a certain outcome. Fines have a strong effect also on the behaviour of the majority of people who have never been fined.

As mentioned above, bridges and barriers can affect **financial resources**. This may affect freedom of movement and **health and physical integrity** as well. Greater financial resources allow lifestyles that include healthier diets, healthier conditions of residence, or healthier ways of moving and travelling. Wealthier people can afford better health care. Health and physical integrity, however, do not depend only on financial resources.

The quality of services and infrastructure offered in a country also affects health and physical integrity. For example, the quality of public transportation has effects relevant to health. Long and uncomfortable commutes to work in noisy, poorly ventilated, or draughty railway carriages may be very unhealthy in the long run. Dark parking garages where muggings take place pose a risk to physical integrity. Routes to school without safe sidewalks and with dangerous pedestrian crossings put the safety of children at risk.

While command and control instruments, economic instruments, and service and infrastructure instruments affect one's freedom of movement, financial resources, or health and physical integrity, we should not forget that informal ways of building bridges and barriers also have effects. Informal social reactions to a person's behaviour can affect the degree of **co-operation and voluntary service** that the person receives from others. If somebody behaves in a way others disapprove of, he or she may receive less help and support in situations where help is needed. A poorly groomed person in shabby clothing may be less frequently helped when trying to get a baby carriage into a bus or when out of coins for a parking meter. Organised promotion of disapproval for certain behaviours can be used as a powerful instrument of inducing behaviour change. It makes a difference whether most people just look the other way when they see someone littering or if they address the person and criticise his or her behaviour.

2.2. *Shaping the sphere of experience*

Shaping stimulus conditions can, as discussed above, affect behaviour consequences. There is, however, a second aspect of shaping stimulus conditions that is also of great importance. The stimuli in a person's environment also shape the sphere of possible experience. When parents take their children to a wild mountain landscape to backpack and tent in a marvellous setting, they are allowing them to have specific experiences. When organisers of a scout camp allow young people to sit around an open fire on a summer evening, where people are playing the guitar and singing, this creates opportunities for physical and social experiences that may affect their values and ideas about the world.

Service and infrastructure instruments, but also communication and diffusion instruments, are – among other things – means of allowing experience.

2.3. *Creating behaviour products*

The sphere of experience depends on more than natural settings. In the modern world, the physical environment is shaped to a large extent by human behaviour. The surface of the earth consists increasingly of cultural landscapes, human settlements, and infrastructure. The construction of buildings, roads, railway lines, electric power lines, the channelling of rivers, and the deforesting of landscapes all have a considerable effect on the stimulus conditions for human individuals. Stimulus conditions are especially affected by service and infrastructure instruments. The same goes for all the technical objects created by human behaviour that shape the stimuli that we perceive in daily life. In this context, we must not leave out art products, particularly music (see also communication and diffusion instruments). The relevance of linguistic products of human behaviour will be discussed in section 2.6.

Even the traces and products of others' behaviour have an effect. Behaviour traces allow us to draw conclusions about behaviour. And thus – like directly observable behaviour – they allow us to draw conclusions about **social norms**. For instance, litter on the ground tells us that others have littered here before. If it looks as if many have done so, we may find it normal to litter. Therefore, at littered spots the chances that other people will litter becomes higher (cf. Cialdini et al., 1990). Behaviour traces have a behaviour diffusion effect, whether intended or not. In the case of littering, the occurrence of littering behaviour can be reduced by frequent clean ups, that is, by eliminating the traces of the undesired behaviour. The same holds for graffiti and other forms of vandalism, where the most effective strategy to reduce the problem is fast removal. In both cases, a norm is established about what is socially desirable. The removal of undesired behaviour traces is a service instrument strategy.

Of course, socially desired behaviours leave traces as well. In fact, all human products are traces of human behaviour. For instance, the presence of flowers in gardens and in window boxes along a street may have an effect on the perception of passers-by and indirectly affect their own motivation and behaviour. The meaning that traces have in the perceiver's eye depends on his or her goals. People searching for paths through the snow in winter will be glad to follow others' footprints. However, a person who wants to be alone will probably avoid frequented tracks.

2.4. *Presenting images and sounds*

As discussed in section 2.3, the mere creation of behaviour products and the resulting shaping of the environment affect others' behaviour. However, the products of human behaviour can be presented deliberately to others. This can be done either by pointing out these products or specific sections of the natural environment directly or by using media, such as photography, television, video, and sound recording technology. Thus, people can be shown sections of their environment that they would not perceive directly. Of special importance among the stimuli that are deliberately presented to others are photographs and music. Both pictures and the sounds of music have a high potential to trigger emotions within the target persons. Music is often utilised to accompany images, statements (see 2.6.), and requests (see 2.7.) as a means of providing clues about their **relevance** and **positive or negative valence**.

2.5. *Presenting model behaviour*

An important aspect of our environment is the **behaviour of others**. Observation of others' behaviour influences us strongly. A large proportion of our personal behavioural repertoire is not the product of our own trial and error. We profit from the experiences of others by observing what they do, in what context, and to what degree of success. From the observed situation we draw conclusions as to the model's goal. We observe the behaviour the model uses as a means toward achieving that goal. And we notice what effects the goal-mean-pattern causes under the observed conditions. Observational learning (Bandura, 1986) is an important way of extending an individual's scope of action and forming behaviour. This is true for both desirable and non-desirable action, as in the case of real-life consequences of media violence (cf. Comstock and Paik, 1991; Carlsson and von Feilitzen, 1998; Centerwall, 1992; Drabman and Hanratty Thomas, 1974; Grossman and DeGaetano, 1999). On the one hand, model behaviour has an effect on learning processes and leads to long-lasting effects. On the other hand, model behaviour can also influence a person's behaviour within the observation context. As mentioned above, if we see people discarding trash somewhere, we are more likely to do so ourselves (cf. Cialdini et al., 1990). Or if in an emergency situation people just watch and do not intervene, it becomes likely that other people will also remain unresponsive and not offer their help. This is because they conclude from the inactivity of the others that the situation can be interpreted as harmless (cf. Latané and Darley, 1970). Model behaviour is an important source of social influence, independently of whether the model intends to have an effect on others or not. It is a powerful means of social influence that can be used deliberately. But it is also a means of unintended behaviour diffusion that works automatically. If there are scenes of overcrowded airports before Christmas on television, it seems quite normal that people fly to warmer climates at that time of the year. In contrast, the presentation of socially and ecologically desirable goals, and the behaviour strategies to achieve them, is an important possible use of communication and diffusion instruments.

Presenting model behaviour allows the observer to learn behavioural means to achieve goals. However, observation need not be direct. By listening to or reading reports, we also learn people's goals, means, and effects. Telling others about what people intended to achieve, what they did, and how they succeeded is a presentation of behaviour related knowledge that affects behaviour.

2.6. *Making statements*

Making **statements about reality** – whether true or false – may have an effect upon others. Statements may refer to persons, animals, plants, inanimate objects, physical states, subjective phenomena, or other things. Theories about the functioning of the world, the relations between actions and their effects, knowledge about how things are going to develop – all this affects the planning of behaviour and later routine behaviour. This implies that by making statements about the functioning of the world we have a powerful means of influencing behaviour. Both the deliberate withholding of information and deliberate providing of true or false information shapes behaviour. This is why the information media are so essential in modern societies.

2.7. *Requesting actions*

Beyond displaying behaviour, producing, leaving or removing traces, and providing knowledge, there is an additional way to influence behaviour. People can be **requested** to behave in a certain way. This can be done more or less politely and in a casual or imperative style, as speech act theory reveals (cf. Searle, 1969). Command and control instruments always contain an appeal to action. In addition, however, they also contain the threat of punishment if people are not willing to comply with the appeal. Communication and diffusion instruments, in contrast, are suitable instruments for distributing action appeals without sanction threats.

There are two main types of action appeals, appeals **without arguments** and appeals **with arguments**. For instance, a sign in a public restroom asking people to "keep this room clean" expresses an action appeal without supplying any justification. Action appeals without arguments are only effective if they can activate a motive that the addressee already has. This motive does not need to be identical with the goal of the appeal. If observers are present, the appeal may be obeyed because of a motive to fulfil the expectations of others.

Successful action appeals that provide arguments are built upon presuppositions that the addressee is expected to agree with. These may be theories about the functioning of the world ("cancer endangers human life") or presupposed goals ("human life must be protected"). The appeals then make an additional statement containing some new information, such as scientific findings ("micro-dust from diesel vehicles promotes cancer") or a topical political statement ("the law up for vote will lead to increased emissions of carcinogenic substances"). Action appeals then conclude with a logical consequence that results from the presuppositions and the new information ("micro-dust emissions from diesel vehicles should be reduced" or "the law should be rejected by the voters").

3. Psychological effects of interventions

3.1. Modified perceptions, categorisations, appraisals, knowledge, and goals

Knowledge and goals influence behaviour. In order to influence behaviour we should therefore influence knowledge and goals. But as discussed above, we cannot modify these directly. By modifying stimuli, we can only change perceptions (see Figure 1, page 82). This, however, can have significant effects, since the perceived phenomena trigger categorisations and appraisals (Figure 4). People always try to make sense of subjective experience. Automatically or deliberately, we classify the phenomena we experience. In other words, we attribute concepts to them. Thus we categorise objects as large, fast, round, loud, and so on. Attributions of concepts of subjective valence (good, bad, useful, useless, pleasant, unpleasant, etc.) or relevance (important, unimportant) are a subset of categorisations. They can be called appraisals. Appraisals are experienced both verbally and emotionally. Verbally, we experience appraisals as written or spoken words (e.g., "good", "bad"); emotionally, appraisals are experienced as felt emotions (the phenomenal experiences of joy, fear, anger).

Categorisations and appraisals depend on prior knowledge and goals within the target individuals. On their part, categorisations and appraisals can cause lasting changes in knowledge and goals. What people learn from stimuli not only depends on the stimuli themselves, but also on prior knowledge and goals.

Perceptions, categorisations, and appraisals on the one hand, and knowledge and goals on the other, are interrelated with each other. Goals and prior knowledge control perceptions and appraisal interests. But it is also true that perceptions and appraisals influence our goals and knowledge. To put it simply, we perceive and interpret what is relevant to our current goals, and we do this on the basis of our current knowledge. Conversely, we set our goals and adapt our knowledge depending on what we perceive and how we interpret the current situation. If we are hungry, we want to eat. If we want to eat, we look for food, and so forth. The relevance of the motivational state for cognitive processes has been discussed by many cognitive psychologists (see, for example, Neisser, 1976).

Since all perception and appraisal occurs in the light of pre-existing goals and knowledge, whenever we change a person's environment, this change is perceived and interpreted on the basis of prior goals and knowledge. If we want to motivate someone to behave differently, we have no other choice than to offer **stimuli** that are **relevant** in the light of the target person's prior goals and **understandable** on the basis of that person's prior knowledge. This is the fundament on which we can try to build or modify goals and knowledge, which will modify perceptions, categorisations, and appraisals.

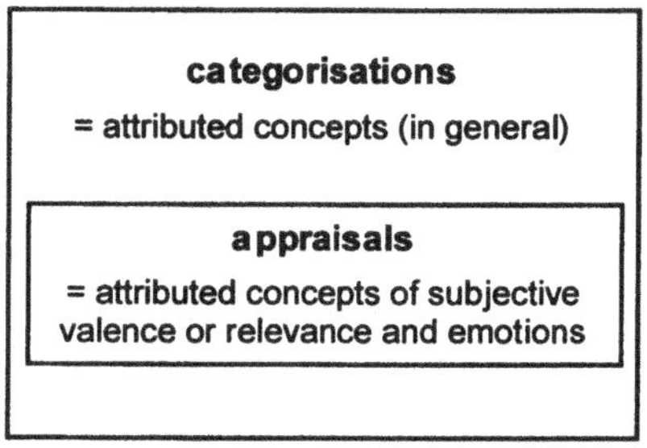

Figure 4: Categorisations and appraisals.

3.2. *Arousal*

Target persons do not always notice the modifications of stimuli that are created by command and control instruments, economic instruments, or service and infrastructure instruments. If the changes are not obvious to them, they must be made salient by using communication and diffusion instruments. For instance, new laws, new price offers, or changes of services must be communicated to those whose behaviour is to be influenced by them.

Communication and diffusion instruments are supposed, in a first step, to arouse **attention**. In a second step, they must show the addressee the relevance of the messages offered. For this, they must be – consciously or not – related to pre-existing goals of the target person. Demonstrating the relevance of the message to those motives will evoke emotions in the target persons. Telling people that the achievement of crucial personal goals is endangered can arouse fear. Communicating to people that certain action possibilities will allow them to achieve their goals and to fulfil their wishes can arouse their curiosity and let them experience feelings of satisfaction or joy.

As we use the term, communication and diffusion instruments are based not only on voluntarily communicated messages, but also on demonstrated meaningful behaviour (actions). Thus the voluntary or unintended diffusion of behaviour strategies by observational learning (Bandura, 1986) is a fundamental mechanism of communication and diffusion instruments. Or as Watzlawick (1967) expressed it: "One cannot not communicate." In Watzlawick's radical sense, informal social reactions are just a kind of communicative instrument.

The **arousal** caused by attention and emotions is a necessary prerequisite if people are to learn the contents of certain messages. Only if messages trigger a certain level of arousal will their content be remembered and become able to influence later behaviour.

3.3. Disruption of well-established goal achievement strategies

3.3.1. Creating maladaptation

When we build bridges or barriers, people's well-established goal achievement strategies come under pressure. These strategies are no longer well adapted to the changed environment. With the new barriers people find that successfully achieving certain goals using previous methods has become more difficult or – in the case of absolute barriers – even impossible. Suppose the non-stop train which brings you to the workplace at half past seven in the morning is no longer offered. Then your old habit of catching that train near your home is no longer possible. There may be alternatives, such as a train that stops at every station along the way and leaves half an hour earlier. Not always do new barriers affect the actions we are used to performing. They can also affect behaviour alternatives that we have not chosen so far. Some behaviour alternatives become even less attractive than before, which may foster our conviction that our choice of behaviour was right. Also, new bridges can have either a direct effect on our present behaviour or they can concern alternatives. In the first case, an action suddenly becomes easier than it was before, which will promote selection of this alternative. In the second case, when an alternative becomes more attractive, the present choice becomes relatively less attractive.

3.3.2. The motives involved

The changed behaviour effects caused by environmental change may be relevant to a number of specific motives. We suppose that we have a motive to eat, or a motive to feel comfortable, or a motive to earn money, and so on. The motives an individual has are – as discussed above – organised within a hierarchical system, insofar as some motives are more general and abstract and others are more specific and refer to concrete behaviour. The more specific motives can be regarded as means in order to achieve the contents of the more general motives, i.e., the more general goals. For example, earning money serves the more general purposes of being autonomous and having many behaviour choices. One possible taxonomy of motives has been suggested by Bischof (1985: 331). We hypothesise that on the top levels, the goal-hierarchy is universal. However, on the medium and lower levels, the structure of an individual's goal system is highly idiosyncratic, susceptible to environmental influence and – especially on the lowest levels – changing. On the most general motivational level, we strive towards **obtaining pleasant experiences** and **avoiding unpleasant experiences**. Below these super-ordinate motives there is an extensive and individually structured set of more specific motives. However, there are motives that many people share, because they are the product of shared cultural influence (e.g., the motive to own a powerful car), and these are crucial to instruments for inducing behaviour change.

The functioning of command and control instruments is based upon our autonomy motive, especially our freedom of movement motive (particularly in the case of imprisonment) and our motive to secure our financial resources (in the case of fines).

The effectiveness of economic instruments is also based on our autonomy motive, mainly on the motive to secure financial resources. This means that the motivational bases of command and control instruments and economic instruments overlap to some extent.

Service and infrastructure instruments base upon different motives than the first two classes. Securing physical and psychological health and integrity and a number of related motives stand at the centre. These are, for instance, the motives to minimise efforts and to experience convenience and comfort, but also motives to protect personal privacy. Here also, the autonomy motive may be involved, with its subordinate motives of saving time and securing financial resources.

The involvement of pre-existing motives is not only crucial to an understanding of adaptations to changed environments, but also in order to explain why certain meaningful, communicative stimuli have an effect upon behaviour. The effectiveness of communication and diffusion instruments can be built upon a broad variety of motives. Of particular interest here are the motives of maintaining self-esteem and maintaining social reputation. Related to both are altruistic motives, such as the motive to leave a healthy environment to future generations and other species. However, all the motives mentioned above may be addressed by communication and diffusion instruments. Using communication and diffusion instruments, we can try to focus on the relevance of certain action alternatives to autonomy, financial resources, health and integrity, and so on.

The effect of informal social reactions, finally, is based predominantly on their relevance to the goals of maintaining social reputation and self-esteem.

3.4. New behaviour programmes

Finally, through instruments of influencing behaviour target persons can also be motivated to learn new behaviour programmes. The disruption of well-established goal achievement strategies leads not only to modification of goal strengths or the forming of new goals, but it can also lead people to learn new behaviour programmes. For example, if high petrol prices make it more difficult and finally impossible to use a motor boat as a strategy for having fun at the weekend, people might become motivated to learn to sail.

4. Conditions of efficacy and acceptance

Not all the instruments discussed in this book are equally effective. And not all of them are equally accepted by those who want to influence the behaviour of others or by those whose behaviour is to be influenced. The efficacy and acceptance of instruments should be examined separately. An instrument may be extremely difficult to implement because of poor acceptance by those who decide on its use, e.g., public voters on a referendum.

But the same instrument may be highly effective when it is actually applied. For instance, the building of a high-speed railway line may run up against strong voter resistance. But as soon as it is in service, it will have a considerable effect on the mobility behaviour of many.

4.1. Acceptance

Acceptance of instruments depends, among other factors, on the **cost** of achieving the changes of stimuli each instrument is based upon. Acceptance depends not only on the total cost of a measure, but especially on the costs for various actors. The enforcement of certain laws may be relatively inexpensive for the state and thus for the taxpayers as actors, but it may be very costly for specific target groups, such as motorists, homeowners, farmers, or shopkeepers.

The acceptance of external changes also depends on how they affect the success conditions for non-financial motives. External changes may have an effect on the individual **autonomy** of certain groups. Speed limits reduce the autonomy of motorists. Or some measures may affect the individual **safety** of specific groups. If a railway line removes conductors from late night trains that were formerly staffed, this reduces passenger safety. On the other hand, lower speed limits increase the safety of all traffic participants.

4.2. Efficacy

If an instrument has overcome the difficulties of its implementation, which is a matter of acceptance, its **efficacy in achieving internal effects** depends on internal, i.e., psychological preconditions within the individuals whose behaviour is supposed to be influenced. The better an instrument fits the motivational preferences of the target persons, the more efficient it will be. If a person is strongly motivated to save money, every intervention that reduces the price of desired behaviours and raises the price of undesired behaviours has a good chance of succeeding. The efficacy of an instrument, however, depends not only on the strength of an isolated motive within the target individuals, but also on the relative strength of the motive and individual patterns of solving conflicts among different motives. Suppose that a young woman has a strong goal to save money. In addition, she also has a strong desire to be admired by others, and she thinks that owning an expensive car serves that purpose. Her behaviour will depend strongly on how she deals with the goal conflict between saving money and gaining admiration. If her pattern of conflict solving strongly favours the second goal, making changes in monetary environmental conditions will have only a weak effect. The fact that the volume of road traffic is not always smaller in countries where fuel prices are comparatively high may be a consequence of widely shared individual patterns of solving conflicts among incompatible goals.

The initial effects of all instruments are based upon the goals of the target persons prior to the intervention. If after raising fuel prices, people use public transportation more frequently and drive less, this can be explained by motives and coping strategies that already existed before the change (e.g., "always use the cheapest means of transport possible"). The changed environmental situation caused by an instrument also affects the motivational pattern itself within all people who live in this environment. For instance, higher petrol prices may also create or strengthen the goal to buy more fuel-efficient vehicles. In the short term, higher petrol prices may cause people to use their cars less frequently. In the long term, they may move people to invest in more economical cars. The oil crisis in the 1970s had the effect that people who previously had not cared about the fuel efficiency of cars began to evaluate cars with regard to fuel consumption.

With instruments that mainly target external changes, changed motivational preferences are a consequence of the adaptations of individuals to a changed environment. But instruments that mainly target internal changes can be used to change motivational preferences – and also appraisal and categorisation processes – directly. The efficacy of these instruments depends on how well they are adapted to the motivational state of the target persons at every stage of the process. In an initial phase, the user of a communication instrument must address the target persons differently, talk about different contents, and use different language than in an advanced stage of the process. If low energy consumption is not a goal for people, it does not make sense to start communication by addressing this topic. But it may make sense to talk about saving money or, if the target persons are parents, about the state of our children's environment in the future. The instrument has to speak the addressee's language and build upon his or her needs.

4.3. *Time effects*

When evaluating the different instruments of inducing behaviour change, it is also important to consider the **stability** of their effects. In general, the effects of the instruments that are targeted mainly at external changes tend to disappear after the external changes are revoked. This does not at all diminish their initial efficacy, but it implies that the measures have to be maintained. Promotion of public transportation, for example, cannot be based upon a few temporary special offers, but requires permanent quality and price standards. For as soon as formerly good public transportation deteriorates, mobility behaviour will change.

Instruments directed at internal changes, however, may lead to effects that persist even after the intervention. This is of utmost importance, because the possibilities of controlling external conditions are limited. Command and control depends upon a certain level of supervision. In situations where this supervision is not possible, behaviour depends on the functioning of internal control. Anti-littering laws are effective only where the risk of being caught is sufficiently high. Whether or not a person will litter when he or she is out hiking alone depends on internal motives that are connected neither to command and control, nor to costs (such as the motive to keep the environment unspoiled). Through the implementation of communication and diffusion instruments, we can address such motives specifically.

The persistence of communicative influence depends on how convincing the messages are. And this depends on how well a message fits into the pre-existing motivational patterns of the addressees. If a message seems to be consistent with and relevant to personal goals, people are motivated to pay attention and to think about the arguments provided.

4.4. *Potential of self-diffusion*

The initial diffusion of any mental content or behaviour by instruments to induce behaviour change affects the target persons of those instruments. In a second step, the changes in the behaviour of those target persons can have effects upon the behaviour of persons who were not involved in the initial diffusion. Each target person can have, voluntarily or not, an effect on the behaviour of others. Whether or not intended by the model, some model behaviour will be copied by others. These secondary effects are based upon **self-diffusion** processes that are not under the control of the original instruments or their originators. However, instruments of inducing behaviour change can deliberately promote the self-diffusion of behaviour.

Bandura (1986) discusses a number of factors that determine the likelihood and speed of behaviour diffusion. Diffusion processes are enhanced if the conditions of the former behaviour alternative are altered. Economic instruments and service and infrastructure instruments can enhance the diffusion of behaviours by improving the success conditions of desired behaviours and deteriorating the success conditions of undesired ones. Of course, command and control instruments also affect the success conditions of selected behaviour alternatives. Since the diffusion of behaviours depends to a large extent on observational learning of model behaviour, the factors that promote observational learning also promote diffusion. An important factor is the **functional value** (Bandura, 1986) of the model behaviour. The more successful a model seems to be when pursuing a goal with a certain means, the more attractive the observed means is. The perception of the functional value of a behaviour depends strongly on the subjective relevance of specific goals in the eyes of the observer. To influence the propagation of behaviours, we can stress the relevance of certain goals and play down the relevance of others. A campaign, for example, can stress the relevance of child safety and play down the relevance of travel speed.

To put it more generally, the self-diffusion of any cultural content – of behaviours or mental contents – depends on the benefit for its bearer. Cultural contents that are more useful or pleasant for their bearer are more likely to be transferred to others and are thus more likely to proliferate. Dawkins (1989) adapted Darwin's (1964) biological theory of evolution to memory contents. He called them – in analogy to genes – memes. Like genes, memes underlie the processes of variation and selection. A behaviour or mental content that brings its bearer reward is more likely to be reproduced than one that brings no reward or one that causes inconvenience. Some behaviours or mental contents are evolutionary stable, whereas others will be selected out by their bearers. How useful and attractive a certain behaviour will be to the bearer depends on his or her goals. In order to promote certain behaviours it makes sense to influence those goals by communication and diffusion instruments. The most effective way to do this consists in **combining measures** that affect action conditions – command and control instruments, economic instruments, service and infrastructure instruments – with communication and diffusion instruments in order to enhance the acceptance of those measures and the attractiveness of the desired behaviours.

5. Summary and conclusions

To induce behaviour change, we have to take into account that individuals are motivated by goals. All interventions to transform the behaviour of individuals have to fit into the present structure of active goals or they have to transform this structure so that the desired behaviour becomes – in the particular situation and from the perspective of the target individual – the self-evident option.

To achieve this, we must transform either the objective situation (external structure) or the individual patterns of perceiving, appraising, and coping with the situation (internal structure). In general, a combination of strategies is more effective than trying to change just one parameter within the functional circuit of the situation-person interaction. Mental change enhances people's willingness to accept situational change and the effects of situational pressures upon behaviour. On the other hand, situational change stabilises new mental patterns, if they become more convenient for the person. As long as the discrepancy between subjective goals and situational reality is not reduced by changed preferences, an individual will try to change the situation as soon as the opportunity arises. For this, constellations with preference-situation accordance are more stable. The exclusive use of one class of instruments is less effective and leads to less stable results than the well co-ordinated use of all available instruments.

References

Ajzen, I. and Fishbein, M. (1980) Understanding attitude and predicting social behaviour. Prentice-Hall, Englewood Cliffs, New Jersey.

Baars, B.J. (1988) A cognitive theory of consciousness. Cambridge University Press, Cambridge, MA.

Bandura, A. (1986) Social foundations of thought and action: A social cognitive theory. Prentice-Hall, Englewood Cliffs, New Jersey.

Bargh, J.A. and Barndollar, K. (1996) Automaticity in action: The unconscious as repository of chronic goals and motives. *In*: The Psychology of Action: Linking Cognition and Motivation to Behaviour, Gollwitzer, P.M. and Bargh, J.A. (eds.), Guilford, New York.

Bischof, N. (1985) Das Rätsel Ödipus: Die biologischen Wurzeln des Urkonfliktes von Intimität und Autonomie. Piper, München.

Bronfenbrenner, U. (1979) The ecology of human development: Experiments by nature and design. Harvard University Press, Cambridge, MA.

Carlsson, U. and von Feilitzen, C. (eds.) (1998) Children and media violence. *Yearbook from the UNESCO International Clearinghouse on Children and Violence on the Screen.* Nordicom, Göteborg.

Centerwall, B.S. (1992) Television and violence: the scale of the problem and where to go from here. *The Journal of the American Medical Association (JAMA), 267*: 3059-3063.

Cialdini, R.B., Reno, R.R. and Kallgren, C.A. (1990) A focus theory of normative conduct: Recycling the concept of norms to reduce littering in public places. *Journal of Personality and Social Psychology, 58/6*: 1015-1026.

Comstock, G. and Paik, H. (1991) Television and the American child. Academic Press, London.

Darwin, Ch. (1964) On the origin of species: A facsimile of the first edition. With an introduction by Ernst Mayr. Harvard University Press, Cambridge, MA.

Dawkins, R. (1989) The selfish gene. New Edition. Oxford Press, Oxford

Drabman, R.S. and Hanratty Thomas, M. (1974) Does media violence increase children's toleration of real-life aggression? *Developmental Psychology, 10/3*: 418-421.

Foppa, K. (1988) Über Möglichkeitsräume von Handlungen. *Psychologische Beiträge, 30*: 248-254.

Frey, B.S. and Foppa, K. (1986) Human behaviour: Possibilities explain action. *Journal of Economic Psychology, 7*: 137-160.

Frey, B.S. (1990) Ein ipsatives Modell menschlichen Verhaltens. *In*: Ökonomie ist Sozialwissenschaft: Die Anwendung der Ökonomie auf neue Gebiete, Frey. B.S. (ed.), Vahlen, München.

Geller, E.S., Winett, R.A. and Everett, P.B. (1982) Preserving the environment: New strategies for behaviour change. Pergamon, New York.

Grossman, D. and DeGaetano, G. (1999) Stop teaching our kids to kill: A call to action against TV, movie and video game violence. Crown, New York.

Gutscher, H., Hornung, R. and Flury-Kleubler, P. (1998) Das Transaktionspotentialmodell: Eine Brücke zwischen salutogenetischer und pathogenetischer Sichtweise. *In*: Gesundheits- oder Krankheitstheorie? Saluto- versus pathogenetische Ansätze im Gesundheitswesen, Margraf, J., Siegrist, J. and Neumer, S. (eds.), Springer, Berlin.

Latané, B. and Darley, J.M. (1970) The unresponsive bystander: Why doesn't he help? Appleton-Crofts, New York.

Luce, R.D. and Raiffa, H. (1990) Utility theory. *In*: Rationality in Action: Contemporary Approaches, Moser, P.K. (ed.), Cambridge University Press, New York.

Neisser, U. (1976) Cognition and reality: Principles and implications of cognitive psychology. Freeman, San Francisco.

Ryan, R.M., Sheldon, K.M., Kasser, T. and Deci, E.L. (1996) All goals are not created equal. An organismic perspective on the nature of goals and their regulation. *In*: The Psychology of Action: Linking Cognition and Motivation to Behaviour, Gollwitzer, P.M. and Bargh, J.A. (eds.), Guilford, New York.

Schank, R. and Abelson, R. (1977) Scripts, plans, goals and understanding: An inquiry into human knowledge structures. Erlbaum, Hillsdale, New Jersey.

Searle, J.R. (1969) Speech acts. Cambridge University Press, New York.

Stern, P.C. (2000) Toward a coherent theory of environmentally significant behaviour. *Journal of Social Issues, 56/3*: 407-424.

Tanner, C. (1998) Die ipsative Handlungstheorie: Eine alternative Sichtweise ökologischen Handelns. *Umweltpsychologie, 2/1*: 34-44.

Tanner, C. (1999) Constraints on environmental behaviour. *Journal of Environmental Psychology, 19*: 145-157.

Tesser, A., Martin, L.L. and Cornell, D.P. (1996) On the substituability of self-protective mechanisms. *In*: The Psychology of Action: Linking Cognition and Motivation to Behaviour, Gollwitzer, P.M. and Bargh, J.A. (eds.), Guilford, New York.

Ulli-Beer, S. and Kaufmann-Hayoz, R. (1998) Ipsatives Handlungsmodell. Grundzüge und Diskussionsansätze. Arbeitsbericht Nr. 5 aus dem Projekt "Veränderungshindernisse" im SPPU. IKAÖ, Universität Bern.

Watson, J.B. (1970) Behaviourism. Norton, New York; *first edition 1924.*

Watzlawick, P. (1967) Pragmatics of human communication: A study of interactional patterns, pathologies, and paradoxes. Norton, New York.

Empirical Studies on Selected Instruments and Instrument Combinations

Changing Things – Moving People.
Strategies for Promoting Sustainable
Development at the Local Level
R. Kaufmann-Hayoz/H. Gutscher (eds.)

Between Efficiency and Sufficiency. The Optimal Combination of Policy Instruments in the Mobility Sector towards Sustainable Development

Ursula Mauch, Nicole North, and Raffael Pulli

There is a consensus within environmental science that sustainable development can be achieved only through a combination of varying instruments, including environmental policy instruments. The present contribution examines so-called sufficiency and efficiency strategies in the mobility sector.

Efficiency, or "eco-efficiency", strategies are linked to technological innovations, such as the catalytic converter or the three-litre engine, but also to price increases on petrol, which all ultimately aim towards an increase in resource efficiency. In the main, efficiency strategies are implemented through economic or command and control instruments. In contrast, sufficiency approaches always contain an element of renunciation of resource use, such as when people give up the use of private vehicles. Sufficiency means more intensive utilisation or shared utilisation of goods, as in car sharing schemes, for example. Sufficiency strategies are frequently implemented through service and infrastructure instruments and through communication and diffusion instruments.

In this chapter, we analyse such service and infrastructure instruments as car sharing ("Mobility Switzerland") as well as approaches based on communication and diffusion instruments, such as Global Action Plan for the Earth (GAP) and the Model Bicycle City of Münsingen (Velomodellstadt Münsingen: Veloville). We compare the ecological effects of these efforts to the effects that can be achieved through economic and command and control instruments. The focus of the inquiry is the issue of the optimal combination of instruments to reduce the volume of traffic and, thus, traffic's serious environmental impact.

1. Introduction

1.1. *Background*

1.1.1. *A debate on concepts of sustainability*

In the framework of the recent scientific and sociopolitical discourse on models, concepts, and strategies of sustainable development, two trains of thought stand out (see, among others, Hofmeister, 1999):

- The conceptual-strategic approach using the efficiency principle. This is known as the "harder" sustainability strategy. Methodologically, the strategy is manifested in approaches that seek to measure and increase resource productivity and resource efficiency. The challenge is directed principally towards producers of goods.
- The conceptual-strategic approach emphasising the "sufficiency principle" as the "softer" sustainability strategy. At the level of resource management, however, this strategy has basically the same aims. The challenge is directed in the main to consumers to alter their life styles and shift to more sustainable patterns of consumption.

The two approaches share the goal to alleviate environmental problems by means of a reduction in resource and energy turnover. In this way, both approaches are quantity-oriented. While the efficiency strategy seeks to boost the productivity of raw material resources, the sufficiency strategy targets a change in consumption behaviour. To overstate the case somewhat, we can say that both approaches prescribe restriction and renunciation (Hofmeister et al., 1999a).

Researchers today agree that sustainable development can be achieved only through a combination of efficiency and sufficiency strategies. From a political perspective, however, there are fundamental contradictions relating to the scientific consensus. Representatives of industry and liberal economists take the efficiency strategy approach and demand that corresponding economic instruments be implemented to provide incentives (via the price mechanism) for increasing efficiency (BCSD, 1994). Others conclude that in addition, or even principally, a real change in sufficiency has to occur for environmental and resource consumption not to exceed the carrying capacities of ecological systems. Environmental policy measures that follow the sufficiency approach are mostly service and infrastructure instruments or communication instruments.

1.1.2. *Individual mobility as a core problem*

The struggle for environmentally sustainable solutions is most evident in the mobility sector. In Switzerland, transport is the main producer of CO2 emissions, is a major source of noise pollution, and consumes approximately 33% of the total final energy consumption. Compared to 1980, transport-related energy consumption has increased by 50%. Between 1980 and 1995 the energy demand for private transport rose by 20%. The total distance travelled by the Swiss per year is 100 billion kilometres (billion = a thousand million, or 109). The growth in the sector is mainly due to private motorised transport (BUWAL and BFS, 1997). The greatest part of the total distance travelled is leisure time travel; of 100 kilometres travelled, 50 are travelled in leisure time, while only 25 make up the commute to work or school. The remaining 25% are split equally between shopping and business trips. On average, each person travels 34 kilometres a day, travelling 20% of this distance by public transportation. Three-quarters of all households have one or more cars (BFS, 1999). There is thus a great need for action in the mobility sector to reduce energy consumption, particularly in view of national and international policies on the impact on climate.

Various solution attempts to make transport more environmentally sustainable involve efficiency and sufficiency approaches. For example, technological innovations, such as the catalytic converter or the three-litre engine car, as well as the implementation of economic instruments, such as fuel price increases, are measures to increase efficiency and reduce resource use. Conscious renunciation of personal vehicle ownership with a shift to public transportation or a car-sharing organisation are examples of the sufficiency approach.

1.2. Guiding questions and aim of this contribution

Until a few years ago, it seemed unlikely that the political process would approve the implementation of government incentives through economic instruments to increase resource efficiency.[1] This led to increased promotion and implementation of instruments that integrate a sufficiency strategy and, thus, a voluntary component. The main force behind this movement were private actors and local authorities rather than the cantons or the federal government.

Our main interest in this chapter is directed to instruments that show the characteristics of the sufficiency strategy. On the one hand, the goal of these instruments is to reduce traffic levels and thus traffic's impacts on the environment; on the other hand, the instruments aim to effect a change in the goals and knowledge of the target group towards more sustainable patterns of mobility behaviour. The main issues to be examined are:

- What are the ecological effects of these instruments with regard to energy savings and emissions reduction (outcome)?
- How can the ecological effects of these instruments be reinforced by combining them optimally with economic and command and control instruments of the efficiency strategy type (implemented by the government)?

---

[1] This might have changed had the proposals for energy taxes been approved by voters in September 2000. The CO_2 Law that passed in mid 2000 makes a provision for a CO_2 tax by 2004, should CO_2 target values not be reached by then. The CO_2 Law would raise the price of petrol by CHF 0.50 per litre.

Following a detailed presentation and evaluation of the efficiency and sufficiency strategies in section 2, section 3 investigates selected sufficiency strategy instruments in the mobility sector with regard to ecological effects as well as to success factors and barriers. We will look at car sharing (Mobility CarSharing Switzerland), the Household EcoTeam Programme of the Global Action Plan for the Earth (GAP), and the Model Bicycle City of Münsingen (Veloville). Following a comparative analysis of the ecological effects of the different instruments, we draw conclusions regarding the optimal combination of efficiency and sufficiency strategies and the corresponding instruments in section 4.

2. Two approaches for sustainable development

2.1. *More prosperity – less resource consumption: The efficiency revolution*

How can you raise the standard of living and at the same time lower resource consumption? This is possible only if each economic activity results in more advantages and less environmental impact, which means that fewer resources are implemented more efficiently. This is the strategy that is associated with the "efficiency revolution".

Pointing the way in 1995, the Wuppertal Institute proposed an increase in resource efficiency, or resource productivity, by a factor of four to ten in the next 30 to 50 years in industrialised countries. The book *Factor Four: Doubling Wealth, Halving Resource Use* (Weizsäcker et al., 1997) presents fifty examples of at least quadrupling resource productivity (energy use, materials use). Since then, the goal of factor four to ten has found tremendous resonance at the international level. Austria incorporated the concept into its environmental policy planning, and Finland and Sweden are discussing it at the programme level. In 1997, the special session of the General Assembly of the United Nations took over the Factor Four goal.

The resource efficiency of the total economy can be measured via the ratio of gross national product and the quantity of primary materials used. It is important to take into account that this efficiency measure is not a direct expression of the absolute level of resource use (both can increase with strong economic growth). However, it is just this level of resource use that must be reduced for actual alleviation of environmental impact (Bringezu, 1999). Efficiency increases are quickly negated by growth effects ("rebound effect"). The best example is cars. Through technological improvements in fuel efficiency, the average fuel consumption of cars per 100 kilometres dropped 10% from 1980 to 1998, while at the same time the total fuel consumption by cars increased by 30%. A rise in fuel efficiency from 10 kilometres per litre to 33 kilometres per litre (= 3 litres per 100 kilometres) is simply not enough. Ultimately, polluting carbon fuels will have to be replaced. There is not enough to be gained ecologically by constructing ever more efficient internal combustion engines, because fossil fuel reserves will inevitably be depleted, even if at a later date. What is more, environmental degradation has not been halted. From this perspective, the efficiency approach is not innovative (Hofmeister et al., 1999).

All in all the efficiency revolution, as described in *Factor Four*, is not yet taking place or is progressing very slowly. The main reasons for this lie in incorrect incentive structures or a lack of them in society as well as distortion and failure of the market, for example through subsidies. Moreover, investments to improve efficiency compete against other investments that can return yields of 15% per annum. Conserving resources would have to produce similar yields, and with today's structure of the economy, that is possible only in small sub-sectors (Schumacher, 1997). The call for the efficiency revolution is thus in the main linked to implementation of economic instruments that, via the price mechanism, create incentives to increasing efficiency and that are held to be cost-efficient altogether.

2.2. *New models of prosperity: The sufficiency revolution*

The strategy of sufficiency goes back to the debate in the 1970s on self-limitation or – practically unanimously rejected – restriction through coercion. Sufficiency always entails the foregoing of personal material goods and consumption. It points to a new conception of affluence that is not measured in income or expenditures, but focuses on qualitative goals (fulfilment, solidarity, community, clean environment, and so on) that are to be implemented in a new lifestyle and new means of production. The new models of affluence attempt to make a simple, modest lifestyle socially acceptable that relieves the strain on the environment and hurts no one. Of all the goals, the highest priority is given to protection and conservation of the natural environment.

Granted, implementation of this guiding model is doubtful in the face of the lack of positive resonance in society. This idealistic approach, which involves modest consumption and the kind of wealth that is based in a stronger orientation to moral values, does not have much of a chance to gain a political majority. This is most certainly the case on the global scale (see here, among others, Fischer, 1995).

Sufficiency also means, however, that individuals and the community more frequently ask themselves if material possessions are truly necessary to well being. And they ask whether certain goods might be utilised more intensively. Telephones, lawnmowers, photocopy machines, televisions, computers, and cars are things that many people only wish to use. Under the strategies for more intensive use, there are numerous options for community use of goods through renting, sharing, or borrowing. Common (simultaneous) use is already the norm in many areas: from the "invisible" utility systems for water, electricity, telephone, and sewage; to public roads, parks, forests, and lakes; to aeroplanes, trains, buses and trams, concert halls and cinema theatres; to company car pooling. Divided use (in temporal succession) of goods is also practised in many areas: part-ownership of company planes; taxi companies; video rentals; launderettes; time-sharing through car sharing associations (see here section 3), washing rooms in multiple-family homes; or sharing of tools among neighbours (compare Bringezu, 1999 and Stahel, 1999).

Unnoticed by many economists, large markets for divided, or shared, use have already developed today. For example, time-sharing of holiday flats has become a worldwide business with a turnover of approximately CHF 6 billion. In addition, many routes to more intensive use of resources, such as sharing or borrowing, are taken by self-help groups and co-operatives, which means that they take place outside the monetary economy (Stahel, 1999).

More intensive, shared use of resources avoids unnecessary waste and can bring economic gains to the users. Shared and more intensive utilisation goes hand in hand with the principle of "using with care", and it bases principally on trust. Misuse may result in expulsion from the "community". Shared utilisation thus supports a shift in participants' values towards sustainability.

2.3. *From efficiency and sufficiency towards consistency?*

The discussion on efficiency and sufficiency has paved the way for a third strategy, called the consistency strategy. Figuratively, consistency means compatibility, acceptability, coherence, agreement. Consistency does not focus primarily on reducing resource use through efficiency gains or renunciation, but rather aims at sustainable management of resources, in very large volumes, through a comprehensively conceived, closed economic cycle. Here the consistency strategy aims at the development of basic innovations that open up fundamentally new paths of technological and product development (compare, among others, Huber, 1995, or Hofmeister et al., 1999a). An example of innovation of this kind is the fuel cell, as opposed to traditional car engines or combustion plants; fuel cells produce no or low emissions. A step further is taken with the general transition from fossil fuels to hydrogen, in the form of solar hydrogen, for instance (Hofmeister et al., 1999a).

Establishing consistency while at the same time increasing efficiency is also the goal of integrated environmental protection according to ISO 14000 standards. This route is no longer a matter of "end of pipe" measures for aftercare of the environment. These standards dictate integrated, proactive protection of the environment. Here possible negative environmental impact is taken into account from the very start of the planning of a product (Schumacher, 1997).

Through the aid of any of the three strategies of efficiency, sufficiency, or consistency alone the goal of environmental acceptability does not seem achievable. A meaningful combination of all three strategies is required. The instruments examined in detail in section 3 and the instruments looked at for comparison purposes in section 3.3 all – in varying ways – integrate efficiency and/or sufficiency strategies.[2]

---

[2] Instruments that follow a pure consistency strategy were not investigated.

3. Instruments with characteristics of a sufficiency approach

3.1. Introducing the case studies

The instruments examined show the following features:

- They develop on private initiative without pressure from above (that is, not upon set instructions from federal or local authorities) and from the bottom-up.
- They are closely related to environmental consciousness and sustainable practices. That is, they aim towards a direct or indirect effect on the environment and an ecological revolution in daily living.
- They integrate environmental policy goals in their activities, or do so at least at the start.
- Participation is on a voluntary basis.
- Participants accept higher costs and/or changes in lifestyle.
- They thus contain the fundamental elements of a sufficiency approach.

The instruments showing sufficiency characteristics that we examined fall in the typology under service and infrastructure or communication and diffusion instruments. A sufficiency strategy can be followed and implemented by individuals, households, or communities. At the level of the individual, we will look at car sharing (Mobility Switzerland), at the level of households we examine the Household EcoTeam Programme of Global Action Plan (GAP), and at the community level we turn to the Model Bicycle City of Münsingen (*Veloville*). The following describes the most important aspects of these instruments as well as their goals (sections 3.1.1 to 3.1.4) and assesses their ecological effects and market potential (section 3.2.). A comparison to the ecological impacts and market potentials of economic instruments and command and control instruments follows in section 3.3. Finally, success factors for raising the market potential are outlined (section 3.4).

3.1.1. Reducing traffic by sharing cars (Mobility Switzerland)

Fifteen years ago, the idea that several persons could share one car in order to save on costs and to reduce environmental impact gained organisational impetus with the forming of two associations, Sharecom and ATG (*Autoteilet Genossenschaft*). The idea continued to grow and develop, and since 1993, it has been promoted and supported, financially and conceptually, by the federal government's energy action programme Energy 2000.[3] In 1997 the two organisations merged to form Mobility CarSharing Switzerland. By the end of 1999, 1,200 cars were available for use by approximately 30,000 participants in the scheme.

Customers, both individuals and companies, use car sharing vehicles in different ways. Private individuals (about 83%) tend to use the vehicles mainly for leisure-time activities and also for

3 See also the study by BÄTTIG AND BALTHASAR.

goods transport or big shopping trips. Companies use the cars for business trips, but for leisure-time activities and shopping as well. At present, Mobility CarSharing's offer of cars for holiday travel, daily shopping, and the commute to work have not found much response. This will change in future, however, particularly with regard to holiday trips, as a result of collaboration with the Swiss Federal Railways and other public transportation providers. Participants will have at their disposal car sharing vehicles, which are often stationed at the railway stations, all over Switzerland (Muheim, 1998).

Theoretically, there are 1.7 million potential customers for car sharing. They range in age from 18 to 74, are in the possession of a driver's license, and live in developed zones of municipalities of over 2,000 residents. In a survey of this group, 613,000 people stated that they could well or very well imagine participating in the scheme. This corresponds to 9% of the population of Switzerland (Muheim, 1998).

3.1.2. Reducing household related environmental problems (Global Action Plan)

GAP was founded in 1989 as a non-profit organisation in the United States. GAP's main goal is to promote sustainable lifestyles through educational instruction in easy and concrete resource-efficient practices. In 1990, GAP set the target to reduce water use by a third, household waste by 75%, and CO_2 emissions by 20% by the year 2000. Programmes were developed for communities, households, schools, and companies. The GAP strategy is based upon Roger's Diffusion of Innovations theory (Rogers, 1995), which is a model of the gradual acceptance of new ideas in society.[4]

In GAP's Household EcoTeam Programme, five to eight households form a team to adopt resource-efficient practices in the areas of waste, water, energy, mobility, and consumer behaviour with the aid of a workbook of easy-to-use suggestions for practice.[5] In order to demonstrate the results of the programme and their own actions, participants measure their usage at the start and at the end of the household programme. The programme is built on three pillars: specific suggestions for practice based on sustainability goals, active support and advice to teams, and feedback by means of collecting results and reporting back to the teams.

In the mobility sector, GAP suggests a number of measures to reduce mobility-related environmental impacts. They include giving up driving your own car and switching to public transportation; giving up driving in part; fuel-saving driving practices and adherence to speed limits; the

---

4 The model identifies five different phases of diffusion (innovators, early adopters, early majority, late majority, laggards) and describes the groups within a population that tend to accept an innovation at each phase. As soon as an innovation has been adopted by enough individuals within a particular group, it achieves the "critical mass", and the innovation's further rate of adoption becomes self-sustaining. GAP was designed to facilitate fast development of this critical mass in "early adopter" groups, neighbourhoods, and communities (Market Street Research Inc., 1996).

5 For example, weather-strip doors and windows, turn hot water boiler temperature to 55 to 60 degrees, replace lightbulbs with energy-saving bulbs, separate trash, install a flush-interrupter to toilet tank, calculate commuter expense, experiment with using public transportation to work, choose durable products, and so on.

forming of commuter car pools; purchase of a more fuel-efficient vehicle or electromobile; more rational and efficient planning of business trips; and foregoing flights to vacation spots and long holiday trips by car.[6]

3.1.3. Promoting bicycle mobility by a community based action (Veloville Münsingen)

Münsingen, the regional centre of the Aare Valley between Bern and Thun, has a population of more than 10,000. With a volume of 20,000 vehicles, traffic jams the centre of town every morning and evening. As a consequence of the large proportion of traffic in the town itself, alternatives within the town were required.

In the framework of the project "Traffic Management in Energy Towns", initiated by the sponsor association *Energiestadt*[7] in collaboration with the federal energy action programme Energy 2000, an energy-saving campaign was started in Münsingen. The conception and planning of the campaign was turned over to a broad-based committee (representatives from town administration, business, local citizens). The committee, in co-operation with local businesses, decided to hold a special bicycle stand event.

On the day of the event in May 1995, bicycle stands were set up in front of local shops and stores. Münsingen was christened "*Veloville*" (bicycle city), and new, blue *Veloville* markers were added to the official town signs. Information was distributed to all households, and local clubs and associations held numerous special bicycle events. The bicycle stand campaign was a success with the public and was received with great interest by the national media. The results of the campaign were evaluated in an interview survey of participants and external observers (INFRAS, 1997).

---

6 Within the framework of the Swiss Prority Programme Environment (Swiss National Science Foundation), the Centre for General Ecology IKAÖ, IKF, University of Zurich, and INFRAS collaborated in 1997/1998 on a written survey of GAP participants, whereby the present authors investigated the topic of mobility. In addition to questions about everyday practices, the participants were asked about their choice of transportation, travel distances, political attitudes, and attitudes towards GAP measures in the area of mobility. See here also BRUPPACHER AND ULLI-BEER and VATTER ET AL.

7 *Energiestadt* (Energy City) was founded in 1989 and is sponsored today by NGOs and the Swiss Federal Office of Energy.

3.1.4. Summarising the case studies

The case studies can be summed up as follows:

Table 1: The action programmes at a glance.

Case study	Sponsor or provider	Goals	Implementation
GAP	Independent NGO	Aims to promote the development of environmentally sustainable lifestyles	Simple and easy-to-do instructions for practices that can be applied in groups
Mobility CarSharing	Private association	Aims at cost savings and reduction of environmental impact by having many persons share a small number of vehicles	Nation-wide fleet of vehicles and combined mobility through collaboration with public transport organisations
Veloville Münsingen	Broad based working committee of administrators, businesses, and citizens	Aims to improve the traffic situation within the municipality	Numerous events in a bicycle campaign and a commitment from the municipality to make the centre more bicycle-friendly

3.2. Possible ecological outcome of instruments incorporating a sufficiency approach

3.2.1. Mobility CarSharing

While initially mainly people who did not own cars joined the car sharing organisation, there have been an increasing number of new memberships in connection with the purchase of a car. Total mobility in kilometres has not changed significantly as a result of CarSharing membership. Whether members or non-members of CarSharing, people travel a distance of between 13,000 and 16,000 kilometres by private and public transportation per year. Only when people give up their cars as a result of joining the CarSharing scheme is their travel reduced by any notable amount (see Figure 1).

By increasing the proportion of mobility by public transport, people who have joined CarSharing have reduced their energy use (fuel) on average by 4,200 megajoules (MJ) per year (from 14,000 MJ to 9,800 MJ). With about 30,000 active customers at the end of 1999, this resulted in total energy savings of approximately 130 terajoules (TJ) per year, which is more or less equal to the annual power consumption of a small municipality with a population of 5,000 residents.

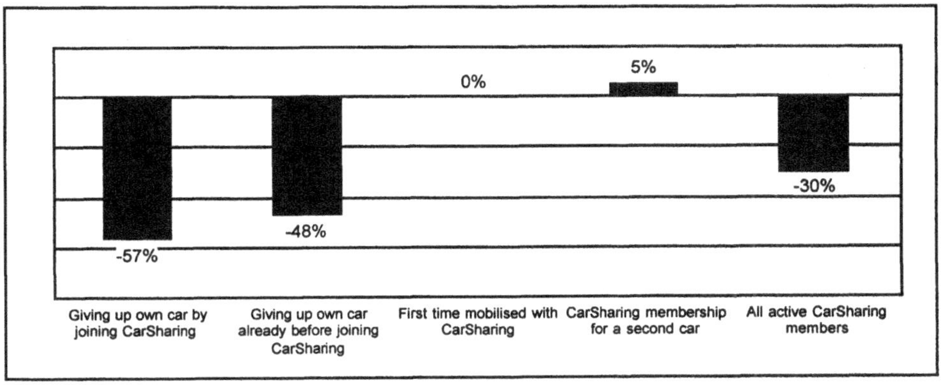

Figure 1: Energy savings according to customer category (source: Muheim, 1998). Through joining CarSharing, the average personal energy requirements for private motorised and public transportation can be greatly reduced (100% = without CarSharing).

With about 22,000 MJ fuel requirements per year, persons using their own private automobiles have a higher energy consumption than participants in CarSharing. If these potential customers join in, the average energy savings increases, because the overall impact of CarSharing depends on the number of users and the proportion of those who give up car ownership. As the proportion of people with their own cars decreases, the level of customer-specific energy savings increases to an estimated figure of 6,400 MJ per year. If the theoretical customer potential of over 600,000 interested persons were fully exploited, CarSharing could reduce motorised mobility by 1,700 million kilometres (2.3% of the total personal mobility demand). Moreover, distances travelled by individual cars are travelled in a more energy-efficient manner due to the fuel-saving fleet of cars. The specific fuel requirements decrease by 17%. All together, the result is a **reduction of fuel requirements by a total of 3.9 petajoules (PJ) per year**, or about 3% of total fuel consumption by private motorised transport (Muheim, 1998).

3.2.2. Global Action Plan

In Switzerland more than 1,000 households have worked through the GAP Household EcoTeam Programme to date. According to GAP 1996, the results of over 100 GAP teams yielded savings of 18% in trash production, 14% in water use, 8% in energy use, and 13% in CO_2 emissions through motor travel. These figures reflect only the energy savings during the course of the programme, however, and do not include eventual continued savings. Significant improvements can be achieved mainly in the areas of waste and consumption, with partial success in the area of water. On the other hand, most of the suggestions in the areas of mobility and energy can not be put into practice. Problems arose whenever implementation of the measures involved high investment costs, great efforts in terms of time, significant reduction of comfort, and inadequate or lacking infrastructures. Other measures that were difficult to put into practice were as a rule those that required co-operation with other, less interested persons (Känel et al., 1998).

According to our survey at the start of the programme, participants tend to question the effectiveness of the measures proposed by GAP. Participants thus find it difficult to apply the practices suggested for the area of mobility, as they assess the potential effectiveness to be low. Over half of the participants are willing to practice the measures of careful, fuel-saving driving and adherence to speed limits (see Table 2). At the same time, however, these two measures are judged to be the least effective measures with regard to the environment. Participants find measures more effective that involve total renunciation of cars, respectively a shift to the use of public transportation. On total then, the participants demonstrate a moderate degree of willingness to put the suggested measures into practice.

Table 2: GAP participants' assessment of GAP measures in the mobility sector.

GAP activity	What measure do you intend to put into practice in the coming months?	What is your assessment of the environmental effectiveness (impact) of the measure?			
		low	moderate	high	no response
Give up driving altogether, shift to public transportation	40%	8%	11%	75%	6%
Give up driving in part	41%	10%	73%	11%	6%
Careful, fuel-saving driving and adherence to speed limits	51%	29%	54%	11%	6%
Form commuter car pools	0%	8%	60%	17%	14%
Purchase a fuel-efficient car or electromobile	2%	16%	46%	24%	14%
More rational and efficient planning of business trips	6%	19%	40%	29%	13%
Forego flights to vacation destinations and long holiday trips by car	25%	10%	16%	68%	6%

Planned implementation of measures and assessment of their effectiveness (survey of 63 participants in the GAP programme).

No definite prognosis can be made as to how long the measures in the area of mobility remain effective. Due to the problems in implementation, the medium- and long-term effects of the measures are certainly lower than the figures provided by GAP.[8] There is an additional factor besides: according to our survey, GAP participants already drive only half the average Swiss personal travel distance (about 7,000 km per year), and they are more conscious of environmental issues.

If we assume an optimistic amount of energy savings of about 10%, a potential of 10,000 participants,[9] and effects maintained over a longer term for 10 to 20% of all participants, GAP could achieve a long-term maximum of 0.2 to 0.4 PJ savings in energy per year.[10]

3.2.3. *Bicycle and pedestrian model town (Veloville)*

With its declaration of a *Veloville*, Münsingen has committed itself to long-term promotion of the bicycle. Further campaign events are planned, and issues important to pedestrians have also gained weight. The survey interviews showed that the population would be receptive to a discussion and an examination of traffic and the choice of the means of travel.

Despite that, no change in behaviour on the basis of *Veloville* could be established (INFRAS, 1997). Members of the working committee do not see any reduction of traffic on the main road – on the contrary, they believe that the traffic volume has increased. For this reason, no quantitative figures on energy savings caused by the campaign can be derived.

3.3. *A comparison with command-control and economic instruments*

A number of further instruments are being implemented in the mobility sector or are planned for the near future. There are communication and diffusion instruments (propagating ecological driving) on the one hand, and command and control instruments (speed limits), and economic instruments (petrol price increase, energy tax) on the other. The following energy savings can be achieved in comparison to the instruments examined in section 3.2 (see also Figure 2):

- Ecological driving: Another measure supported by the federal energy action programme Energy 2000 is "eco-driving".[11] If appropriate learning contents are included in driver education courses, a clear and lasting reduction in fuel consumption can be achieved (IAP, 1995). Over the long term, in comparison with the traditional style of driving, eco-driving can yield fuel savings

---

8 According to Graf 1997, 40% of participants in a survey report that energy savings have been maintained up to the present date, while 55% do not respond and thus appear to be sceptical about how long the measures remain effective. The figure for mobility measures must be significantly lower.

9 To be achieved around the year 2010 based on the current diffusion rate.

10 Assumed: 7,000 km per year and person, 8.5 l petrol per 100 km.

11 Eco-driving is a fuel-efficient and thus more economical and more environmentally friendly style of driving (correct choice of gear, low engine revolution driving, avoidance of unnecessary braking by means of a defensive driving style, regular tire pressure checks, and so on).

of 10 to 15% (Energie 2000, 1998). With 40,000 new drivers trained in eco-driving annually (in 1999 about half of all new drivers), an energy savings of 200 TJ can be reached. Each year, of course, there are more new drivers, and this will increase the energy savings through eco-driving. If all drivers in Switzerland were trained eco-drivers, over 20 million PJ of fuel energy could be saved (at 10% of the fuel requirements).

- Speed limits in residential neighbourhoods: A comprehensive reduction of the speed limit in residential neighbourhoods from 50 to 30 km per hour could result not only in increased traffic safety, but also in fuel savings (INFRAS, 1998). This is based on the assumption that fuel consumption decreases at slower speeds. On the whole, we could reckon with energy savings in the amount of about 0.2 PJ annually.[12]
- Price increase on fuel: A comprehensive introduction of a fuel tax of CHF 0.50 per litre would lead to a reduction in distance travelled by car or a change in driving behaviour. A reduction in energy consumption of 10 PJ per year could be achieved (INFRAS, 1998).
- CO_2 Law: The effects of the CO_2 Law on the transport sector are assessed in the framework of the energy perspectives of the Swiss Federal Office of Energy (Prognos, 1996). In comparison with the reference scenario (implementation of all approved measures), the CO_2 Law can achieve energy savings of about 41.5 PJ per year by 2030 (-12% compared to the reference scenario).

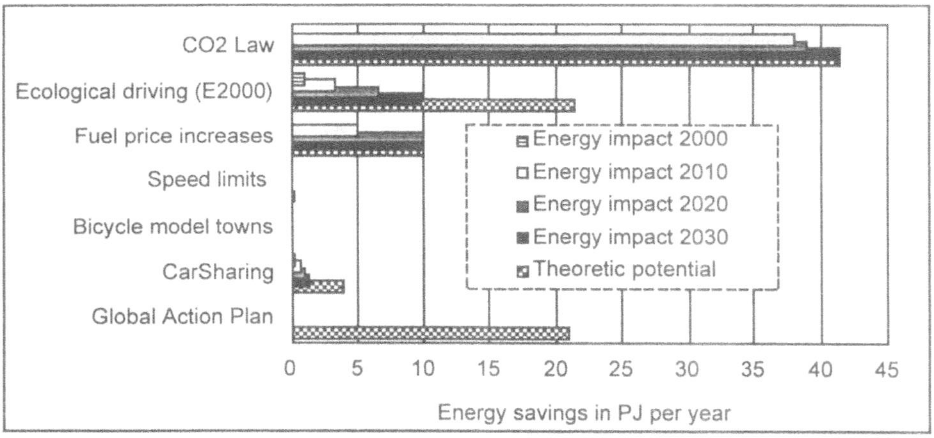

Figure 2: Long-term effects of measures in the transport sector if the potentials are fully exploited.

[12] In the spring of 2000, both the Federal Council and the National Council voted against a general reduction of speed limits in residential neighbourhoods.

Figure 2 shows that the effects in terms of energy reduction that can be yielded by federal government or cantonal economic instruments or command and control instruments (fuel price increase, implementation of the CO_2 Law) as well as communication instruments (ecological driving) that follow an efficiency strategy are considerably greater than the effects of the instruments with characteristics of a sufficiency strategy that we examined above.

3.4. *Success factors and barriers for further implementation*

For the goals set by GAP to be achieved through the measures they propose, two factors are extremely important. First, a change in lifestyle must be sustained over the long term. However, in the mobility sector in particular, the GAP measures are very difficult to put into practice. The barriers appear to include the high investment costs, high expenditure of time, and great sacrifice in terms of comfort. Whether the measures would remain effective in the long term has not been examined in a comprehensive manner. It is significantly easier to apply the measures once and in a limited fashion than to maintain the behaviours through time.[13] It is also probable that a change in lifestyle is more difficult for a person who is not very conscious of environmental concerns than for a person who has a high level of awareness. For this reason, we hardly expect to see a comprehensive change in lifestyles.

As the second important factor, a widespread diffusion of the programme is essential. According to GAP, a critical mass of 10 to 20% of the population is required for successful implementation of the programme, and this could be achieved in three to four years. However, this has not taken place so far. It appears to be difficult to find a sufficient number of interested persons. This was apparent also in our own survey. GAP participants make only cautious predictions in this regard, and they tend to see a high enough potential for participation only in their own circles of friends. The critical mass could hardly be reached in that way. As long as the two factors – changes in lifestyle and diffusion – can not be better implemented, the ecological contribution will remain quite modest.

The situation is somewhat different with car sharing. Car sharing can contribute significantly to environmental sustainability in the mobility sector. For this, the attempt has to be made to tap the full customer potential (600,000 persons). An analysis of present-day users' reasons for participating in the Mobility CarSharing scheme shows a shift from ideological to more pragmatic and cost-oriented arguments. Mobility behaviour changes most with participation if, at the same time, members give up the use of their private cars. This means that car sharing can be made more attractive using two approaches. On the one hand, CarSharing itself can improve the quality and comprehensiveness of its services, for this has a positive effect on the rate of customer growth. On

[13] However, a recent study in Holland indicates that behaviour changes resulting from GAP participation are not only maintained through time, but also become stronger (see here also BRUPPACHER AND ULLI-BEER).

the other hand, positive effects can be achieved through improving the services and status of public transportation and non-motorised means of mobility. A tough barrier to membership in a car sharing scheme is fear of a loss of convenience and comfort. Potential customers also place high value on a wide choice of vehicles and safe, clean, and well-equipped cars. The success of the instruments examined above is thus dependent upon internal, programme-specific and external, programme-independent factors. For campaigns in a small context, such as *Veloville* Münsingen, broad based support and a good information policy seem essential. For national action programmes like GAP or CarSharing, ease of application and the optimisation of framework conditions in the mobility sector must be underlined.

4. Summary and conclusions towards an optimal combination of instruments

The research on strategies for sustainable development has shown that both efficiency and sufficiency strategies have a role to play. Consistency – the economical management of resources by means of a comprehensively conceived economic cycle – must also be taken into consideration, particularly in the development of new technologies. In the face of growing consumption, technological innovations and increased efficiency alone will not suffice, however, to master ecological problems. What is more, efficiency gains are easily negated by growth effects and have in the past led to an increased demand for products and ultimately to an increase in resource use. At the same time, efficient products and services will most likely succeed on the market much more quickly than any shift in values towards sufficiency can take place. In addition, people will continue to show a wide spectrum of individual differences in future. While some people base their self-definition on their material possessions, others tend to seek fulfilment by means of contacts to other people and independent decisions over time. Thus, strategies of the efficiency revolution and the consistency revolution will succeed only if a sufficiency revolution occurs as well. Here community based, democratically organised self-imposed restriction through the sharing of goods can play an important role.

The service, infrastructure, and communication instruments in the mobility sector examined above – car sharing (Mobility CarSharing Switzerland), Global Action Plan's Household EcoTeam Programme, and the Bicycle City of Münsingen (*Veloville*) – show the characteristics of the sufficiency approach. They encourage people to give up their personal cars, to change over to public transportation, or to travel more by bicycle. In comparison to economic or command and control instruments implemented by the authorities that show the characteristics of the efficiency approach, the "sufficiency instruments" yield significantly smaller measurable ecological effects. This is because up to now, sufficiency instruments appeal primarily to people who are already environmentally conscious, while government instruments as a rule apply to the entire population. Only a small part of the population has been reached by sufficiency instruments, which demand a greater willingness to incur costs, seem to restrict the personal freedom to act, and demand a change in lifestyle. If the theoretical customer potential of such instruments were more fully exploited, the ecological effects would be more significant.

A feasible way to strengthen the sufficiency instruments in the mobility sector that we examined would be to implement appropriate economic and/or command and control instruments. Price measures can be seen as the key to changes in mobility behaviour. This has also been acknowledged by earlier studies that concluded that instruments integrating price incentives are much more effective than sufficiency approaches or command and control instruments, especially in the energy and transport sector. These studies, however, also conclude that environmental consciousness and voluntary self-commitments can influence environmental action – last but not least in the sense of an important parameter for political decision-making processes (Diekmann, 1994; Frey and Busenhart, 1994).

Therefore, the energy policy planned for Switzerland that foresees price increases for fossil fuels promises to be an optimal basis for reinforcing "sufficiency instruments". Once the government has established these new framework conditions, participation in Mobility CarSharing or GAP will become more attractive to participants from an **economic** perspective as well.

However, as long as fuel prices remain low, the true costs of mobility are not borne, and private motorised transport continues to receive high priority, the growth in private traffic volume can only be checked by means of a big boost to the attractiveness of public transportation services and through the implementation of measures by federal and cantonal authorities. And so for the meantime, for sufficiency instruments to succeed, it is important that participants continue to show a willingness to pay the higher price of environmentally friendly products and services and a readiness to adopt changes in their lifestyles.

References

BCSD/Business Council for Sustainable Development (1994) Getting eco-efficient. International BCSD/UNEP, Institute for Environment and Development. World Business Council for Sustainable Development. Online in the WWW: <http://www.wbcsd.com>

BFS/Bundesamt für Statistik (1999) Statistisches Jahrbuch Schweiz. Bern.

Bringezu, S. (1999) Aus weniger mach mehr. Über den Zusammenhang von Effizienz und Umweltverbrauch. *Politische Ökologie, 17/62*: 14-17.

BUWAL/Bundesamt für Umwelt, Wald und Landschaft and BFS/Bundesamt für Statistik (1997) Umwelt in der Schweiz 1997 – Daten, Fakten, Perspektiven. Bern.

Energie 2000 (1998) 8. Jahresbericht des Aktionsprogramms Energie 2000. Beilagenband 1. Bern.

Diekmann, A. (1994) Umweltbewusstsein oder Anreizstrukturen? Empirische Befunde zum Energiesparen, der Verkehrsmittelwahl und zum Konsumverhalten. *In*: Kooperatives Umwelthandeln, Modelle, Erfahrungen, Massnahmen, Diekmann, A. and Franzen, A. (eds.), Rüegger Verlag, Chur, Zürich.

Fischer, W. (1995) Nachhaltige Entwicklung – eine Norm für die Gestaltung unserer Zukunft. Online in the WWW: <http://www.zam.kfa-juelich.de/tff/tff_sustaina.html>

Frey, B.S. and Busenhart, I. (1994) Umweltpolitik: Ökonomie oder Moral. *In*: Kooperatives Umwelthandeln, Modelle, Erfahrungen, Massnahmen, Diekmann, A. and Franzen, A. (eds.), Rüegger Verlag, Chur, Zürich.

GAP/Global Action Plan Schweiz (1996) Pressespiegel. Zürich.

Graf, E.O. (1997) Energiesparaktionen. BFE/Bundesamt für Energie, Bern.

Hofmeister, S. (1999) Über Effizienz und Suffizienz hinaus. Zur methodischen Weiterentwicklung des Stoffstrommanagements mit Bezug auf Nachhaltigkeit. *Politische Ökologie, 17/62*: 34-38.

Hofmeister, S., Huber, H. and von Gleich, A. (1999a) Wege nach Ökotopia. Kann Nachhaltigkeit ohne Sparsamkeit erreicht werden? Eine Gesprächsrunde. *Politische Ökologie, 17/62*: 8-12.

Huber, J. (1995) Nachhaltige Entwicklung durch Suffizienz, Effizienz und Konsistenz. *In*: Nachhaltigkeit in naturwissenschaftlicher und sozialwissenschaftlicher Perspektive, Fritz, P., Huber, J. and Levi, H.-W. (eds.), Stuttgart.

IAP/Institut für Angewandte Psychologie (1995) Evaluation des Projekts "Eco-Fahrweise: Fahrlehrer- und Expertenausbildung". Zürich.

INFRAS (1997) Evaluation Energiesparaktionen im Verkehr auf Gemeindeebene – Beispiel "Veloville" Münsingen. Energie 2000, Ressort Treibstoffe, Bern.

INFRAS (1998) Kosten-Wirksamkeiten von Umweltschutzmassnahmen im Verkehr. Forschungsauftrag 41/96 auf Antrag der Vereinigung Schweizerischer Verkehrsingeniere, SVI, Zürich.

Känel, E., Magun, B., Öhri, R. and Sanchez, A. (1998) Umweltverantwortliches Alltagshandeln beim Global Action Plan: die Bedeutung sozialer Netze. Schriftenreihe Studentische Arbeiten, Nr. 11. IKAÖ, Universität Bern.

Market Street Research Inc. (1996) The market potential for the household ECOTEAM Programme, prepared for GAP, Northampton, MA.

Muheim, P. and Partner (1998) CarSharing – der Schlüssel zur kombinierten Mobilität. Evaluation der Unterstützung des CarSharing durch Energie 2000. BFE/Bundesamt für Energie, Bern.

Prognos (1996) Energieperspektiven der Szenarien I bis III, 1990 – 2030. BFE/Bundesamt für Energie, Bern.

Rogers, E. (1995) Diffusion of innovations. Fourth Edition. The Free Press, New York, London, Toronto.

Schumacher, L. (1997) Strategien zur Erreichung einer nachhaltigen Entwicklung. Online in the WWW: <http://www.ph-freiburg.de/wirtscha/schumac4.htm>

Stahel, W.R. (1999) Geteiltes Notebook – doppelte Freude? Nutzenorientierung als Strategie für eine ressourcenschonende Gesellschaft. *Politische Ökologie, 17/62*: 63-66.

Weizsäcker, E.U., Lovins, A.B. and Lovins, L.H. (1997) Factor four. Doubling wealth, halving resource use. Earthscan Publications LTD, London.

Changing Things – Moving People.
Strategies for Promoting Sustainable
Development at the Local Level
R. Kaufmann-Hayoz/H. Gutscher (eds.)

Voluntary Collective Action in Neighbourhood Slow-Down – Using Communication and Diffusion Instruments

Heinz Gutscher, Hans-Joachim Mosler, and Jürg Artho

We will show that the implementation of communication and diffusion instruments can successfully motivate a large number of people to change behaviour relevant to the environment. The goal of the intervention was to reduce neighbourhood driving speeds in a district of 10,000 residents quickly, without coercion, and with no investment in costly construction measures.

A combination of various communication instruments was used. We asked people to sign public, written statements of self-commitment; model behaviour was made visible; reminders and feedback were given. The presenting of and persuading about facts, appraisals, and goals as well as sending appeals also played important roles.

Non-media diffusion served as one type of diffusion instrument. This based upon direct personal contact between multipliers and target persons or target organisations. A simple diffusion task was also applied. As a second type, mass media diffusion instruments were implemented: Television and radio announcements, newspaper reports, printed materials, and various forms of posters, as well as dispensers and reminder objects. More than 1,000 people – approximately one-fifth of all registered drivers – committed themselves in writing and publicly to drive more slowly during the campaign phase of five months. The resulting reduction in driving speed was comparable to that achieved elsewhere through conventional command and control measures, i.e., through the implementation of speed limits, police controls, and fines.

1. Introduction

If we want to change the behavioural patterns of large groups of people or entire populations, we have the option, in addition to the classic command and control instruments and economic instruments, to implement communication instruments. Whereas classical instruments influence behaviour patterns in the population through modifying the chances of success of that behaviour, communication instruments target individuals' goals, and the means they use to achieve them, and aim to change individuals' knowledge (compare here the chapter by Flury-Kleubler and Gutscher). In the present study, the local authorities wanted to reduce driving speeds in residential areas quickly – in the sense of an initial measure – and without having to resort to coercion or expensive construction measures. With these goals in the forefront, we recommended the implementation of communication and diffusion measures.

The success of an action campaign depends upon solving two tasks. When communication instruments are implemented, intervention must be followed by diffusion. As the aim is not to produce effects in single individuals, but to affect entire populations, the communication instruments have to be "transported" to the target groups through the aid of diffusion instruments.

The goal of **diffusion instruments** is to

- gain the attention of the correct target persons and to expose them to the psychological effects of the communication instruments used.

The goal of **communication instruments** is to present the contents or messages in such a way that they are

- attended to,
- understood,
- accepted,
- remembered over time,
- and, via the formation of intentions,
- expressed successfully in behaviour (see the typology chapter by KAUFMANN-HAYOZ ET AL., section 3.5.4).

The research goal of the project was to test the use of communication and diffusion instruments in the context of a real-world problem relating to the environment and to investigate the functioning of the instruments in large groups. A suitable task, with a large number of people involved in the problem, was found in the municipality of Münsingen in the Canton of Bern. The goal of a driving speed reduction in residential areas can be conceived of as a new kind of *public good* (see Gutscher et al., 2000). To create this public good, the co-operation of a large number of residents must be secured.

Therefore, the goal of the intervention was to stimulate enough co-operative behaviour to clearly reduce average driving speeds. Co-operation consisted of making a public self-commitment in writing to reduce driving speed to 30 km/h in residential neighbourhoods during the five months of the experimental phase, despite the legal speed limit of 50 km/h.

A reduction of average driving speed contributes to an improvement in the general quality of life. Noise is abated, there is a slight reduction of emissions, and most of all, pedestrian and bicycle traffic becomes safer and more attractive. This latter improvement is a prerequisite if walking and bicycling are to replace private motorised transportation in residential neighbourhoods.

A core issue on the way to sustainable development is the aspect of *collectivity*. A great many problems, particularly environmental problems, are the consequence of the aggregation of a large number of individual "contributions" to the problem. Successful solutions, therefore, require the aggregation of a large number of "contributions" to the solution. Individuals who face a demand to change their behaviour, as induced from the outside and entailing manifold "costs", very often place their behaviour options and corresponding appraisals into the context of others' behaviour. In a concrete sense, this means that people compare the extent of necessary personal behaviour changes to what others do or seem prepared to do. In this situation, a typical response is uncertainty. People are not sure that the goal, achievable only through an aggregation of the behaviour effects of very many other people, can ever be reached. They wonder whether enough people are also willing to make a contribution towards creating the public good. This uncertainty proves to be a core barrier to all collective action (compare the chapter by VATTER ET AL.). The frequently cited fear of the 'free rider' seems hardly significant in comparison (see Gutscher et al., 2000).

A broadly effective new behavioural norm to "slow down" when driving can only be established through *collective action*. Due to the uncertainty that people experience, it is highly unlikely that spontaneous self-organisation would establish that kind of new norm. We thus directed our attention to the central question of whether advantageous conditions for successful collective action can be created through implementing communication and diffusion instruments.

2. The slow-down campaign in Münsingen

The aim was to mobilise enough persons to put the speed reduction goal into practice in their daily driving during a five-month long experimental phase. The plan was to first find a sufficient number of participants and then to have them begin the slow-down in a co-ordinated way (simultaneously). The entire course of the project had four phases. In a preparatory phase, the concept was designed and financial support secured; in a first main phase, participants were sought; the second main phase saw the new behaviour put into practice; and in a post phase, we conducted measurements and performed evaluations.

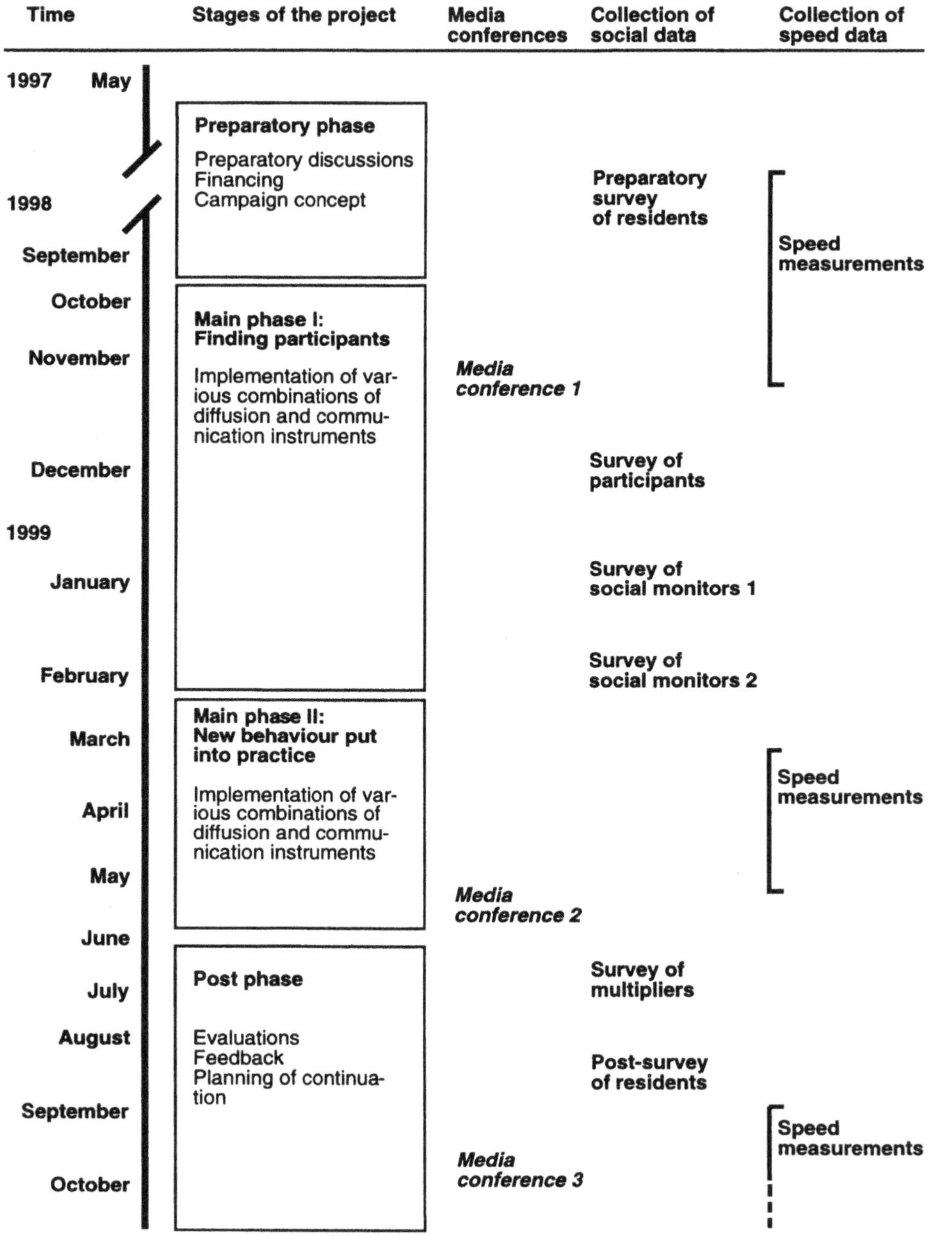

Figure 1: The entire course of the action campaign.

2.1. *Preparatory phase*

The combination of basic research, applied research, and actual implementation posed great difficulties in this case study. Because the object to be studied was the campaign that would be conducted, or the processes that the campaign based upon, we first had to secure financing for the "creation" of the object of the study. On the project side, financing had been granted only for scientific design and conception and consulting during the campaign. The scientific work in this phase had to be restricted to theoretical conception, further development, and theory-based concrete design of the communication and diffusion instruments, as well as preparation for the collection of empirical data. This preparatory phase proved to be enormously expensive in terms of time and lasted from May 1997 to September 1998 (see Figure 1) – almost a year and a half. During this period, scientific results could hardly be produced – a fact that tried the patience of both the research team and the responsible parties within the Swiss National Science Foundation. All those involved became increasingly nervous.

This process of trying to market the project was lengthy and stressful, as failure would put a halt to the project, notwithstanding the basic grant for the scientific part of the work. Looking back, however, we see that the parallel processes of designing the campaign concept and establishing a local sponsor using funds "brought in" could not have been cut shorter. The mutual learning process, the establishment of trust, the mustering of courage to act, and the development of optimism about the success of the campaign – for all of these, the local sponsor required time. Having to take the time yielded a crucial advantage: the sponsor ultimately conducted "its own" campaign and in future will be in a position to itself further develop the impetus created by the project.

Early on, the project secured a central factor in the success of any such campaign: a central, key figure who was willing to act and who served as the core person in the formation of a local sponsor. In Münsingen this key person was an employee of the local planning department and building control office who had been given a green light by his superiors to follow-up with us on the idea of a campaign. As the goal of the campaign lay in the mobility sector, it was natural to include an existing, informal traffic committee in the preparations and to foresee a role for the group as a campaign committee. The town then turned this informal committee, in no small part due to its new and active role in the preparations for the campaign, into a democratically legitimised Traffic Commission and integrated it into the official structures of the municipality, which increased its influence. With this turn of events, both the members and chairman of the committee changed. For us, this meant that we had to expend efforts once again to convince the new members of the value of the campaign. The Commission held a number of meetings that focused on finding financial resources and on intensive discussion and development of the campaign concept and the accompanying research. The research team took part in eight of the meetings, which were held every one or two months. Quite early on, an external communications consultant was charged with designing the campaign materials as well as with public relations work.

2.2. *Main phase I: Finding participants*

As soon as financing for the campaign was secured through contributions from various institutions, it was time to realise the plans. In this phase, from October 1998 to the end of January 1999 (see Figure 1), diffusion and communication instruments were utilised to find as many people as possible who were willing to take part. The initiative and responsibility for the basic concept of the campaign had been ours; now the local working committee took on the task of recruiting participants. This working committee was made up of the Traffic Commission and a co-ordination office.

Co-ordination tasks were taken over by the municipality's planning department and building control office. Their main duties were to make suggestions about the form of the campaign to the Traffic Commission and to put them into practice. The co-ordinators also collected the self-commitment sign-up sheets, lanced the campaign in the press, and conducted speed measurements during the campaign.

2.2.1. *Use of diffusion and communication instruments*

Media and non-media *diffusion instruments* were used to gain participants in this phase I. As a form of non-media diffusion, traffic commissioners acted as multipliers, in that they propagated the goals of the campaign through personal contacts with others in their social circles. Another form of non-media diffusion consisted in a simple diffusion task. Diffusion via the mass media was a further key element. Media conferences were held on radio and television, and regular reports appeared in local newspapers. We also published advertisements on the campaign in the papers. A core element of the diffusion was a flyer, printed in colour, that served as a vehicle for various communication instruments. The flyer was distributed personally by the multipliers and could also be picked up from dispensers that were placed in a number of local businesses and in the town hall. In addition, posters were used, both in the conventional form and also in the form of a highly visible tower erected at a central location ("barometer of success").

The *communication instruments* used in this recruiting campaign phase included presenting facts, presenting appraisals, goals and norms, presenting options, persuading about facts, options, appraisals and goals, and sending appeals. The most important communication instrument implemented was stimulating self-commitment. The purpose of this instrument was to move as many drivers as possible to commit themselves, in writing and publicly, to participate in the slow-down. To support diffusion, various local clubs and organisations were presented as models to the rest of the community by becoming known as members of a "supporter club" participating in the campaign.

2.2.2. *Combinations of diffusion and communication instruments in practical application*

In the participant recruiting phase, Traffic Commission members acted as multipliers. Within the framework of a non-media diffusion strategy, multipliers used verbal communications, both in the form of dialogues (personal conversations with individual persons) and monologues (talks, short announcements to clubs, and so on). The multipliers held personal conversations with individuals who themselves had access to larger groups, such as business owners or club presidents. These individuals were typically personal acquaintances or work colleagues of the multipliers. The multipliers received aids in the form of a fact sheet and a set of informative overhead projector transparencies to use for their talks at club meetings. In all of these efforts, the multipliers implemented a whole range of various communication instruments: they presented facts on the 30 km/h speed limit, demonstrated the option of the new driving behaviour, and introduced the appraisals and goals. In doing so, the multipliers attempted to convince their conversation partners of the validity of the facts, the new behaviour options, the appraisals, and goals. Finally, they appealed to people in an attempt to activate them to make a self-commitment to participate in the slow-down.

Prior to the public phase of recruiting participants, these efforts were directed to the forming of a "supporter club". The goal of this public self-commitment was to gain the agreement of political parties, clubs, associations, and businesses to have their names publicised as active supporters of the campaign. In this way, a certain degree of successful diffusion could be communicated to the public at a very early stage. These institutions and organisations, as reference groups, took on an important orientation function for their own members as well as for greater social circles. Members of the supporter club thus served as models. Moreover, the fact that very diverse groupings, even contrasting political parties, all supported the same campaign created a norm-setting effect: Any cause supported by so many could hardly be controversial. This communicated a feeling of security as well, for from the start, no participant felt alone. The members of the supporter club made sponsor contributions and gave other forms of support. In return, their names were cited in almost all the information materials distributed for the campaign. A total of 20 businesses, 11 clubs and associations, and 6 (of 8) local branches of the political parties were gained as supporter club members.

As compared to media diffusion instruments, the non-media form of diffusion has the crucial advantage that it is flexible and can be adapted specifically to the recipient. The 12 Traffic Commission members, coming from diverse social and political groupings, made successful use of their knowledge of local conditions and the peculiarities of each conversational partner in their face to face attempts to convince others to participate. It was evident that social relationships and commitments played an important role here, for the most successful arguments proved to be the call to "take an active role in our community" and to "make our roads safe". Environmental issues were a less effective argument. We estimate that each multiplier appealed to between 50 and 100 persons. All of the multipliers concluded that they had played a positive and useful role, and they stated that they would be willing to play an active role again in future campaigns of this kind.

The main communication instrument of the campaign was the call to sign a written self-commitment. The direct diffusion instrument was printed information in the form of a flyer that served as the vehicle of the communication instruments. The appeal to self-commitment belongs here, and facts, appraisals, and goals were also presented in the flyer. Signing the self-commitment sign-up sheet in the flyer committed participants to drive more slowly and to keep to a speed limit of 30 km per hour in the neighbourhoods beginning in February, the start of main phase II. With their permission, the names of participants were publicised on posters and in newspaper advertisements in so-called "team player" lists. In this way, those already signed up to participate were presented to others as models. The names of the organisations taking part in the supporters club were also posted. This served mainly to demonstrate the broad based social support that the campaign was receiving in the local population. The flyer was available in dispensers that were placed in various shops, at the town hall, in businesses, and in clubrooms. Self-commitment sign-up sheets could be turned in on the spot or sent in by postage-paid mail.

On three occasions, members of the Traffic Commission and town selectmen manned information booths to gain support for the campaign from passers-by. The booths were set up at the annual Christmas Fair and, for a half-day each, in front of the two most frequented shops in the area. The people manning the stands acted as multipliers, who by establishing direct personal contact with the passers-by, implemented the various communication instruments. This form of recruiting proved to be very successful (see Figures 3 and 4).

Information was released to the media before, during, and after the campaign (see Figure 1). Press releases, here media conferences, are a means to stimulate the mass media to aid diffusion of communication instruments. Mass media utilise well-established, traditional mechanisms of distribution, and they are in a position to direct attention to local events. Media coverage reported in the main on the facts, the behaviour options, and the appeals that were sent out. Due to the local, or regional, significance of the campaign, the media conferences found a response mainly in the reporting of the regional and local media. However, a short report was also shown on national television in the evening news (SF DRS, 25.5.1999, 19:30).

The local press was also used for blanket coverage to produce effects through persuasive communication instruments. We placed regular advertisements in the local papers presenting facts and belaying arguments against the collective action project that became known to us through the preparatory survey of residents prior to the action phase. The key element of the ads was always an appeal for participation. Appeals also appeared on 15 posters, in the form of slogans supporting the campaign.

A month prior to the action phase, registered committed participants received a letter that asked them to pass along information flyers to others and to convince them to participate in the campaign. This form of non-media diffusion is a type of diffusion task (see the typology chapter by KAUFMANN-HAYOZ ET AL., section 3.5.3.) These persons' prior commitment to participate certainly played a crucial role in their motivation to convince others to join in. The success of this type of participant recruiting became evident in a further rise in the number of participants (see Figure 4).

At a roundabout in the town centre we set up an eight-meter high tower as a "success barometer". A moveable counter showed the current number of people signed up to participate in the slow-down. Figure 2 shows that this spectacular method of giving feedback to the community on the status of collective action willingness received a lot of attention, particularly also from non-participants. Due to its function, an installation of this kind can be seen as a special form of the "poster" diffusion instrument (see the typology chapter by KAUFMANN-HAYOZ ET AL., section 3.5.3.). Through the aid of the barometer tower, participants were not only given feedback on the current status of participation, but – once a certain level of participation had been reached – a kind of participation norm was presented.

2.3. *Main phase II: Putting new behaviour into practice*

Once the application phase began, it was time for participants to express their self-commitment through behaviour. Here again, both diffusion and communication instruments were implemented. This phase lasted for five months from 1 February 1999 to the end of June 1999 (see Figure 1).

2.3.1. *Diffusion and communication instruments implemented*

The *diffusion instruments* used in the application phase included coloured flags and electronic boards that displayed the speed of passing cars. Both of these instruments can be seen as particular forms of the poster type. At the start of this phase, all registered participants also received a letter, printed materials in the form of bumper stickers, and a key chain as a reminder object. The local media followed the entire application phase in their reporting.

Serving as *communication instruments* in this phase were appeals, feedback, reminders, prompts, and models. Moreover, the effect of the previous public self-commitments in writing now came to fruition.

2.3.2. *Combinations of diffusion and communication instruments in the application phase*

Just prior to this phase II, all registered participants received a letter thanking them for their commitment to participate. This is a form of feedback. The letter also appealed to those registered to put their self-commitment into practice in their driving, starting in February. Enclosed with the letter were reminder objects. A key chain showing the campaign logo was to serve as a daily reminder. Bumper stickers showing the campaign logo would allow other drivers to identify participants in the slow-down. This made those participating models as well. Through the fact that the participants could now be identified publicly, they also became subject to the effects of social control. This increased the pressure on them to put their self-commitment into practice.

The effectiveness of self-commitment in writing is due in the main to the pressure to be consistent, in intention and behaviour, within one's person. Where self-commitment is also public, there is a pressure towards personal consistency that as a rule leads people to fulfil the promises that they make.

In order to mark the start of the application phase clearly and visibly to all, 120 coloured flags were hung in the town showing the campaign logo and the slogan "voluntary 30 km/h slow-down in Münsingen". These functioned as reminders to those participating and as prompts to others.

Further reminders and prompts were provided by conventional posters, special shop window displays, the success barometer, and intensive press coverage.

Throughout the entire phase, three mobile units measuring driving speeds were moved around from place to place within the municipality. Clocked speeds posted on the electronic boards gave drivers feedback on their actual speeds and served to remind drivers about the campaign.

2.4. *Post-phase*

Following the official end of the campaign, we conducted various speed measurements and conducted social scientific surveys. Moreover, this was the time to analyse the data and evaluate the experience acquired in the interaction with the municipality.

2.5. *Measurement concept*

A comprehensive concept for measurement and data collection was developed for the entire campaign (see also Figure 1). A representative sample of 1,412 residents between the ages of 18 and 65 received a written questionnaire by mail both in the preparatory phase and in the post-phase. The response rates were 49% (n=688) and 31% (n=430), respectively. The first survey provided information on how much people knew about the topic of a 30 km/h speed limit and how they appraised it. This allowed us to develop optimal argumentation to support the goal of the campaign. In main phase I, all registered participants received a card with a brief questionnaire to tap the paths of diffusion. Here the response rate was 29% (n=286). Twice during phase I, we surveyed so-called social monitors, who were residents selected by us for this purpose (n=17 and n=18). During the campaign, we questioned the social monitors by telephone about signs of the campaign within their social circles. This information allowed us to have continuous checks on the spread of the campaign. In the post-phase, the multipliers (n=12) were surveyed with regard to their campaign experiences.

In order to obtain "hard" data with regard to the goal of reducing driving speeds, speed measurements were conducted at ten sites in the municipality prior to the application phase from September to December 1998. The measurements were taken by two different means: mobile radar instruments and invisible induction strips installed in the roads. The measurements were repeated again in main phase II, from February to May 1999. In the post-phase from September 1998 to March 2000, a third measurement took place.

3. Effects

Figure 2 presents the quantitative results of asking participants and non-participants how they had become aware of the campaign. Important diffusion instruments proved to be the flyers placed in dispensers in local businesses (28% of the responses), the information booth at the Christmas Fair (32%), and personal conversations (25%). As attention-getters, the feedback tower, the posters and the reports in the newspapers were clearly the top instruments.

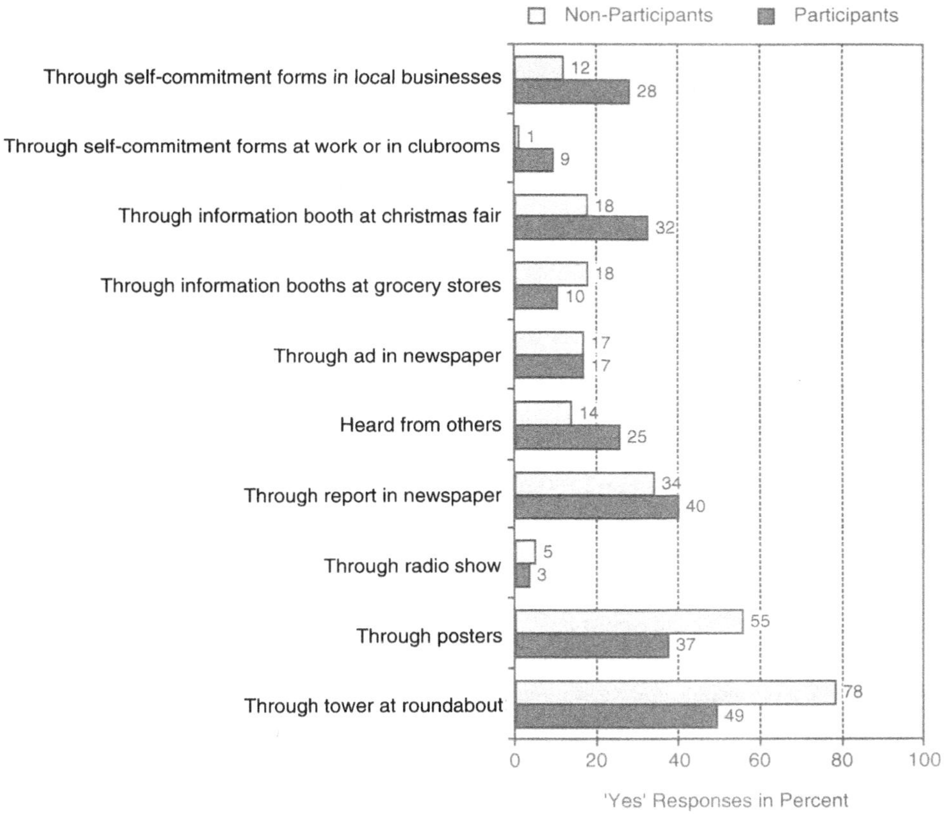

Figure 2: How participants and non-participants became aware of the campaign (multiple answers were possible; data source: post-survey of residents).

The data on the paths of diffusion of the self-commitment sign-up sheets stem from the survey of participants conducted in main phase I. Figure 3 reveals the paths by which the flyers containing the tear-off sign-up sheets for self-commitment reached the participants. Here again we see the overwhelming importance of the diffusion by direct personal contact at the Christmas Fair and in local businesses and shops: Almost two-thirds of the participants were reached in this manner.

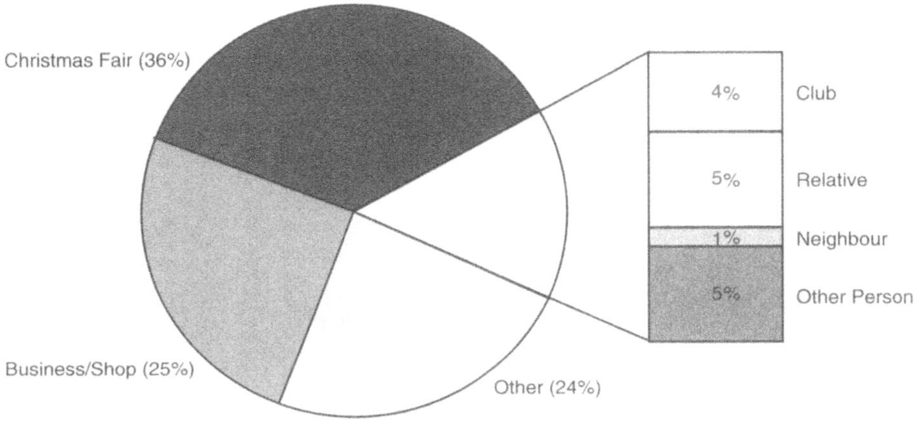

Figure 3: Diffusion paths of the sign-up sheets for self-commitment (data source: survey of participants).

Similar conclusions can be based on Figure 4. The steeply rising curve of participation following each instance of personal contacting through information booths attests to the great success of this method of diffusion.

The information booth at the Christmas Fair gained the participation of about 350 people. Between the Christmas Fair and New Year's, another 250 persons joined in. The information booths in front of local grocery stores recruited another 150 participants. An appeal to registered participants to fulfil a diffusion task by recruiting others brought in another 100 participants. Once main phase II began, there were no particular attempts to recruit additional participants, but another 50 persons joined in as a result of self-diffusion (see the typology chapter by KAUFMANN-HAYOZ ET AL., section 2.5.1.). A total of 1,015 participants were gained. With a population of 10,000 residents, or approximately 4,000 registered drivers, this means that every fifth or sixth car on the road was driven by a person participating in the slow-down.

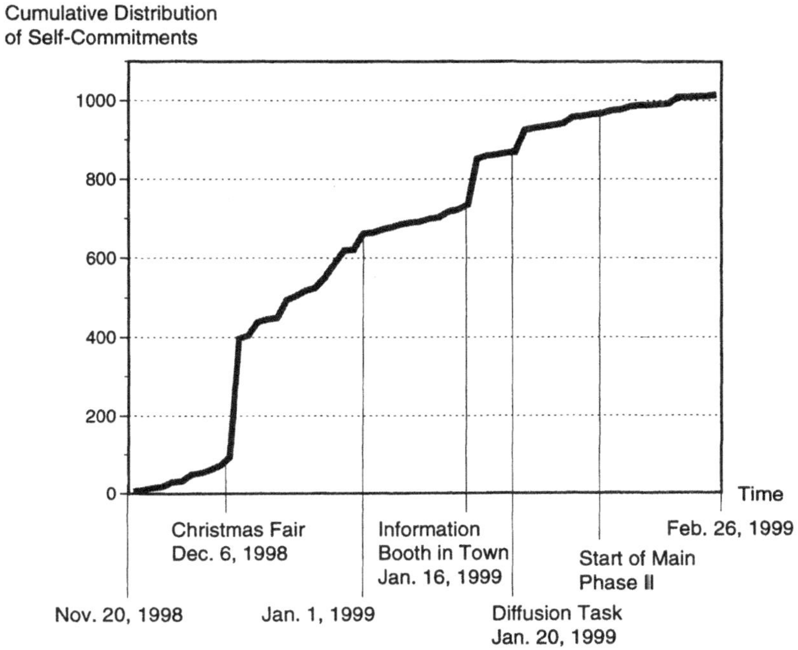

Figure 4: Cumulative diffusion chart (data source: daily count of incoming self-commitment sign-up sheets).

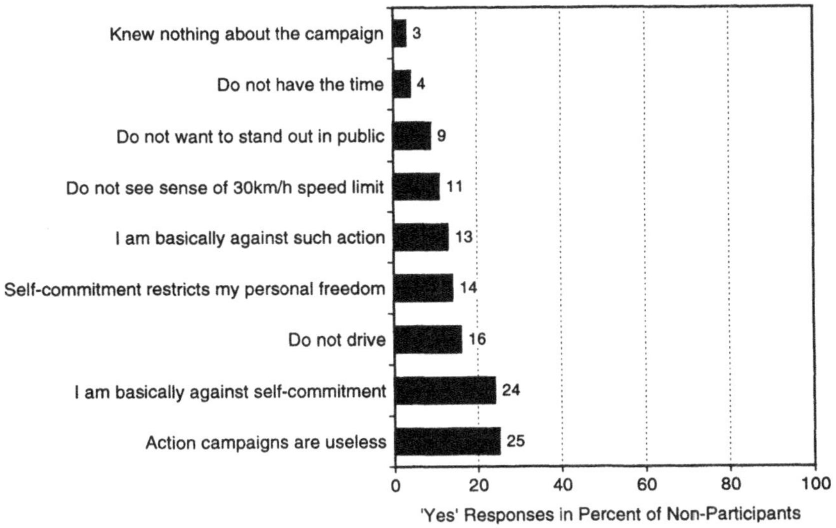

Figure 5: Reasons for not participating (multiple answers were possible; data source: post-survey of residents).

Examination of the reasons for not participating (see Figure 5) shows that people most frequently stated that they did not expect the campaign to be effective. Among the non-participants in the post survey (n=298), 25% stated that they did not take part because such campaigns are useless and 24% were basically against self-commitment. Also important was a reluctance to restrict personal freedom (14%). 13% expressed a fundamentally negative attitude towards this kind of collective action. A tenth revealed that they did not see the sense of a speed limit of 30 km/h.

The reasons given for participating, on the other hand, are shown in Figure 6. These data from the post-survey of residents (n=118 participants) provide an interesting correction to the findings of the multiplier survey that environmental arguments had little influence: The participants did indeed come from circles that routinely support environmental causes (noise, pollution, emissions). Another important factor proved to be the quality of the information materials provided.

Figure 7 shows the effectiveness of the various types of reminders in alerting people to the reduced speed campaign. The coloured flags along the streets received the most mention (71% of participants, 57% of non-participants), followed by the roadside speed measurement board and the posters.

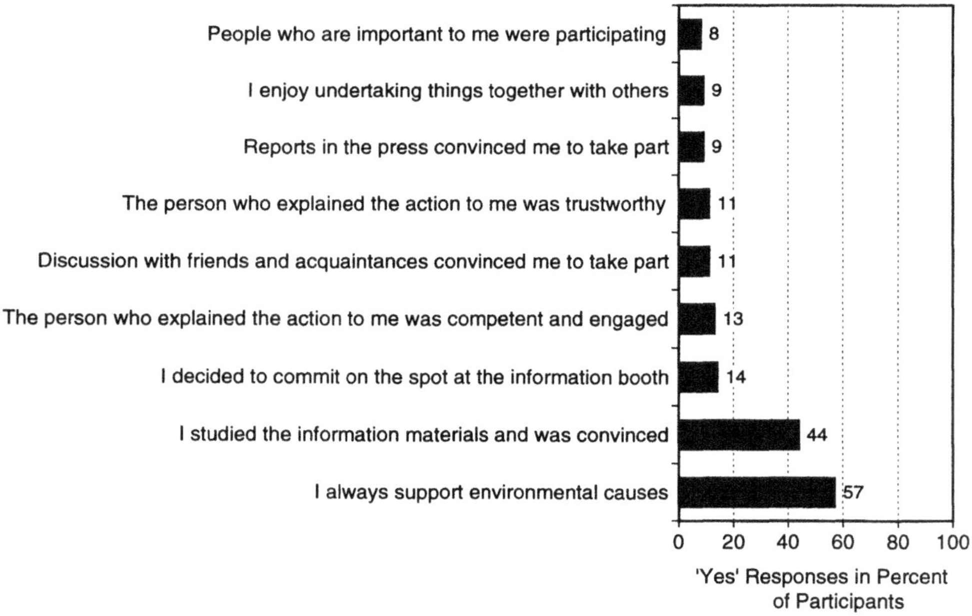

Figure 6: Reasons for participating (multiple answers were possible; data source: post-survey of residents)

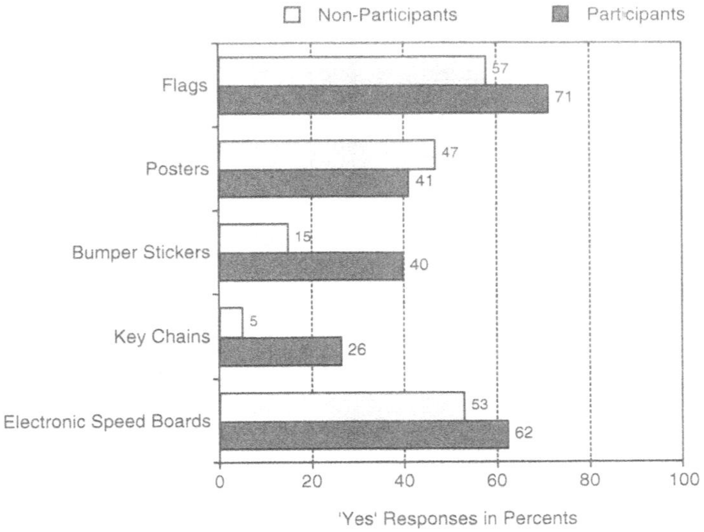

Figure 7: Reminders: What alerted you? (multiple answers were possible; data source: post-survey of residents).

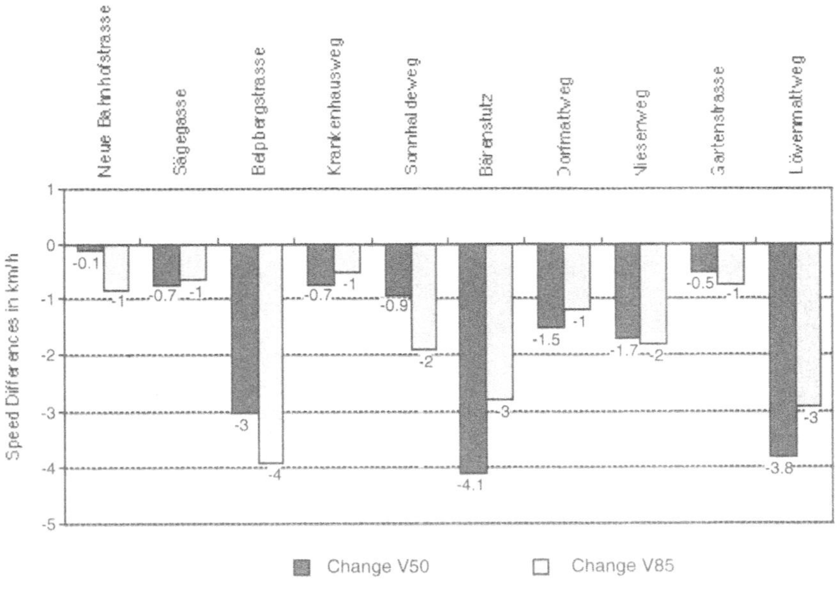

Figure 8: Changes in driving speeds measured (V50 and V85 stand for the speeds that are maintained and not exceeded by 50% and 85%, respectively, of vehicles; data source: speed measurements during preparatory phase and main phase II).

What effects did the slow-down campaign achieve? Figure 8 presents the figures, broken down for the individual streets in the neighbourhoods. V50 and V85 are standard measurement values that are used in the field of traffic science. They stand for the speeds that are maintained and not exceeded by 50% and 85%, respectively, of vehicles. Figure 9 presents a similar picture: The percentage of vehicles that did not exceed 35 km/h during the campaign is clearly higher than before the campaign. Evaluation of driving speeds measured four months after the conclusion of the slow-down campaign reveals a persisting, though somewhat less marked, speed reduction effect.

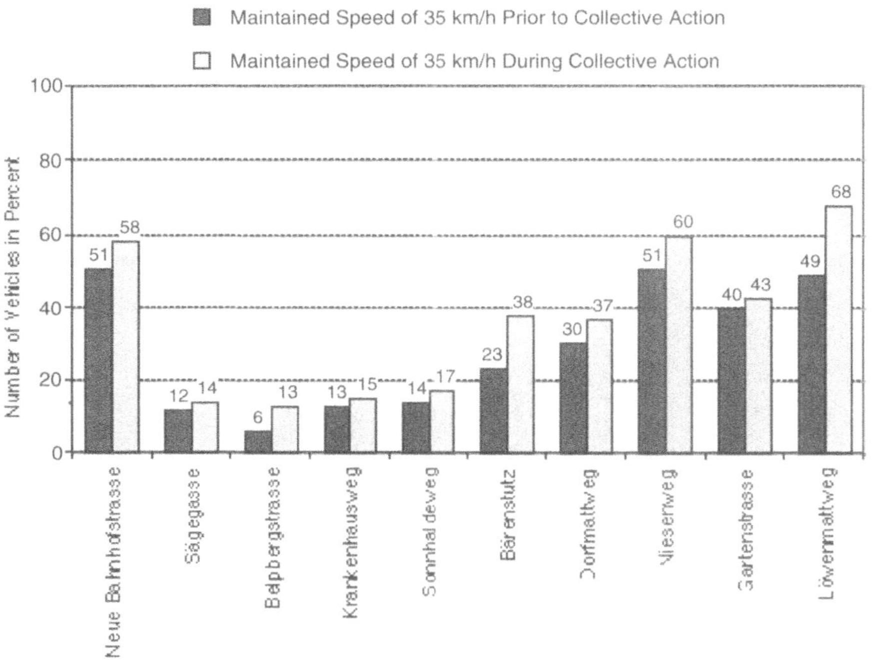

Figure 9: Percentage of vehicles not exceeding 35 km/h (data source: speed measurements during preparatory phase and main phase II).

4. Discussion

How should the speed reductions achieved be rated? It is important to emphasise that our voluntary collective action campaign for a 30 km/h driving speed on roads that have a legal speed limit of 50 km/h should not be compared to a target reduction of 20 km/h. First of all, a speed of 50 km/h is still legally allowed; and secondly, average driving speeds on the roads we studied were already – depending on construction and width of the roads – clearly lower (compare here also Figure 9). Based upon values achieved with the introduction of 30 km/h speed limits in other cities, our written goals before beginning main phase II were set at the range of –1 to –5 km/h (V50). For five of ten streets, this range was successfully achieved; for the rest of the roads, the speed reductions were somewhat smaller.

A relevant example for comparison is the city of Graz in Austria, which lowered the speed limit on neighbourhood roads from 50 km/h to 30 km/h by law following an information campaign. The new legal limit was also implemented by means of police controls. The average speed reduction achieved in Graz by means of these command and control instruments was –1.9 km/h (Kuratorium für Verkehrssicherheit, 1998: 2). Although this speed reduction may seem very small, it illustrates that even small changes in speed can have massive results: The number of drivers involved in accidents in Graz following the introduction of the new 30 km/h speed limit decreased considerably. The number of severely injured dropped by 24%, while the number of mild injuries fell by 12%. There was no change in emission levels except for a slight reduction in NOX; a perceptible drop in the noise level was registered. No changes were found in fuel consumption (Pischinger et al., 1994).

The evaluation of the success of the diffusion and communication instruments in attracting public attention shows overwhelmingly that informing people at the local level through direct personal contact is effective. The data suggest also that our main communication instrument, public self-commitment in writing, functioned well. The effects documented by the speed measurements show that people fulfilled their self-commitments – in spite of the absence of external social sanction mechanisms or threats. The measured effects can not be explained in any other way.

This leads us to the question of the psychological determinants of participation or non-participation. What was the crucial factor in the decision to participate? The important differences that distinguish participants from non-participants lie in the areas of attitudes and expectations of effectiveness. Those who chose to take part had a much more positive attitude towards voluntary collective action campaigns as well as clearly higher general expectations of collective action in terms of successful effects than non-participants. This receives additional support from the finding that non-participants very frequently explain that they did not participate because they had a negative attitude towards collective action and, furthermore, did not expect collective action of this kind to succeed. For a person to make an individual behavioural investment in collective action in order to provide a public good, it seems crucial that the public good in question be important and positive and that the individual expect that the attempt to provide the public good will be successful. If the probability of success seems too low, there is too great a risk that the personal contribution will be lost – it is just not worth the effort.

For individuals facing a decision for or against participating in a voluntary collective action campaign, their own store of information may be decisive in the end. As part of this knowledge, the extent of their information about the current status of the process providing the public good is important. Another crucial aspect may be people's expectations of the significance of making a personal contribution early on or only later as a campaign progresses.

Our own research group – on the basis of social psychological concepts and with the aid of simulation tools – has begun to focus on the central problem of how the two levels, the individual decisions to participate and the collective, interact. Within the framework of a microanalytical, aggregative simulation approach, we have succeeded in designing an actor model that on the basis of simulation experiments yields very concrete recommendations for the conception, planning, and organisation of collective action campaigns (Mosler and Tobias, 2000; Mosler and Brucks, in prep.). Based on the theory of planned behaviour (Ajzen, 1985; Ajzen and Madden, 1986), the model outlines the influences on a person's intention to participate. These include attitude towards participation, subjective norms regarding participation, perceived behavioural control, and – as an expansion of the theory – the return that the person expects to receive as a result of collective action. In accordance with the elaboration likelihood model (ELM; Petty and Cacioppo, 1986; for computer simulation of the ELM, see Mosler et al., 1998; Mosler et al., 2001), attitude is moreover conceived of as dependent upon the persuasive influence of other actors. Independently of these components, the model also calculates the intention to participate as affected by the benefit the person expects to gain after reckoning the use of the public good, the costs of contributing, and possible additional incentives. The research results presented here stand as initial, rough indications of the importance of these concepts.

5. Summary and conclusions

The study shows that scientifically-based combinations of communication and diffusion instruments can be implemented with success. This is particularly the case where it is desirable to produce initial, relatively rapid effects and to prepare the public for further, more extensive interventions. Through the course of our work, we have become increasingly aware that many more such investigations will be needed before a reliable inventory of instruments based on social scientific knowledge can be built. To be able to offer guidance and direction to steps towards sustainable development on a secure scientific basis, researchers will need to conduct a great deal more participatory, transdisciplinary, and applied research in future. For this, promotion of scientific research will have to include the providing of more mental and financial venture capital, for research of this kind, at least in part, forces us to deviate from the "results guaranteed" paths of traditional research operations.

References

Ajzen, I. (1985) From intentions to actions: A theory of planned behavior. *In*: Action-control: From cognition to behavior, Kuhl, J. and Beckman, J. (eds.), Springer, Heidelberg.

Ajzen, I. and Madden, T.J. (1986) Prediction of goal-directed behavior: Attitudes, intentions, and perceived behavioral control, *Journal of Experimental Social Psychology, 22*: 453-474.

Gutscher, H., Keller, C. and Mosler, H.-J. (2000) "Roads as new common pool resources, speed reduction as a public good – Two case studies in organizing large-scale collective action." Available CD: 8th Biennial conference of the International Association for the Study of Common Property (IASCP), Indiana University, Bloomington, Indiana. Online in the WWW: <http://129.79.82.27/IASCP/Papers/gutscherh041300.pdf>

Kuratorium für Verkehrssicherheit (1998) Geschwindigkeitsmessungen Tempo 30/50 in Graz. Author, Graz; *tech. rep.*

Mosler, H.-J. (2000) Computersimulation sozialpsychologischer Theorien. Studien zur Veränderung von Umwelteinstellung und Umweltverhalten. Psychologie Verlags Union, Weinheim.

Mosler, H.-J., Ammann, F. and Gutscher, H. (1998) Simulation des Elaboration Likelihood Model (ELM) als Mittel zur Entwicklung und Analyse von Umweltinterventionen. *Zeitschrift für Sozialpsychologie, 29*: 20-37.

Mosler, H.-J. and Brucks, W. Integrating resource dilemma findings in a simulation model; *submitted to European Journal of Social Psychology*; *in prep.*

Mosler, H.-J., Schwarz, K., Ammann, F. and Gutscher, H. (2001) Computer simulation as a method of further developing a theory: Simulating the Elaboration Likelihood Model (ELM). *Personality and Social Psychology Review; in press.*

Mosler, H.-J. and Tobias, R. (2000) Die Organisation kollektiver Aktionen durch Beeinflussung der individuellen Teilnahmeentscheidung. Eine Simulationsstudie. *Kölner Zeitschrift für Soziologie und Sozialpsychologie, 52/2*: 264-290.

Petty, R.E. and Cacioppo, J.T. (1986) Communication and persuasion: Central and peripheral routes to attitude change. Springer, New York.

Pischinger, R., Sammer, G. and Schützenhöfer, A. (1994) Tempo 30/50 in Graz. Ergebnisse der wissenschaftlichen Begleituntersuchung zum Modellversuch "Tempo 30 für alle Grazer Wohngebiete – Vorrangstrassen ausgenommen". Kurzfassung. Stadtbaudirektion, Graz; *tech. rep.*

Changing Things – Moving People.
Strategies for Promoting Sustainable
Development at the Local Level
R. Kaufmann-Hayoz/H. Gutscher (eds.)

Taxation of Residential Energy Use

Myriam Garbely and Alastair McFarlane

The chapter focuses on different new and "classical" economic instruments, i.e., taxes and grants, and on some command and control instruments aimed at improving the energy efficiency of the residential building sector. A justification is given why energy conservation investment in the building sector is essential on the way towards sustainable consumption. The functioning and the particular implementation problems of the proposed policies are discussed.

The analysis and the simulation of the economic and environmental impacts require a model that includes technical knowledge about the energy efficiency of buildings as well as behavioural relations from economic theory. The methodological framework is discussed for this model, where buildings are assumed to be heterogeneous with respect to their energy efficiency. We discuss further some failures in the working of economic instruments and show how they could be reinforced and complemented by communication and diffusion instruments.

1. Introduction

One of the most important goals of medium and long-term energy and environmental policy is a substantial reduction of polluting and non-renewable energy consumption. There are several reasons, both environmental and economic, for this goal: A first is the conservation of non-renewable resources, a crucial challenge for sustainable development as well as for the independence of nations that import energy. A second reason is to limit the environmental damage done by the pollution of burning fossil fuels. A third reason is to overcome certain barriers that lead to inefficient use of resources. There may be no unique instrument that will reduce energy consumption to the desired level. Therefore, it is worthwhile to analyse, for different sectors and applications, what instruments are the most appropriate, if new instruments might be required and how these various instruments could be combined together to be more efficient.

The residential sector is the source of a significant amount of energy use and pollution. In Switzerland, heating is responsible for more than a third of energy consumption and nearly half of total CO_2 emissions. In OECD countries, energy used in buildings accounts for nearly half of the total energy consumed and, with fossil fuels being the primary energy source, the consumption of buildings produces 22% of total CO_2 emissions (OECD/AIE, 1991).

There exist many technically feasible measures to reduce energy consumption in the built sector. Furthermore, the residential sector is not exposed to international competition. Unilateral measures can be introduced without being a threat to the competitiveness of the Swiss manufacturing sector. The taxation of real estate is within the competence of the cantons, and the Energy Law[1] allows them to introduce specific policies for residential buildings. This has in particular the advantage that local differences, of which there are many in the residential sector, can be taken into account by policy makers.

2. Residential energy demand

2.1. The determinants of residential energy demand

The demand for residential energy can be thought of as the demand for an input into the production of thermal comfort (Berndt and Wood, 1979). Fuel is important because it, along with the structure itself, allows residents to "produce" housing services. Energy demand depends on three factors:

The existing stock of equipment, i.e., all heated living surfaces. The current housing stock is the result of past investment decisions. The heated living surface per capita depends on both the average volume of an apartment and the size of the household. These variables are linked to the life style, the socio-economic environment, and wealth. The long-term evolution of the heated living surface per capita depends on social, political, and cultural choices with respect to housing policy and land zoning policies. From the point of view of sustainable development one might ask the question of the desirable living space for a household and for an individual in the future. Is an individual house socially optimal for everybody? These questions have to be thought about in the long term, in sociological and political terms as well as in ethical ones.

The intensity of use of a residential building. This depends on the desired comfort, i.e., the indoor temperature and the number of heating hours, and the climate, particularly the difference between indoor and outdoor temperature. This intensity of use is the only element that can vary in the short term. However, except in case of wasteful behaviour, there are not a lot of possible interventions as there are no substitution possibilities as, for instance, are in the transportation sector.

---

[1] See Art. 9 of the *Loi sur l'énergie*, 730, Recueil systématique du droit fédéral, January 1999.

The energy efficiency of a residential building. This is represented by an indicator of specific energy consumption for heating purposes which assesses the thermal quality of the building and is measured in MJ/m^2year and calculated for a standard heating temperature of 20°C and for mean climatic values. The indicator is determined by the physical characteristics of the building, which depend on technical progress and how fast it is adopted by investors. The indicator does not depend on the behaviour of the inhabitants and on climate variations. It is an indicator of demand and not an indicator of actual energy expenses.

The energy efficiency of a building depends on its building year, its retrofitting cycle, and the quality choices made at those moments (standards in use, insulation costs, costs for solar installations). The indicator of energy efficiency does not change between two renovations, as long as a correct maintenance is guaranteed.

In order to achieve a reduction of resource consumption, Agenda 21 focuses on changing consumption towards sustainable levels of consumption as well as on changing consumption patterns (Pearce, 1994). Changing consumption in housing requires modifying the individual available room, i.e., a change of lifestyle that goes far beyond energy policy purposes. Changing consumption patterns means turning to less resource-intensive products. Thus, policies to substantially reduce energy consumption in housing in the medium and long term ought to be concerned with improving energy efficiency.

2.2. *Barriers to energy efficiency in residential housing*

For the most part existing buildings are far from energy efficient. Furthermore, when constructing new buildings or when retrofitting, not all investors choose the most efficient thermal quality, even if it appears to be profitable.

There are many possible reasons for this behaviour. As is the case for all environmental investment, the expected repayment period is quite short. This comes from the risk due to uncertain future energy prices. Investors want their investment to be profitable within a few years, even if the payback period and the lifetime of the building are much longer. In addition to that, there are some other obstacles that are specific to the buildings sector:

- Economic and financial barriers. These come from the large proportion of tenants (about two-thirds of the population in Switzerland). Investment costs have to be paid by the owner who can pass only a part of it on the tenant. The tenant, however, benefits from a lower energy bill. This barrier is in particular effective for retrofitting where the legal restrictions prevent the owner from including the whole investment costs in the rent, as is currently the case in Switzerland. One might argue that in new buildings there can also be a gap between the decision maker and the beneficiary of energy efficiency. The decision affecting energy efficiency of the building shell and heating equipment is often made by the builder or architect, and not by the purchaser. Sutherland (1992) shows that this does not produce economic inefficiency, as builders attempt to construct marketable buildings, which mirror consumer preferences. But this does not guarantee optimal energy efficiency.

- There is a gap between construction standards and technical evolution. Furthermore, standards cannot involve too strict requirements, as they have to be applied to the whole sector.
- Investors have to make decisions in reliance upon partly uncertain data. They have to place bets on future prices, in particular on energy prices and interest rates in order to calculate the discounted profitability of an energy investment. The future return (the avoided energy costs) is highly uncertain.[2] This risk does not affect all investors in the same manner. Households and small private companies are particularly affected, as they do not have a sufficient portfolio of earnings assets to diversify this risk. Further investments in a building shell are illiquid without marketability. Thus households require higher discount rates on energy-efficiency investment than publicly held corporations.
- Information about retrofit costs and performances is often incomplete, sometimes even conflicting, and in any case expensive to obtain. Again the problem is particularly acute for households, as they must recover their search and information costs with a single, very infrequent purchase.
- Some practical problems may also handicap an efficient retrofit, such as, for example, the sharing of retrofit cost between adjacent buildings and security and access problems to occupied renovation sites.

3. Policies to encourage energy-conserving investment in buildings

3.1. General description

Various kinds of instruments may influence investors' choices. Some of these instruments are applied in all of Switzerland or only in some cantons, while others are currently being discussed as legislative proposals or in initiatives.[3] In addition, we will consider innovative policy instruments that merit consideration. The policies aimed at encouraging investment in energy efficiency of buildings,[4] in which we are in particular interested, are the following:

---

2 Hassett and Metcalf (1993: 710) show that the so called energy-paradox ("very attractive investment opportunities in energy efficient capital, opportunities with high *ex ante* rates of return are routinely passed over by investors") is not due to irrational behavior but to the uncertainty of future prices.

3 The most important projects currently under debate in Switzerland are: The introduction of taxes on CO_2, energy or fossil energy; an ecological fiscal reform; the "solar initiative" including a tax on non renewable energy the income of which is mainly devoted to solar energy; the "energy and environment initiative" taxing non renewable energy and electricity from large hydraulic plants in order to encourage energy saving and curb wasting; the initiative "to guarantee the AVS (Swiss old age pension) – tax the energy and not the labour".

4 As the aim is to reduce the demand for polluting energy, a substitution of fossil energy by renewable energy will be assimilated to an increase in energy efficiency.

Incentive taxes: In Switzerland, some examples of energy-related taxes, although not specific to the building sector, are under current proposal by the government and the Parliament. These consist mainly in taxes on energy, taxes on fossil or non-renewable energy, and taxes on CO_2, but they have not yet been implemented. An innovative instrument specific to the building sector would be a tax based on the indicator of specific energy consumption for heating purposes.

Investment subsidies: Another approach to encouraging energy-saving investments and investments in renewable energy is to provide subsidies, as does the investment programme Energy 2000 in Switzerland. Subsidy policies can be implemented in various manners, including income tax credits or low interest rate loan programmes. Subsidies can be allocated for any energy-conserving investment, such that the investment achieves a level of efficiency that exceeds the minimal legal requirement. The subsidy level may also depend on an index of energy consumption.

Standards: Regulation is the most frequently applied policy tool. Construction standards set a minimum level of quality for buildings, i.e., for the indicator of specific energy consumption (SIA norms currently in force in Switzerland). Such standards are defined for new buildings as well as for retrofitting. They define the maximum level of specific energy consumption for heating purposes, measured in MJ/m^{2}year. Equivalent norms defined for building components also exist for partial renovations. The standard of the European Community (EN 832)[5] defines the thermal performance of buildings in a similar way.

All of these instruments, even the innovative ones, i.e., taxation and subsidy schemes based on the thermal quality of the building, belong to classical regulatory and economic instruments. They rely on command and control mechanisms or on the reactions of landlords to economic incentives. There are other measures that may influence, directly or indirectly, the thermal quality of a building. Among them are:

- information and consciousness raising campaigns;
- research and development projects;
- private initiatives, in particular those supported by local communities, such as pilot projects;
- individual calculation of heating costs;[6]
- quality labelling, such as the "Minergie" label introduced by the canton Zurich in Switzerland or the home-energy rating system (HERS), which provides standardised comparisons between the energy consumption of a home against a reference one.

---

5 The European calculation method based on EN 832 will be used in the next version of the Swiss norm.

6 It some cases already the threat of the introduction of such a measure has an effect on energy efficiency. Thus the canton of Geneva has renounced introducing individual calculation of heating costs for all buildings having an index of energy consumption for heating purposes lower than 600 MJ/m^{2}year. The effect was that owners preferred to improve energy efficiency, rather than to install the possibly more expensive individual calculation system.

The more individual character of these instruments makes it difficult to evaluate their impact through standard micro-economic simulations. In section 5 of this chapter we discuss how communication and diffusion instruments that are realised in such measures may contribute to more effective implementation and functioning of economic instruments.

3.2. *Actors and target groups*

The target group for energy efficiency improvement decisions are property owners themselves. Without the proper incentives, property owners will be reluctant to invest in energy efficiency improvements. Barriers to investment are capital costs and uncertainty regarding many factors, including the movement of energy prices, the cost of investment, how long they will own their homes, and the resale market value of efficiency improvements. All residents of an apartment building are also directly affected. Their energy bills for heating are reduced, but even if they are tenants, they may have to bear a part of the investments costs. Energy efficiency policy affects the construction sector directly. All economic sectors can then be affected indirectly by the modified prices of the construction sector as well as by an eventual change of the energy consumption of households.

Energy conservation ordinances and standards, as well as taxes on energy, have to be implemented by the government at the national level. A tax paid by the homeowner would depend on the legislation of the cantons. Financing programmes, such as grants and zero and low-interest loan programs, can be offered by the federal government as well by cantons and local municipalities. Thus energy efficiency policies concern all the key players identified by JEGEN'S Study.

3.3. *Problems in implementing energy efficiency policies*

3.3.1. *Incentive taxes*

Taxes on (fossil or non-renewable) energy and tax on CO_2: All empirical studies show that there is a rather low price elasticity; hence to be efficient for the construction sector the tax should be quite heavy. Such taxes affect the whole economy. Therefore the problem of their acceptability and of an eventual international co-ordination has to be faced. The incentive for an owner of a rental building remains quite small, as he can only pass on to the tenant a limited part of the investment cost. For a country like Switzerland, which imports all its fossil energy, the implementation is easy to combine with existing custom taxes.

A tax based on the indicator of specific energy consumption for heating purposes: The relevant authorities for the taxation of real estate are the cantons. This allows the taking of local and regional characteristics into account. There is an incentive for the owner, who may reduce his fiscal burden by improving his building. Tenants have to support a part of the investment cost, but their residential energy costs will be reduced.

The initial implementation may be complicated, as the fiscal authorities would have to make an inventory of the indicators of specific energy consumption of all residential buildings. Note, however, that already this indicator has to be calculated and declared for any construction or renovation permission request. In Geneva, for instance, real estate management companies have to declare the indicators of specific energy consumption of all buildings they are in charge of to the Energy Office of the canton. These data ought to be cross-checked with the tax returns of the owners and completed by those data concerning single-family houses.

In following periods, renovation permits would allow, if suitable, justification of improvements to the indicators of specific energy consumption. Otherwise, the tax return would not change from one taxation period to the next. The great weakness of such a taxation scheme is that it is too novel to gain acceptance (see the contribution by JEGEN).

3.3.2. Subsidies

There should be no special problems when implementing grants except their funding. Nevertheless existing subsidy programmes, like the Energy 2000 programme in Switzerland, have shown important differences in the propagation speed of these measures among regions. Thus, regions with the most energy inefficient buildings do not necessarily benefit the most from such grants, and thus they can become less cost effective. The implementation of subsidies with a level depending on indicators of specific energy consumption does not imply particular procedures, as the indicator has to be declared for obtaining a retrofit or construction permit.

3.3.3. Standards

Standards have the advantage of being largely accepted (see the contribution by JEGEN), and their functioning has been experienced over many years. Professionals govern adaptations of the standards in order to make sure that the sector can apply them easily. In Switzerland, the standards are calculated on plans, and control is part of the construction permit allocation procedure. As for any standards, the difficulties are twofold. Their adaptation to technical progress might be quite slow. They do not ensure economic efficiency, i.e., that the improvement of energy efficiency with the lowest cost is carried out first.

4. Analysing the impacts of energy policies

There are a few sophisticated approaches to analysing the impacts of energy policy on the housing sector. The first is empirical, the best examples of which are analyses of an American tax credit for energy-conserving investment in the residential sector (Hassett and Metcalf, 1995; Walsh, 1989). Estimating the effects of a specific policy on consumer and investor behaviour would yield a statistical model that can be used to predict the effects of different programmes in slightly different circumstances. However, this approach requires that the policy be in effect. To predict the consequences of introducing a new policy, one could estimate a statistical model of energy demand to forecast the impacts of a new tax (for examples, see Quigley, 1984a; 1984b). A final alternative would be to draw from empirical research done in other countries. However, this approach may not be the most appropriate one for Switzerland, because the housing market in Switzerland is different than in other OECD countries. For example, most studies of the energy tax credit in the United States concern homeowners, whereas in Switzerland, two-thirds of the people are tenants. When an empirical approach is not appropriate either because the policy does not yet exist, or the data are incomplete, or studies of other countries are not applicable, then we have to resort to more abstract methods.

Theoretical approaches differ according to the accuracy of the data used to calibrate the model, the assumptions underlying the economic model, and the use of computer simulation. The advantage of a theoretical approach is that one can choose a model according to the data available and the interests of the researchers. An additional advantage is that one can compare different policy proposals and analyse their effects under a wide range of scenarios. One disadvantage of this approach is that it is based to a large extent on guesswork. A second disadvantage is the credibility of the results. An elegant mathematical model, which can be solved with a pen and paper, is often seen as too simplistic. However, a more complex and computable model, which may involve a system of many equations, can be seen as a "black box". Thus, it is important to choose models that are already theoretically well developed so that the results are transparent and thus easier to interpret and communicate.

4.1. The technical framework and data

To carry out our analysis, we thus look for existing theoretically well-founded models, i.e., for economic models of retrofitting behaviour that are centred on the profits to be had from investing in energy efficiency. The goal of these models is to quantify the change in the profitability of retrofitting that an incentive tax would introduce (see Hassett and Metcalf, 1993; Jaffe and Stavins, 1994). The two works cited are theoretical in nature. Empirical studies are rare because in all countries, not just in Switzerland, there is a lack of information on the profits of retrofits. The only large-scale study of which the authors are aware was completed in the United States (Goldman et al., 1984, 173 observations). To analyse the impacts of different energy policies on the housing market, we need data on energy-saving investment and the corresponding change in energy consumption.

To gain more information on the profits of retrofits, we take two approaches: first, using data from case studies of actual renovations and second, engineering simulations of energy conservation. The former empirical approach involves identifying the cost of energy-saving investments during major renovations. These data allow us to estimate a functional form for the relation between the cost and the energy efficiency.[7] We can simulate the effect of a tax. However, there are some problems: a small amount of observations, a possibly biased choice of buildings, and the fact that renovations will have other goals than energy efficiency. A second approach is what has been called a "technical-economic" approach, which is based on methods used by engineers to specify a cost function for improving energy efficiency based on the attributes of the building.[8] Adding a tax would shift the cost curve. In this way, we can simulate its effect on retrofit activity.

Despite the accuracy of the above two approaches in explaining the supply side, they do not account for consumers' decisions and how the market values energy-saving investments. This may be especially tricky in the case of rental housing, where incentives are far more complex than for homeowners. Thus, for a worthwhile economic model, we must integrate the demand side. As mentioned, this requires some creativity, because market data on retrofitting do not exist. Finally, despite the lack of economy-wide data on retrofits, the data on the housing market itself is plentiful. Thus, we are able to have fairly good context data.

4.2. *The economic model*

4.2.1. *A simple model of investment in energy saving equipment*

Before building a computable equilibrium model, we considered a more manageable model of investment in housing with which we could derive some comparative statistics. A preliminary version of the model is described in Dupraz et al. (1998). The model is simple in that we treat conversions as independent of one another. This simplification allows us to construct a model to focus on the effect of a variety of policies on the timing of development and renovation. Property owners decide when to invest in energy-saving equipment in order to minimise their fuel bills. Since lower fuel costs translate to higher rents, the profit-maximising decision of the landlord is also an energy cost-minimising one. Thus the goal of the property owner is to minimise the present value of energy costs in choosing the optimal timing and intensity of the retrofit.

Formalising this optimisation problem and deriving the first order condition for timing, we obtain the result that the marginal benefits of investing now (paying lower energy bills) as opposed to later must equal the marginal opportunity cost of investing now (paying interest on the physical cost of the investment). Inspecting the second order condition for timing we see that if fuel prices are not increasing, then a retrofit will not be a cost-minimising strategy and will not take place. The first order condition with respect to the intensity is that the value of the marginal product of capital must equal the unit cost of capital. The second order condition is that the marginal product is di-

---

7 This function is similar to Conserved Energy Supply Curves defined by Stoft (1995) that are based on a sequence of packages of conservation measures, each of which provides additional conservation at a higher marginal cost than the previous packages.

8 A detailed description of the model based on Roulet (1987) is given in Dupraz (1999).

minishing. These are all standard conditions in economics. Our contribution is to use these conditions to develop insights into the effects of energy taxes and investment subsidies.

First, we consider an **excise tax**, paid by the owner of the building, on the consumption of energy. Such a tax will hasten development under certain conditions, because it increases the financial benefits of retrofitting. One of the insights is that the tax will have no long-run effects on retrofitting if energy prices are declining. In that case, there would be a rush to retrofit by a few property owners, but afterwards there would be no retrofit activity. Here we see the difference between a regulation and a tax. If we want to be certain that energy consumption is reduced, regulations may be necessary. A second insight is that the impact of the tax will be greater the smaller the growth of prices. This can be explained by considering the "hurdle price", which is defined as the critical level of the energy price that triggers investment. A change in the tax will change the hurdle price, and when growth is slow, it takes more time to adjust to the new hurdle price. A third insight is that a tax can be designed to ensure retrofitting as long as the sum of the price of energy plus the energy tax increases over time.

The second policy that interests us is a **subsidy to energy conserving investments**. An investment subsidy accelerates conversion by lessening the marginal cost of investing now. A subsidy to the cost of energy-conservation will have a greater effect on timing when the capital cost of development is greater. Thus, higher interest rates augment the impact of a subsidy. One insight is that the size of the effect of the subsidy appears to be sensitive to more factors than an energy tax. Thus, it is likely to be easier to calculate the correct tax than the correct subsidy. Comparing energy taxes and the investment subsidy, we see that the ratio between a tax and a subsidy that have the same effect on timing will vary by building. Thus, there is not a universal equivalence of the two policies when buildings are heterogeneous.

Our model becomes more difficult to solve when we consider the initial development of the building along with the retrofit. We arrive at another insight: a policy that affects the initial construction will also affect renovations. Energy taxes appear to have a greater impact on energy efficiency at the time of construction than on renovations. Finally, imposing energy standards will delay both development and retrofitting and as a result reduce the housing supply. A more complex version of this model that includes the impact of a variety of housing policies, such as rent control, on retrofit activity is currently under development.

4.2.2. A more complex model of investment in energy conservation

The difficulties in modelling housing markets are well known by economists. According to Anas and Arnott (1991: 3) these stem from the characteristics of housing, the most important of which are "durability, complexity and multidimensional heterogeneity, indivisibility in consumption, adaptability, non-convexity in production, spatial fixity, the importance of transactions costs and the presence of asymmetric information". For a complex model of the housing market and investor behaviour, we draw on the work of Anas and Arnott (1991; 1997). When we say "complex", we mean a model where both housing and residents are heterogeneous. The benefit of such a model is that we are able to see how a tax on a landlord is different from a tax on a resident. We can also model a larger set of conversion decisions, so that there are richer supply dynamics. In addition, we can see how a policy affects different classes of people. The challenge that we faced is how to model energy consumption.

First, we start with a brief description of the model of Anas and Arnott (1991). The elements of the model are as follows:

- the model is competitive so that everyone is a price-taker;
- the housing market in question is treated as a part of a larger economy;
- the model is in discrete time, the time period is one year; all variables may vary over time;
- there are random components of the model, and thus the choices of consumers and landlords are modelled using probabilistic discrete choice theory.

The demand side is characterised by the following:

- All households are tenants.
- Households are mobile: they can leave the city to obtain the desired housing.
- Multiple types are allowed. In our present models, they differ only by income.
- Households vary by tastes, which are idiosyncratic and follow a Gumbel (double exponential) distribution.

The supply side is as follows:

- An apartment possesses a vector of distinguishing characteristics. In this study, we consider only two: energy efficiency and quality (comfort).
- Each owner has only one property. Their objective is to maximise the value of their property by choosing the profit-maximising conversion strategy.
- Every year a number of investors decide to convert their apartments. There are different types of conversions: construction, quality upgrading, and retrofitting.
- The costs of a conversion are idiosyncratic (vary by landlord) and follow a Gumbel (double exponential) distribution.

Energy is modelled as follows:

- Energy consumption is inelastic with respect to energy prices.
- Only energy efficiency of the apartment has an effect on energy consumption.
- The energy efficiency of an apartment is set at its construction but can be upgraded in later periods.

The random variables in this model are the preferences of households and the costs of conversion for landlords. Thus, we cannot predict with certainty what will happen in each period. We can only compute the *expected* effect. This model was chosen for this stochastic element. In every period, the market for housing services generates a "temporary stochastic equilibrium". Competition between households for different types of apartments generates a vector of rents. Landlords will respond to these rents by making the type of conversion that maximises profits. When landlords buy their land, they do not know exactly what the costs of conversion will be. The equilibrium is set where the *expected* demand for each type of housing is equal to the *expected* supply. Thus, the equilibrium is stochastic, because the actual demand may be different than the actual supply. Also, this "equilibrium" is only temporary since the economy is non-stationary, and year after year a new equilibrium will occur.

The goal of this model is to predict the proportion of the housing stock that has been converted to energy-efficient buildings. Since more energy-efficient buildings are in general more valuable (Neels, 1982), the objective of the property owner would be to choose the optimal conversion strategy to minimize total energy costs (conversion cost plus expenditures on energy). From there, we can analyse the impact of different policy instruments on the energy efficiency of the entire housing stock.

4.3. Computer simulation model

Computer simulations are helpful whenever we have a large number of equations, and they are necessary when these equations are non-linear. For our simplest model of investment decisions, a computer is not required for us to make statements about the effects of an energy tax. However, as soon as we complicate the model by treating initial construction and renovation as joint decisions, computer simulation is required for finding a solution. Our second and more complex model was designed for computer simulation from the beginning. It involves two major additional complications. The first is that the model is probabilistic. Thus, an equation can be very difficult to solve depending on the probability distribution that we choose. The second complication is that we consider multiple types of housing, consumers, and time periods. As mentioned previously, computable models are not a perfect substitute for economic theory. However, they allow us to build more realistic models of the economy and thus can give us insight into difficult questions.

4.4. Preliminary results from computer simulation

In this section, we illustrate the impacts of different policies with the simulation model described in the previous section. Before discussing specific policies, we introduce the results of our model in the absence of any policy. The dynamics of the supply of housing and built land are illustrated in Table 1 below.

Table 1: Conversion probabilities (%) with no policy.

	convert to:									
	vacant land		bad comfort/ low efficiency		bad comfort/ high efficiency		good comfort/ low efficiency		good comfort/ high efficiency	
convert from:	T1	T2	T1	T2	T1	T2	T1	T2	T1	T2
vacant	36.1	20.8	23.0	12.2	11.2	12.3	19.7	27.3	10.0	27.4
bad/low	---	---	41.6	19.1	18.3	17.3	26.6	31.8	13.5	31.8
bad/high	---	---	---	---	62.4	39.8	---	---	37.6	60.2
good/low	---	---	---	---	---	---	66.4	50.0	33.6	50.0
good/high	---	---	---	---	---	---	---	---	100	100

Conversion includes construction, quality upgrading, and retrofitting.
T1: Beginning period, with fixed housing stock.
T2: Intermediate period, with flexible conversion and stock variables.

Housing is characterised by either *bad* or *good* comfort (utility) with *high* or *low* energy efficiency. First, the symbol "---" marks conversions that were assumed not to take place: Housing may not be downgraded nor demolished. In the third and final period, there is supposed to be no more conversion.

We see that in the first period, for all types of housing including vacant land, the largest portion is kept in its current use and is not converted. Given the uncertainty of conversion costs, this result in one that is intuitive. The goal of an energy policy would be to encourage a larger proportion of owners to convert towards high-efficiency housing (either *bad/high* or *good/high*). In the second period, which is the period before the last, a higher percentage of housing and land is converted towards higher efficiency. Also, most of the housing that was built as low efficiency in the first period will be upgraded to higher efficiency in the final period. Another striking fact is that in the second period, there is a sizeable increase in conversion rates to other housing types. However, we see that although not all housing is converted towards high-efficiency housing, a large proportion of housing is. We start the simulation with half of the stock energy efficient and finish with four fifths. Thus, given our present assumptions, the market seems to work. Certainly, different parameters may give us different results. However, our goal is not to focus on the impact of varying all of the parameters, but only the tax. Table 2 presents the effects of a 100% energy tax on the quantity of energy consumed.

Table 2: Difference in conversion probabilities from a 100% energy tax.

	convert to:									
	vacant land		bad comfort/ low efficiency		bad comfort/ high efficiency		good comfort/ low efficiency		good comfort/ high efficiency	
convert from:	T1	T2	T1	T2	T1	T2	T1	T2	T1	T2
vacant	+0.1	+0.6	-0.1	-0.2	+0.1	+0.0	-0.1	-0.5	+0.1	+0.1
bad/low	---	---	-0.1	-0.3	+0.2	+0.3	-0.1	-0.4	+0.1	+0.4
bad/high	---	---	---	---	+0.0	+0.0	---	---	+0.0	+0.0
good/low	---	---	---	---	---	---	-0.3	-0.7	+0.3	+0.7
good/high	---	---	---	---	---	---	---	---	+0.0	+0.0

The values indicate the differences in conversion probabilities from those without a policy given in Table 1.
T1: Beginning period, with fixed housing stock.
T2: Intermediate period, with flexible conversion and stock variables.

For the model in question, there is no great effect on the rates of conversion. However, even a small change in the rate of conversion will add up over time. At this point, the results should not be taken as definitive, as there is still some work to do on estimating the values of a few of the parameters. Yet, they are indicative of the direction of the effect of a tax. First, the probability that a conversion will be towards high efficiency increases and that it will be towards low efficiency decreases, for all housing types.

Second, more land is kept vacant. Thus, one of the effects of an energy tax will be to limit the supply of housing by a small amount. The tax imposes a cost on households in the form of higher rents in the intermediate and final periods. For example, for a 50% energy tax, tenants would pay approximately 6% of the tax in the form of higher rents in the final period (for details, see McFarlane, 1999). It is an old rule of public economics that the burden of the tax is not dependent on who nominally pays the tax but on who cannot escape the tax. Given that the housing supply is fairly inelastic and consumers are mobile, most of the burden will fall on the landlords. When the total stock of housing is perfectly inelastic, as in the first period, then there is no change in rents from the tax. It is also important to note that tenants of both high- and low-efficiency apartments will share the same burden, since the increase in rents due to a decrease in the supply has an effect on all housing sub-markets. Thus, there are no special advantages to being a tenant of high-efficiency housing when landlords pay the taxes.

Third, the tax has no effect on the conversion between different types of energy efficient housing. Once housing is energy efficient, it stays that way.

Many of these conclusions are intuitive and would not require a computer simulation. However, the interactions between variables and comparison of different policies are less obvious. An important issue is how the tax revenue is spent. In the previous simulation, the tax was equivalent to an increase in the costs of maintenance. In reality, the effects of taxes are more complex, because the way the revenue is spent should have effects of its own (McFarlane, 2000). We consider the following possibilities:

- the absence of any fiscal or energy policy;
- an energy tax but no spending of the revenue;
- an energy tax and a lump sum transfer to citizens;
- an energy tax and lump sum subsidies for investment in energy conservation; and
- an energy tax and a lump sum subsidy to the annual revenue of landlords.

The energy tax rate is equal to 100%. We make the assumption that there will be no government debt by the final period. We have already investigated the first two cases. The third involves a transfer from landlords to consumers. The effect of such a policy would be to encourage more people to live in the city to take advantage of the fiscal benefits. However, such a use of the tax revenue may not be warranted, since it is horizontally inequitable (between landlords and consumers) and would not necessarily encourage energy conservation. A more useful policy would be to redistribute the tax revenue as subsidies to conversion towards energy efficient housing, either at the time of construction or renovation. Effectively, this redistributes from those that do not invest in energy-efficient housing and those that do. The only inequity is that those that already have high efficiency housing cannot benefit from the tax revenue. However, in reality, there exists a continuum of levels of energy efficiency, and even those at the higher end would be able to receive subsidies. A final policy is to divide the revenue among the landlords. Thus, owners of energy efficient housing receive a net transfer, and owners of energy-inefficient housing pay a net tax.

Our immediate concern is the evolution of the stock of high-efficiency housing. We use the stock of energy-efficient housing at the final period to measure the success of a policy. At the same time, we are interested in the effect of policy on other aspects of the housing market, such as rents, asset prices, and the material welfare of the residents. We certainly do not want to support a policy that increases the stock of energy-efficient housing at any cost. To understand the effects on welfare, we can use a commonly used indicator of the welfare of a community, the total value of built and vacant land. This is a good measure, because it reflects the future value of the income of all residents. We will use the value of built and vacant land in the first period as a measure of the efficiency of the policy. The results of the simulations of the different fiscal policies are as follows:

Table 3: Impacts of policies on high-efficiency housing and on the value of housing and land.

Policies	Variation of the stock of high-efficiency housing (last period)	Variation of the total value of built and vacant land (1st period)
none	+0.0%	+0.0%
tax only	+1.2%	-5.4%
tax and transfer to residents	+1.2%	+0.8%
tax and investment subsidy	+1.3%	-5.2%
tax and transfer to landlords	+1.2%	+0.8%

The values are differences from a situation without any fiscal policy.

It appears that the most effective policy for encouraging an increase in the stock of energy-efficient housing is to spend the tax revenue on subsidies. One reason for the effectiveness of this policy is that all of the tax revenue is targeted at a fairly small group: those considering building energy-efficient housing. However, measuring from the effect on the sum of all land values, it is not the most efficient policy. In other words, although it accomplishes our goal, it is not the least costly. Certainly, redistributing the revenue as investment subsidies is better than not spending it, but it is not superior to the third and fifth policies.

There are two noticeable aspects of the transfer policies. First, transferring tax revenue to landlords or residents has the same effect on the economy. Distributing the energy tax revenue to landlords or to tenants are equivalent policies, because if given to tenants, the tax transfer will be capitalised into rents, and if given to landlords, the annual net rent will increase by the amount of the transfer. Whereas a transfer to landlords raises their income directly, a transfer to residents raises landlords' incomes indirectly by the same amount, because residents are willing to pay higher rents to live in a city where they receive tax breaks. Second, the result of these two policies is a slight increase in total land value. The primary explanation is that the transfers will benefit the majority of the residents (or landlords). To make a judgement as to which is the best policy choice, the investment subsidy or income transfer, would require a better understanding as to the financial benefits of lower energy consumption.

In this model, we have discovered how a tax imposed on landlords might impose a cost on tenants by lowering the supply of housing and raising rents. We have also seen that conversion towards energy-efficient housing is encouraged by an energy tax without reducing the housing stock by a significant amount. To make broader statements on the optimal policies to reach our energy goals the comparison of tax policy with different types of building regulations would have to be furthered refined.

5. Combining economic measures with communication and diffusion instruments

Improving energy efficiency of housing is a condition *sine qua non* of reducing energy consumption. It has been shown (Neels, 1982) that the quality of building plays a greater role in reducing energy consumption than the behaviour of the residents. To simulate possible consumption reduction one has to rely on the hypothesis of rational consumer behaviour with a given set of preferences. Cross section analysis of household consumption (Haas et al., 1998) has shown that a "rebound-effect" can outweigh conservation to a quite large degree. The major argument is that consumers worry more about the utility they obtain from the service than about energy demand per se. Hence, consumer service demand will mainly depend on how easily and how cheaply a certain level of service can be provided. Consumers choose a different service level (e.g., indoor temperature) as prices become cheaper.

Other studies show how the role of the user might be decisive. Marked fluctuations have been observed in the consumption of heating energy in one-family houses of the same size that are equipped similarly. These variations have been found to be due to great differences and changes in usage patterns. Thus, in spite of the importance of an efficient technology and of incentives for its adoption, consumer attitudes and behaviour have to be taken into account by energy policy implementation. Beside economic and command and control instruments, **communication and diffusion instruments** are advocated.

The importance of the construction domain for the national economy as a whole and the large share it has in the energy demand is undisputed. Nevertheless decision-makers are shown to be very cautious or even hostile to new incentives (see the contribution by JEGEN). Hence information about how practical implementation problems can be overcome is a prerequisite to introducing new instruments or altering existing ones.

One obvious problem affecting the adoption of efficient technologies decisions is the lack of information about them. Jaffe and Stavins (1994) argue that it is costly for people to learn of an innovation's existence and to learn enough about it, to find out if it is profitable, and to learn how to use it. Further, if others' use of the technology is an important source of information, then adoption creates a positive externality, because it generates information that is valuable to others. Thus, information programmes presenting facts and options designed especially for property owners and builders will reinforce the impact of already implemented economic instruments (see also the contribution by BRUPPACHER AND ULLI-BEER). Among these instruments, pilot and demonstration projects, as forms of providing experience, and educational courses for specialists are particularly valuable.

The importance of combining incentives and information is also advocated by Stern (2000), who refers to an evaluation study showing that financial incentives for investment in home energy conservation were necessary but not sufficient. Apparently consumer response to subsidies depends on how the subsidy programme is made known to the householders.

Investors meet another obstacle if they cannot credibly present the energy efficiency of the new or renovated building to potential buyers. Then the sale price may not reflect efficiency attributes. Quality labelling, like the "Minergie" or the Home-Energy Rating System[9], could help to overcome this obstacle.

Last but not least, communication and diffusion instruments are necessary to reduce or to avoid the rebound effect in energy efficient homes, i.e., in homes where energy-saving investments have been realised. All kind of communication and diffusion instruments that aim to influence the individual behaviour of inhabitants and housing usage patterns will then prove useful.

6. Summary and conclusions

We have shown that the energy efficiency of residential buildings is an important step towards a substantial reduction of polluting energy consumption. First, one seeks to identify the obstacles to the choice of more energy efficient investments and the determinants explaining the behaviour of investors. Second, the importance and the difficulties of combining economic and technical knowledge to understand, analyse, and simulate such investment decisions are emphasised.

In most computer simulations, we find that the imposition of an energy tax on landlords will increase the stock of energy-efficient housing by a small but not insignificant percentage of the total housing stock (McFarlane, 1999: 4). An energy tax will increase the rate of conversion from low to high energy-efficiency housing by making it *relatively* more profitable to do so. However, the negative effect of such a tax is that the total housing stock will be relatively smaller than without the tax, since developing becomes less profitable. The result of a smaller housing stock is clear: rents will be higher for all tenants, including those living in high-efficiency housing. When the housing supply is elastic, tenants will share the burden of the tax with landlords. Thus, we should be careful to weigh the benefits of any energy policy against the costs that it would impose on the rest of the economy. Measuring the costs and benefits would be difficult, but simply being aware of such costs would aid policy decisions.

Some economic instruments are proposed to overcome the identified obstacles. Their functioning and implementation are discussed. An important outcome is that even for classical instruments like taxes or subsidies, their impact may be rather more effective if they are combined with communication and diffusion instruments. This finding is also underlined by Stern: "There is strong evidence that incentives and information interact, with the combination sometimes being much more effective than the sum of the two interventions" (Stern, 2000: 419).

---

[9] This programme has become effective in large parts of the United States since the Home Energy Rating System (HERS) was included in the Energy Policy Act in 1992. In 1993 a national HERS Council was established with the mission to increase residential energy efficiency and affordability nationwide by advancing uniform home energy ratings, energy-efficiency financing, education, and research through collaborative efforts.

References

Anas, A. and Arnott, R.J. (1991) Dynamic housing market equilibrium with taste heterogeneity, idiosyncratic perfect foresight, and stock conversions. *Journal of Housing Economics, 1*: 2-32.

Anas, A. and Arnott, R.J. (1997) Taxes and allowances in a dynamic equilibrium model of urban housing with a size-quality hierarchy. *Regional Science and Urban Economics, 27*: 547-580.

Berndt, E.R. and Wood, D.O. (1979) Engineering and econometric interpretations of energy – capital complementarity. *American Economic Review, 69*: 342-354.

Dupraz, F. (1999) Spécification technico-économetrique de la fonction d'offre de la qualité thermique de bâtiments renovés. Travail de diplôme en économétrie. University of Geneva; *mimeo.*

Dupraz, F., Garbely, M. and McFarlane, A. (1998) Ecological taxation and energy conservation equipment. Intermediate report to the Swiss National Science Fund, University of Geneva; *mimeo.*

Goldman, C.A., Greely, K.M. and Harris, J.P. (1988) Retrofit experience in U.S. Multifamily buildings: Energy savings, costs, and economics. *Energy, 13/11*: 797-811.

Haas, R., Auer, H. and Biermayr, P. (1998) The impact of consumer behaviour on residential energy demand for space heating. *Energy and Building, 27*: 195-205.

Hassett, K.A. and Metcalf, G.E. (1993) Energy conservation investment: Do consumers discount the future correctly? *Energy Policy, 21*: 710-716.

Hassett, K.A. and Metcalf, G.E. (1995) Energy tax credits and residential conservation investment: Evidence from panel data. *Journal of Public Economics, 57/2*: 201-217.

Jaffe, A.B. and Stavins, R.B. (1994) The energy paradox and the diffusion of conservation technologies. *Resource and Energy Economics, 16*: 91-122.

McFarlane, A. (1999) Even more on equilibrium II: An analysis of energy taxes with an endogenous housing supply, multiple housing types, multiple household types and three time periods. University of Geneva; *mimeo.*

McFarlane, A. (2000) Spending energy tax revenue. University of Geneva; *mimeo.*

Neels, K. (1982) Reducing energy consumption in housing: An assessment of alternatives. *International Regional Science Review, 7/1*: 69-81.

OECD/AIE (1991) Energy efficiency and the environment. OECD, Paris.

Pearce, D. (1994) Sustainable consumption through economic instruments. Symposium on sustainable consumption, Oslo 19-20 January.

Quigley, J.M. (1984a) The production of housing services and the derived demand for residential energy. *Rand Journal of Economics, 15/4*: 555-567.

Quigley, J.M. (1984b) Residential energy standards and the housing market: A regional analysis. *International Regional Science Review, 9/3*: 201-16.

Roulet, C.-A. (1987) Énergétique du Bâtiment, I et II. Presse polytechniques Romandes, Lausanne.

Stern, P.C. (2000) Toward a coherent theory of environmentally significant behavior. *Journal of Social Issues, 56/3*: 407-424.

Stoft, S.E. (1995) The economics of conserved-energy "supply" curves. *The Energy Journal, 16/4*: 109-137.

Sutherland, R.J. (1992) Market barriers to energy-efficiency investments. *The Energy Journal, 12/3*: 15-24.

Walsh, M.J. (1989) Energy tax credits and housing improvements. *Energy Economic, 11/4*: 275-284.

Changing Things – Moving People.
Strategies for Promoting Sustainable
Development at the Local Level
R. Kaufmann-Hayoz/H. Gutscher (eds.)

The Acceptance of Instruments among Energy Policy Key Players

Maya Jegen

This article deals with energy policy, a policy domain of high relevance for sustainable development. Its important issues – CO_2, nuclear waste, energy efficiency, renewable energy – require diverse policy solutions and instruments at different societal levels, as illustrated by several case studies in this volume. This contribution focuses on the cantonal and national level – thus complementing the focus on the municipalities – and on key players involved in political processes. The central question is how these political actors evaluate energy policy instruments. The instruments belong to three of the main instrument types presented in this volume, namely collaborative agreements, economic instruments, and command and control instruments. For each type several concrete instruments have been selected, including the tax on the energetic quality of buildings that is elaborated in the contribution by GARBELY *and* MCFARLANE. *To put the key actors' evaluations into perspective this study focuses on the political structure and analyses the power relation in the domain of energy policy. It attempts to give some indications of the barriers and opportunities that instruments for promoting energy efficiency may encounter in the political process. The analysis is based on empirical data from a survey of 240 key actors in the Swiss energy policy domain.*

1. Introduction

Up to the 1960s energy resources seemed to be unlimited, the demand was growing, and prices tended to decrease. This vision of growth and progress came into question by the early1970s, not only due the oil crisis, which affected all industrialised countries, but also by *The Limits of Growth* (Meadows et al., 1972), a report for the Club of Rome's project on the predicament of humanity, which warned of a dark future characterised by a scarcity of fossil energy resources. Although some of the pessimistic conclusions of this report have subsequently been attenuated, the question of energy supply and its environmental limits remains a major concern of energy policy in both industrialised and developing countries, as the Kyoto Protocol to the UN Framework Convention on Climate Change from 1997 illustrates.

In Switzerland, considerations of sustainability have been on the political agenda since the 1970s, though not in the most efficient way. As an immediate outcome of the oil crisis, the security of supply and the independence of the country had priority among energy policy topics, and nuclear energy was regarded as the most promising technological guarantee. Its negative consequences – security of power plants and nuclear waste – were brought up only through citizens' rising protests, which polarised and completely paralysed the domain of nuclear energy legislation.

Parallel to the search for fossil fuel substitutes, however, various efforts in favour of energy saving programmes were made. Within this context, the idea of an energy tax as an effective instrument for promoting energy efficiency emerged on the list of political items in the early 1980s. However, a popular initiative on energy taxation, launched by the Swiss Energy Foundation (SES), an energy-oriented environmental organisation, was rejected in 1984 by 54% of the votes. A second attempt to introduce an energy tax failed in the deliberation phase of the constitutional article on energy. The supporters from the green-left environmentalist faction were not powerful enough. The constitutional article minus an energy tax was finally accepted by popular vote in 1990.

The Swiss debate on energy policy in the 1990s, like that of the member-states of the European Union, was undoubtedly dominated by the liberalisation of the electricity market, though by now the issue of energy taxation has also been included in the political agenda and found broad support across the political parties. A glance at the actual federal legislation shows that several bills focus on taxation issues: the reduction of CO_2 emissions, the "Energy Environment Initiative" and the "Solar Initiative", the federal decree on energy taxation as well as the project of an ecological tax reform.[1] Improving ecological efficiency is, however, not left to environmental taxes, and there are various voluntary and regulatory tools aimed at the same goal. Thus, while the goal of a sustainable development is widely accepted, the conceptions of how to reach it and who is going to pay what price for it differ considerably. In other words, energy policy instruments encounter variable support in the political process.

This chapter focuses on the preferences of key players with regard to energy policy instruments. We suppose, however, that most of the instruments apply to other environmentally relevant policy domains, too. According to the typology of instruments presented above, but concentrating on **instruments relevant to the political decision-making process**, we examine the **likelihood of political acceptance of the instruments**. We attempt to discern the core criteria responsible for the acceptance or the rejection of an instrument, and we analyse the political structure in order to understand which political alliances could favour or obstruct the implementation of a certain instrument. On this basis, we may gain some insight into the probable political trajectory of an instrument and give some background information on political patterns for the other empirical

[1] See GEHRIG AND NORTH for more detailed information.

studies of this book. Even though these deal primarily with cantonal or municipal levels and focus very much on communication and diffusion instruments or the implementation of legal procedures, a closer look at the political process and the actors involved at the federal level is important, because it generates the legal framework in all domains of sustainable policy-making.

2. The key players in energy policy

The focus on the political acceptability of energy policy instruments implies a certain number of assumptions about political patterns and processes. Following Mayntz and Scharpf's actor-centred institutionalism, decision-making is conceptualised as taking place among individual and corporate actors within institutionalised settings. Its output is given by the interdependent choices of multiple actors that have specific perceptions and preferences of a given policy problem (Mayntz and Scharpf, 1995; Scharpf, 1997). The outcomes of public policies are therefore based mainly on intentional actions, depending on subjective preferences and convictions, and framed by a set of institutionalised – formal or informal – rules. Thus, if we want to know something about potential policy outcomes, we should know something about the key players involved, i.e., the constellation of actors in a specific policy domain. Through the existing coalitions we should be able to generate some hypotheses about potential barriers and opportunities for the acceptance of policy options or tools.

By key players we understand all actors who participate directly or indirectly in the political decision-making process of the energy policy domain or who are directly concerned by the outputs of these decisions, i.e. exponents of political parties, trade associations and unions, the federal and cantonal public authorities, science, and energy consulting firms. This study also takes into account board members of the big utilities *[Überlandwerke]* and actors of social movements whenever they have a decisive impact on the decision-making process. The key players are, therefore, individual actors as well as private and public organisations advocating their ideas and interest within the energy policy domain (Sabatier, 1993: 120). Even though individual actors can be of great importance in the formation of policies, we follow Laumann and Knoke, who assume that corporate entities are the key state policy-domain actors:

> "(...) we conceive of the modern industrial polity as a complex of formal organisations in conflict with one another over the collective allocation of scarce societal resources. Neither aggregates of persons nor agents of class interests, the large bureaucratic state organisations are effective instruments of domination for those elites who can command their authority. Using power relations within and interorganisational networks among them, state managers and interest group leaders struggle to mobilise political resources that shape public policies beneficial to their organisations' objectives, including the survival of the power structure itself" (Laumann and Knoke, 1987: 8).

We consider the political system of Switzerland as such a corporatist network that integrates the key players by mechanisms of co-operation and co-ordination, without allowing all actors the same access to power positions.

2.1. *Selecting the key players*

As mentioned above, we are interested in actors who shape the course of energy policy by defining its issues, actors who are in a position to exert power and influence. The problem of measuring power and influence is as old as political science itself. Baumgartner and Leech recapitulate:

> "(…) the most troubling difficulties (…) stemmed from the vituperative ideological and methodological debates about where power was located in the political system and how to recognise when, by whom, and how influence was wielded" (Baumgartner and Leech, 1998: 58).

It was not our claim to have resolved this problem, but we tried to minimise the bias by combining the decisional, positional, and reputational approaches. Following the decisional approach (Kriesi, 1980), we selected the most important energy policy issues over the last ten years (from 1987 to 1998) and listed the most active contributors across the different phases of the political process. This list was completed by the exponents of relevant positions of the public or private energy sector as well as by actors who were mentioned as very powerful in the preliminary interviews with civil servants. From the initial list containing 1'075 actors we selected a core of 300. As selection criteria, we used positional and participatory attributes (frequency of participation, seats in managing boards, amount of represented capital, etc.) in a cumulative fashion. The final core was interviewed face-to-face during the summer of 1998, using a standardised questionnaire. The response rate was about 80%. All data discussed below are based on this survey.

2.2. *Mapping the energy policy coalitions*

To understand the kind of alliances that exist in energy policy and how their effective power may be estimated, we rely on network analysis.[2] This method is, as Wasserman and Faust (1998: 4) point out, a distinct research perspective within the social and behavioural sciences, because it relies on the assumption of the importance of relationships among interacting units. Thus, this approach emphasises the interdependence of actors and their actions, rather than their independence, and assumes that the relational ties are channels for the flow of material or non-material resources. Social, economic, and political structures in network models are regarded as durable patterns of rela-

---

2 We consider policy network as a generic term and not as a particular form of governance that lies between market and hierarchy. In that our network approach differs from the one used by BÄTTIG AND BALTHASAR.

tions that may provide opportunities or constraints on the action of individuals or organisations. Thus, we look for the political patterns within the energy policy domain.

A priori, we ignore whether Swiss energy policy is a homogeneous field, whether the same actors dominate the whole decision-making process, or whether it is a heterogeneous domain composed of diverse specialised actors with very particular interests. Neither do we know whether the coalitions have been stable over the decade under analysis. For these reasons, we make a distinction between three issue areas within energy policy, namely nuclear energy, energy efficiency, and the liberalisation of the electricity market. In addition, we suppose that each issue area is particularly related to a certain period in time, during which it dominated the energy policy debate. Accordingly, we presume that nuclear energy was the crucial topic until the mid1980s, that energy efficiency became relevant in the context of the growing ecological awareness during the 1980s, and that the liberalisation of the market is the major theme of the late 1990s. This contribution concentrates on the **energy efficiency** domain, given the interest in the acceptability of energy policy instruments for promoting energy efficiency. The two other issue areas are taken into consideration as far as there is a structural difference that might alter the interpretation considerably with respect to the acceptance of instruments in other domains of energy policy.

The relational ties in our network under study are defined as close collaboration among the organisations and institutions. The question related to networks in the standardised questionnaire was the following: "Relating to the domain of energy efficiency, please look at this list and specify the organisations and institutions with whom your (name of the concerned organisation/institution) collaborated closely." The expression "close collaboration" was defined as follows: "Close collaboration implies the common elaboration of positions as well as an intense exchange of information and views about the energy domain at stake." The question about close collaboration was asked for all three issue areas. The list submitted to the respondents contained 138 organisations and institutions, arranged by categories (political parties, economic organisations in general, employees' and consumers' unions, energy related organisations, environmental organisations and movements, as well as federal and cantonal ministries and offices). The answers were summed up in an adjacency matrix, where a tie was coded as 1 and an absent tie as 0.

On the basis of this network data, we examine two major proprieties of the network, its centrality and its cohesiveness.

2.2.1. Centrality measures: who are the central actors?

Looking at the location of actors in a policy subsystem implies the assumption that influential actors are "in the thick of things" and that the close collaboration with their allies substantiates their influence. The literature of network analysis makes a distinction between a "local" and a "global" centrality (Scott, 1991; Wasserman and Faust, 1998). Local centrality can be defined as an attempt to measure the importance of an actor within the network. Its basic idea is that a central actor is at the centre of numerous direct contacts with other actors; he is, therefore, central within his neighbourhood. The idea also implies the existence of several local points of local centrality within a network. In contrast, global centrality, which is referred to as centralisation, relates to specific properties of the graph structure as a whole. This general index looks for actors dominating the network, and the larger it is, the more likely it is that there is a single quite central actor, with the remaining actors considerably less central (Scott, 1991).

Different formulae have been developed in sociometrics for measuring centrality. We look at two centrality measures, degree and betweenness. The former is a very basic robust measure that merely counts the number of actors adjacent to a given actor; it takes into account the direct contacts of an actor. A high degree centrality may thus be considered as the actors' "visibility or the potential for activity in communication" (Freeman, 1979: 219). Comparing the degrees of different actors shows how well they are related to their local environments. The second, more complex measure, betweenness, assumes that information flows through the relational channels by the shortest path and takes into account indirect links as well. The central actor is the one who is between a number of other actors, and who can be central, even though its degree might be relatively low (Scott, 1991: 89). The strategically located actors who monitor the largest amount of information can be considered (political) brokers.

All network analyses were effectuated by UCINET V (Borgatti et al., 1999). Table 1 shows the degree (C'_D) and betweenness (C'_B) centrality indices for the 30 most central organisations and institutions in the energy efficiency domain. The global centralisation index at the top of the table gives an idea of how closely the network is organised around its most central actors; the index 0.52 indicates that there is some variability between the smallest and largest actor-level indices, i.e. there are some central actors in the network. On the other hand, the index of the global betweenness centralisation (0.12) is very low, i.e., brokers do not play an important part in the whole network.

Before examining the actor-level centrality indices let us look at the influence measure, reported in the third column of Table 1. The key players were asked to evaluate the influence of the other organised and institutionalised actors.[3] Only eight actors are regarded as very influential in the energy efficiency area, reaching a nominal value over 100 points, which is an arbitrary cut-off point indicating that 42% or more of the interviewed actors think that the actor at stake is very influential. By comparison, 13 are considered very influential in the domain of liberalisation and 23 in the nuclear energy domain. Furthermore, the eight influential actors of energy efficiency are mainly public, left, and environmental actors: the Federal Office of Energy (BFE) and the Federal Department of Environment, Transport, Energy and Communications (UVEK), to which the BFE belongs, the cantonal energy offices and the cantonal energy directors, the Social Democrats (SPS) and the Green Party (GPS), as well as the Swiss Energy Foundation (SES) and the WWF. This supremacy of public and green-left actors is to some extent confirmed by the centrality measures.[4]

[3] The question read: "Relating to the domain of energy efficiency, please look at this list and specify the organisations and institutions that have a high influence within this domain."

[4] The Pearson correlation between the nominal values of influence and the degree centrality indices is 0.749 at a significance level of 0.01.

Table 1: Centrality indices (expressed as a percentage for each vertex) and influence evaluation for the 30 most important energy efficiency actors. For explanation of the abbreviations see Table 5, page 210.

Centralisation	**0.52**	**0.12**	**–**

Public Actors	**C'$_D$**	**C'$_B$**	**Infl.**
BFE	0.63	0.9	163
Ct Energy Directors	0.63	0.7	109
Ct Energy Offices	0.63	0.7	130
BUWAL	0.36	0.2	76
UVEK	0.23	–	129

Political Parties	**C'$_D$**	**C'$_B$**	**Infl.**
CVP	0.67	0.12	73
SPS	0.62	0.12	124
FDP	0.43	0.3	84
GPS	0.28	0.1	114
LdU	0.23	–	24

Electric Power Industry	**C'$_D$**	**C'$_B$**	**Infl.**
Other utilities	0.42	0.3	26
BKW	0.39	0.2	56
CKW	0.37	0.1	41
VSE	0.31	0.1	90
NOK	0.28	0.1	50
EOS	0.27	–	44
EGL	0.22	–	39

Environmental NGOs	**C'$_D$**	**C'$_B$**	**Infl.**
Solar 91	0.40	0.4	39
Greenpeace	0.31	0.2	67
WWF	0.28	0.1	101
SES	0.27	0.1	115
Greinastiftung	0.27	0.1	27
KSU	0.23	–	18
Anti-AKW-Org.	0.22	0.1	44

Business organisations	**C'$_D$**	**C'$_B$**	**Infl.**
Energy Forum	0.50	0.6	98
Vorort	0.38	0.4	58
SIA	0.36	0.1	84
Business promotion	0.33	0.4	16
SGV	0.24	0.1	38
SAB	0.24	0.1	9

For every local measure there is a corresponding – global – centralisation measure. The formulae for the measures used are as follows:

Degree: Normalised Centrality $C'_D = C_D / (n-1)$; Centralisation $\sum (C'_D\max - C'_D(vi)) / (n-2)$.

Betweenness: Normalised Centrality $C'_B = 2C_B / (n-1)(n-2)$; Centralisation $\sum (C'_B\max - C'_B(vi)) / (n-1)$.

The measure varies from 0 to 1. The value 1 occurs if the network is organised like a "star"; the value 0 occurs in a "complete" network without central actors.

Before examining the actor-level centrality indices let us look at the influence measure, reported in the third column of Table 1. The key players were asked to evaluate the influence of the other organised and institutionalised actors[5] Only eight actors are regarded as very influential in the energy efficiency area, reaching a nominal value over 100 points, which is an arbitrary cut-off point indicating that 42% or more of the interviewed actors think that the actor at stake is very influential. By comparison, 13 are considered very influential in the domain of liberalisation and 23 in the nuclear energy domain. Furthermore, the eight influential actors of energy efficiency are mainly public, left, and environmental actors: the Federal Office of Energy (BFE) and the Federal Department of Environment, Transport, Energy and Communications (UVEK), to which the BFE belongs, the cantonal energy offices and the cantonal energy directors, the Social Democrats (SPS) and the Green Party (GPS), as well as the Swiss Energy Foundation (SES) and the WWF. This supremacy of public and green-left actors is to some extent confirmed by the centrality measures.[6]

Considering the **degree** indices of the 30 most central actors we find the **public actors** mentioned above, i.e., the BFE, the UVEK, the cantonal energy directors and the cantonal energy offices. The **political parties** heavily involved in energy efficiency are primarily the Christian-Democrats (CVP) and the SPS, followed by the Radicals (FDP), and the Green Party (GPS), supported by the Independents (LdU). Note that the subjective evaluation for the GPS is much higher than its degree index. To a lesser extent, this is true for the FDP and it is the opposite for the CVP. The fourth governmental party, the Swiss People's Party (SVP), does not show up at all in the list of the top thirty of the energy efficiency domain, nor is it implicated much in the other issue areas. On the other hand, the cantonal energy offices are much more central in the energy efficiency domain than in the two others, which is corroborated both by the centrality and the influence measure. Jams in the legislative process and an undecided division of authority between the cantons and the federal state gave the cantons a scope of action, and some of them developed quite efficient energy solutions, in particular within the construction sector (e.g., educational programmes, MINERGIE label). The high centrality of the cantonal energy offices reflects their activity in this domain.

The **electric power industry**, which takes a key position within the entire energy policy domain with the presence of the big utilities among the 30 most central actors, seems to be a little less central in the energy efficiency domain (two of the utilities are missing, and the indices are somewhat lower). Surprisingly, some of the utilities seem to be more central than the official representatives of the electric power industry, the Swiss Union of Electricity Companies (VSE). The fact that the "other utilities" score high may indicate that there are some cantonal or municipal electric companies quite actively involved in Swiss energy policy.

5 The question read: "Relating to the domain of energy efficiency, please look at this list and specify the organisations and institutions that have a high influence within this domain."

6 The Pearson correlation between the nominal values of influence and the degree centrality indices is 0.749 at a significance level of 0.01.

The **central business organisations** in all three domains are the Energy Forum, the Swiss Federation of Commerce and Industry (*Vorort*), and the Association for the Promotion of Swiss Business and Trade. All three represent, directly or indirectly, energy-related interests, even though their member organisations do not always share the same objectives (e.g., big energy consumers vs. fossil fuel importers). Actors who emerge as central only in the energy efficiency domain are the Association of Engineers and Architects (SIA), which publishes energy guidelines for the construction sector, the Trade Association (SGV), and the Working Group of the Alpine Regions (SAB).

Among the **environmental organisations**, Solar 91 is especially central; this association, which promotes techniques for renewable energy sources, is more central than Greenpeace and the two "established" actors, SES and WWF. On the whole, the presence of environmental NGOs is stronger in the domain of energy efficiency than in the other two issue areas. Even though these NGOs rank among the 30 most central actors, none of them achieves a top position. This suggests that the environmentalists have to rely on the SPS and its role as a broker whenever they want to put an issue on the political agenda. Amazingly, the Swiss Unions are absent from the energy policy field. The Federation of Socialist Labour Unions (SGB) scores only within the liberalisation domain, and even there it is rather peripheral.

As already indicated by the centralisation index of betweenness, there are no strong brokers in the energy efficiency network (they are fairly stronger in the two other domains). The actor-level centrality shows that when there is an intermediary role to play, it is taken by the two parties, the SPS and the CVP. These two actors are also characterised by a high degree of centrality, and they are in contact with the most important players. Though weak, the brokerage of the SPS may nevertheless be crucial for the environmental NGOs, because they lack highly central positions and other direct contacts with strong players.

To summarise, as the centrality and influence measures reveal, there are fewer central actors in the domain of energy efficiency than in the nuclear and the liberalisation domains, and these few core actors interact mainly with one another. Moreover, the public actors – the BFE, the cantonal energy directors, and the cantonal energy offices – and two of the governmental parties, the CVP and the SPS, seem to be in control within this domain. Fewer actors and less opposition may facilitate interaction in the domain, but it might also indicate that fewer non-specialised actors feel concerned about energy efficiency.

2.2.2. *Cohesive subgroups: who are the strongly allied actors?*

We began by looking at the actors who hold central positions in the energy efficiency network. We ignored, however, how these actors are related to each other. So we turn now to the identification of **cohesive sets of actors**, i.e., subgroups of actors with strong and direct ties.[7] Actors of a cohesive

[7] For reasons of space, the technical aspects of cohesiveness are not specified in the text, and we did not insert the generated tables of hierarchical clusters.

subgroup entertain more frequent interactions within their group than outside, and thus we assume that they are more likely to form coalitions in order to defend their policy options.

At the highest level of cohesion, one group composed of the electricity community (the big utilities ATEL, BKW, CKW, EGL, EOS, and the Energy Forum) and the peak economic organisation *Vorort* stand out. The electricity sector is the most cohesive in all three issue areas, but it is somewhat smaller in the energy efficiency domain. Here the VSE, the association of the electric power industry, does not join the cohesive group. At a slightly lower level of cohesion, there are three distinct groups. The first one regroups the governmental parties (CVP, FDP, SPS),[8] the public actors (BFE, BUWAL, Cantonal Energy Directors), and Clima Suisse, the association of heating and ventilation companies. This latter organisation appears only in the energy efficiency domain. The second group clusters the environmentalists (GPS, Greenpeace, SES, WWF, KSU, Anti-AKW, and *Grimselverein*), the SPS, and the Cantonal Energy Directors. The third group contains again some members of the electricity sector (BKW, CKW, EOS, other utilities, Energy Forum), the *Vorort*, and as main public actors the BFE, the cantonal energy directors, and the cantonal energy offices.

Note that the cantonal energy directors are associated with all three groups at the second level of cohesion. This highlights the importance of cantons within the energy efficiency domain, which is also emphasised by recent legislation (sharing of authority regulated by the energy law, and incentives of the sequential program of Energy 2000). The cantonal position is even reinforced by the Cantonal Energy Offices that appear as a central actor only in the energy efficiency domain.

The simultaneous occurrence of the same actors – the cantonal directors, the BFE, and the SPS – in different cohesive groups indicates that the energy efficiency structure is less polarised than the nuclear energy domain. There, the formation of groups follows the right-left cleavage with the public actors (the cantonal energy directors, the BFE, and the UVEK) joining the centre-right and the (electricity) business camp. Likewise, there is a distinct environmental-left cluster in the energy efficiency domain, but it is related to the other actors via the cantonal energy directors and the SPS, which emerged as the main energy efficiency broker in the centrality analysis described above.

3. Energy policy instruments and their evaluation

3.1. Selected instruments

As developed in the introductory chapters of this book, the typology elaborated by the Integrated Project "Strategies and Instruments" consists of five main types, namely command and control instruments, economic instruments, service and infrastructure instruments, collaborative agreements, and communication and diffusion instruments. For our survey among energy policy key

[8] In the analysis of centrality the fourth governmental party, the SVP, is absent.

players we selected **eight instruments of the types command and control instruments, economic instruments, and collaborative agreements**. These types of instruments are typically the objects of political decision-making. For reasons of comparison, all of the evaluated instruments apply, though not exclusively, to the construction sector. As the consumption of polluting fuels within this sector is high (heating accounts for more than a third of Swiss energy consumption), there is a considerable potential for energy efficiency measures and a series of technologies is available. Moreover, the follow-up programme of Energy 2000 focuses particularly on the cantonal construction sectors (NZZ, 1999b). GARBELY AND MCFARLANE'S contribution in the present volume deals with some of these instruments from an econometric point of view.

Within the category of **collaborative agreements** are classed the ISO 14001 standards, the label MINERGIE, and the SIA 380/1 recommendation. The first represents standards for the promotion of environmental management systems, and the second is a quality label for promoting goods and services with respect to criteria of energy efficiency and renewable energy. The third instrument is a recommendation on energy edited by the Swiss Association of Engineers and Architects (SIA) and is used as a planning tool for energy efficient construction with respect to the maximal heating need per surface and the minimal heat rate (Bundesrat, 1987). The category of **economic instruments** consists of three environmental taxes: a tax on CO_2 emissions, a tax on non-renewable energy, and a tax on the energetic quality of buildings. These fiscal instruments share some common characteristics: they leave the choice between preventing pollution or paying for pollution to the producers and consumers, they aim at the internalising of externalities and continuing incentives for improvements towards ecological efficiency, and they spur on innovation (Schlegelmilch, 1998: 2). On the other hand they differ considerably with respect to their minimum level of taxation, the regulation mechanisms, the sectors affected, and the principle of redistribution of the benefits. The evaluated models of taxation on CO_2 and on non-renewable energy correspond to the bills discussed in Parliament in summer 1998. The CO_2 bill proposes, relative to the level of 1990, a 10%-reduction of the CO_2 emissions by 2010. An obligatory tax will be implemented only if collaborative agreements do not prove to be successful, at the earliest in 2004. The tax on non-renewable energy that was at stake during the deliberation on the energy law proposed a levy of 0.6 cents/kWh; one half of the returns would be assigned to the promotion of solar energy, the other half to the promotion of energy efficiency measures for renovating buildings.[9] The tax on the energetic quality of buildings is discussed in detail in GARBELY AND MCFARLANE'S contribution. Its main idea is to levy a charge on buildings with an energetically bad balance, whereas high quality buildings get a subsidy for the amortisation of their investments. The category of **command and control instruments** contains an individual calculation of heating costs (VHKA) and isolation

[9] The tax on renewable energies has since been modified considerably. It has been detached from the energy law, and Parliament reduced the energy tax on a common denominator of 0.3 cents/kWh in 1999. It was submitted to a popular vote in September 2000, simultaneously with two further-reaching popular initiatives on energy taxation. All three issues were rejected. Due to this outcome, it will be difficult to put energy tax issues, including the ecological tax reform, on the political agenda successfully in the near future.

prescriptions. The VHKA is based on the polluter-pays principle, i.e., each tenant of an apartment is held personally responsible for not dissipating resources. This instrument was particularly contested with respect to old buildings because of the considerable investment for monitoring technologies at the homeowner's expense. The insulation prescriptions for buildings are based on a standard decree elaborated by the BFE and the cantons in 1986 (Bundesrat, 1987).

3.2. Evaluation criteria

The key players interviewed were asked to evaluate each instrument on the basis of five criteria: (1) ecological efficiency, (2) compatibility with economic competitiveness, (3) potential for political consensus, (4) facility of implementation, and (5) capacity for diffusion. These criteria include the ecological, economic, political, and informational dimensions of the instruments. Each instrument is evaluated on each criterion on a scale from −10 to +10.

Before evaluating the instruments, the respondents were asked to rank the relevance of the five criteria. The question read: "On this list, you find five different criteria for evaluating energy policy instruments. Please tell me, which is the most important (1), the second important (2)... the least important (5) criterion for you." Table 2 summarises the rankings for the seven principal categories of organisations and for the central actors of the energy efficiency domain.[10]

The first row of the table shows the overall outcome of this ranking: ecological efficiency criterion is put in first place in 37% of the answers, followed by economic compatibility (34%), facility of implementation (15%), and potential for political consensus (12%). Left far behind is capacity for diffusion (5%). The fact that only a few actors give priority to the propagation of an instrument might indicate that the key players do not consider marketing tasks of primary importance. Not surprisingly, we find the main initiators of the MINERGIE label and the exponents of the BFE among the advocates of the importance of the diffusion criterion.[11] Thus, actors that are directly concerned with the elaboration and the realisation of an innovative instrument emphasise the need for broader information.

---

[10] To keep Tables 2 and 4 intelligible, only the first half of the most central actors from Table 1 are shown.

[11] They are supported by energy consultants and some cantonal politicians of the green-left wing.

Table 2: Percentages of first rank for the five evaluation criteria, overall and for each organisation.

	Ecological efficiency	Competitiveness	Implementation	Consensus	Diffusion	N (range)
Overall % first rank	**37**	**34**	**15**	**12**	**5**	223-227
Centre-right	**27**	**50**	**9**	**12**	**3**	34
CVP	˙44	44	11	0	0	9
FDP	33	50	0	8	8	12
Left-green	**70**	**0**	**7**	**13**	**3**	30
SPS	68	0	11	11	11	19
Environ-mentalist	**81**	**4**	**12**	**0**	**4**	25-27
Solar 91	100	0	0	0	0	1
Federal State	**56**	**19**	**0**	**19**	**7**	15-16
BFE	67	17	0	0	17	6
BUWAL	100	0	0	0	0	2
Cantons	**21**	**31**	**37**	**10**	**7**	42-43
Energy Directors	27	31	31	8	4	26
Energy Offices	17	18	58	9	17	12
Business organisa-tions	**42**	**42**	**8**	**4**	**4**	24
Vorort	0	100	0	0	0	2
SIA	50	50	0	0	0	2
Energy Forum	50	50	0	0	0	4
Electric Industry	**8**	**62**	**13**	**21**	**0**	51-53
BKW	0	50	25	25	0	4
CKW	0	100	0	0	0	3
Other utilities	7	50	14	29	0	14

The positive performance of the criterion of ecological efficiency is somewhat reduced, when we look more closely at the differences among actors. The table shows also the classic cleavage between the centre-right parties and the business community, which are clearly competition oriented, on the one hand, and the much more ecology based green-left actors on the other hand. As expected, the ecological criterion is considered more relevant among the environmental organisations and the federal offices. The electric power industry manifestly prefers the competitiveness to the ecological criterion. The other representatives of the business community distinguish less sharply between these two factors, and not until the second rank is taken into account does their economic preference become obvious. The problem of political implementation is of main concern for cantonal actors, especially for the cantonal energy offices, because they are confronted daily with practical questions of implementation. It might be surprising that federal actors are not interested in implementation, but once again, this outcome is levelled off by the second rank. Federal actors attribute equal importance to political consensus and to economic factors, but have a clear preference for ecological aspects. Political consensus is considered more relevant only by the electric power industry. Again, there is a striking discrepancy between the evaluation of the electric power industry and those of the other parts of the business community.

The rankings of the central actors in the field of energy efficiency vary within the main categories. Thus, 44% of the CVP, which has the highest degree of centrality, put the ecological criterion first, which is much greater than the average of the other centre-right players. Likewise, the BFE and the BUWAL seem to be more environmentally oriented than the other federal actors. As already mentioned, the BFE also stresses the importance of the diffusion criterion, which is in line with the strategy of Energy 2000 (see the contribution by BÄTTIG AND BALTHASAR). Among the business organisations the uncompromising preference of the *Vorort* for business concerns is striking.

3.3. Evaluation of the energy policy instruments

Let us now consider the evaluation of the instruments. The different types of instruments can be interpreted as representing different degrees of action constraints imposed by the state. It could be hypothesised that less interventionist instruments, such as collaborative agreements, are more likely to be accepted by the key players than strongly interventionist ones, such as command and control instruments. However, as shown in Table 3, the insulation prescriptions – a command and control instrument – do best. On the other hand, we find the collaborative agreements ahead of the economic instruments, as expected.

On the whole, the insulation prescriptions are evaluated positively, except for the criterion of political consensus. The positive evaluation can be explained by the fact that this instrument is already well known and implemented. By contrast, the heating bill based on individual consumption and, even more, the tax on the energetic quality of buildings, is evaluated pessimistically on all criteria. The label MINERGIE gets an intermediate rating on all five aspects. Both standards – ISO and SIA – are considered as consensual and compatible with economic concerns, but not very efficient from an ecological point of view. Therefore we can generalise that collaborative agreements respond to consensual and implementing claims, but are not considered as sufficient by the key actors with respect to sustainability considerations. On the other hand, the tax on CO_2 as well as the tax on fossil energy are regarded as ecologically efficient, but are criticised for their lack of competitive and implementing adaptability.

Table 3: Average scores across criteria and ranking of the instruments per criterion (based on the score average).

Instrument	Average* (N)	Ecological efficiency	Competi-tiveness	Political implementation	Political consensus	Diffusion
Insulation prescriptions	5.39 (215)	1 (rank)	2	1	4	1
SIA 380/1, 380/4	4.78 (146)	5	3	2	1	5
Minergie	4.66 (159)	4	4	4	3	4
ISO 14000	4.31 (155)	6	1	3	2	7
Tax on CO_2-emmisions	3.64 (225)	2	6	5	5	3
Tax on non-renewable energies	3.06 (225)	3	7	6	6	2
Individual calculation of heating costs	2.79 (222)	7	5	7	7	6
Tax on the energetic quality of buildings	0.04 (208)	8	8	8	8	8

*Each criterion was evaluated on a scale from –10 to +10; this column reproduces the average of the five criteria.

The number of responses varies with each instrument, as the first column of Table 3 shows. In particular, the questions concerning **collaborative agreements** received perceptibly fewer responses, even though they scored encouragingly in general. This may suggest that the instruments at stake lack visibility and are poorly known. On the basis of our empirical data it is unclear to what extent a carefully directed marketing effort might increase the performance of such instruments. We suppose, however, that the broader promotion of these instruments might contribute to their visibility and to knowledge of their ecological efficiency. Contrary to the insulation prescriptions, the individual calculation of heating costs confirms the assumption that **command and control instruments** find less support among the key players. The pessimistic judgement of the individual calculation reflects Parliament's disagreement on this issue during the deliberation of the energy law. The idea behind this individual calculation was that the polluter-pays principle would lead to energy efficiency. It was contested, however, because of its expected costs for existing buildings. The energy law, which was approved in 1998, now prescribes this individual calculation of heating costs for new buildings, but leaves the regulation of old buildings to the cantons. The instrument that scored worst on all five criteria is the taxation of the energetic quality of a building. Unlike all other evaluated instruments, the key players were not familiar with this innovative instrument and had to decide on the basis of the standardised description they received from the interviewers. The lack of knowledge about a completely new tool may partly explain the bad performance of this incentive tax, as modelled by GARBELY AND MCFARLANE. In the interviews the key players strongly criticised the tax with respect to consensus and implementation. Thus, if the tax on the energetic quality of buildings were to have a chance of being put on the political agenda, its implementation would have to be straightforward and non-bureaucratic.

Focusing now on the **central energy efficiency actors,** we note only slight differences of ranking compared to the general rank order of all actors (Table 4). On average, the insulation prescriptions score highest, the tax on buildings lowest. The preference for the prescriptions is shared by about a third of the central actors, in particular the CVP and the cantonal actors. However, more than half of the central actors prefer collaborative agreements, even though the priorities within this category vary quite a bit. The energy efficiency brokers – the CVP, the SPS, and the BFE – share a common preference for the SIA recommendation. As already indicated, the federal actors assign some importance to the diffusion criteria and encourage collaborative agreements. This marketing strategy might be seen within the context of the federal programme Energy 2000. The target groups of this programme, which encourage voluntary-based mechanisms compatible with economic arguments, are private and public actors. In fact, this is an interesting example of how command and control instruments can support or encourage other types of instruments. The acceptance of the ten-year nuclear moratorium in 1990 paralysed energy policy legislation, in particular the nuclear power decisions process, but favoured the elaboration of the energy programme Energy 2000. Thus, existing legal norms can accelerate the development of economic instruments, collaborative agreements, service and infrastructure, and communication and diffusion instruments.

On average, taxes score lower than collaborative agreements, but they find considerable support among the left and environmental organisations and to some extent among the CVP. It is nonetheless difficult to make a general statement concerning taxes, because the variance within the categories of organisations and among central actors is significant. This is probably due to the complexity of the tax issue where the controversy, no longer over the principle, is over the tax rate and its subsequent allocation.

Table 4: Evaluation of the energy policy instruments by the central energy efficiency actors: Ranking of the instruments per actor category (based on the score averages).

Central actors	**N range**	**Collaborative agreements**			**Economic instruments**			**Command and control**	
		Iso 14001	SIA 380/1	Minergie	Tax on energy	Tax on buildings	Tax on CO_2	Individual heating	Insulation prescript.
General rank order	148-215	4	2	3	6	8	5	7	**1**
Rank order of central actors	74-116	2	3	4	6	8	5	7	**1**
Centre-right	16-34	4	**1**	3	6	8	5	7	2
CVP	10	6	2	5	4	8	3	6	**1**
FDP	12	2	**1**	4	7	8	5	6	3
Left-green	16-30	6	**1**	4	3	8	5	7	2
SPS	20	5	**1**	4	3	8	6	7	2
Environmental	11-23	6	5	4	**1**	8	3	7	2
Solar 91	1	5	6	7	**1**	4	2	8	2
Federal State	12-18	**1**	2	4	6	8	5	7	3
BFE	9	4	**1**	6	5	8	3	7	2
BUWAL	2	2	3	4	5	8	7	5	**1**
Cantons	30-42	4	2	3	6	8	5	7	**1**
Energy Directors	28	2	3	4	7	8	6	5	**1**
Energy Offices	12	6	2	3	4	8	5	6	**1**
Business organisations	16-22	4	2	3	6	8	5	7	**1**
Vorort	3	2	–	5	6	–	3	**1**	4
SIA	2	2	4	**1**	5	8	6	7	2
Energy Forum	4	4	3	**1**	7	8	5	6	1
Electric Industry	28-48	3	4	2	7	8	5	6	**1**
BKW	4	**1**	–	3	5	8	4	6	2
CKW	4	2	4	**1**	6	8	5	6	2
Other utilities	14	4	3	2	7	8	4	6	**1**

4. Summary and Conclusions

For analytical purposes we concentrated on the key players within the energy efficiency area and on instruments that target the construction sector. Since this focus might have been too narrow and might have excluded recent developments in the energy policy domain, we now attempt to expand the discussion and project some (hypothetical) trends in realignments and the chances of the acceptance of efficient instruments with respect to a sustainable development.

Our data indicate that the whole domain of energy policy is split into a centre-right and (electricity) business camp on the one hand, and a left and environmentalist camp on the other hand. Throughout, some key actors are of high centrality: The cantonal energy directors and two of the governmental parties, the CVP and the SPS; the peak business organisation, *Vorort*, and the electricity industry are central players. The cohesion data for the domain of liberalisation, however, reveal disintegrating tendencies within the electricity sector.[12] This trend is in line with the current process of radical restructuring in the liberalised European electricity market (Le Monde, 1999). Thus, we argue that liberalisation, as an external shock[13] (Sabatier, 1993; 1998), is contributing to a realignment of coalitions, in that the electricity industry – a united para-state actor that has been powerful for decades and is not very sensitised towards environmental issues – might lose some of its influence on Swiss energy policy.[14] Against a weakening conservative actor, liberalisation might open a **window of opportunity for environmentalists** to put ecological issues successfully through the political process. At the same time, liberalisation might open a window of opportunity for consumers, too, insofar as monopolies break down and consumers may freely choose their – green – electricity supplier (see also the chapter by GEHRIG AND NORTH) In such a scenario, communication and diffusion instruments and collaborative agreements[15] might gain increased importance. Moreover, the few but central advocates for ecology (BFE, Cantonal Energy Directors, Cantonal Energy Offices, CVP, and SPS; cf. Table 2) in the domain of energy efficiency might unite in an influential environmentally oriented coalition.[16]

---

[12] Regroupings of utilities highlight this fact, e.g., the regrouping of CVE-Romande Energie, Ensa (Neuchâtel) and FEW (Fribourg) in the French speaking part of the country (NZZ, 1999a; Handelszeitung, 1999).

[13] The nuclear accident in Tchernobyl in 1986 was a first important external shock that undermined somewhat the polarised camps of nuclear energy advocates and adversaries.

[14] There is an alternative interpretation, which cannot be discussed here at length. The dynamics of the liberalisation of the market in Europe have a considerable impact on the Swiss debate. If the political decision process is not accelerated, there is a great risk that political regulation will come late and will be outstripped by the economic process. In such a scenario, the political alliances would be of little importance.

[15] A label for green electricity and renewable energy was elaborated by the working group "*Arbeitsgruppe Ökostrom*" (Grasser and Kiefer, 1999: 6). If this label gets enough visibility and credibility, it might be important for a sustainable energy policy.

[16] For the first time, the Alpine cantons and the SPS defended a common position on hydraulic interests in 1996. There is a good chance that this coalition will advocate common standpoints in the liberalisation debate, too, when it comes to energy taxes, including the reimbursement of hydraulic power plants.

But there is another side to this ecologically optimistic scenario. Based on our data there is no cohesion among environmental organisations in the domain of liberalisation. The absence of coordination might be explained by the differences in environmentalists' perceptions of the consequences of liberalisation for sustainable development. While some view liberalised markets with lower electricity prices as a threat to ecological objectives, others look forward to the breakdown of monopolies, as mentioned above; others may not have comprehended the issue yet (relating to the survey moment in the summer of 1998). The lack of cohesion will no doubt limit the political strength of the ecologically-oriented actors.

What can be deduced from this energy policy structure that is in motion with respect to other environmentally relevant domains? As we have seen, **liberalisation entails a number of uncertainties, risks, and opportunities** with respect to the (re)arrangement of coalitions and to ecological issues in the energy domain. It may be speculated that the general neo-liberal turn of the Western world has similar effects on well-established political coalitions in other policy domains as well, and that international politics (European Union, World Trade Organisation, Kyoto Protocol) have a crucial impact on national legislation (for empirical studies concerning Switzerland, see Sciarini, 1995; Mach, 1998; 1999). In other words, the scope for political action at the national level is restricted by external factors because of reciprocity clauses. For sustainability, the risks of such constraints are obvious, but there may also be some opportunities, e.g., the advancement of the EU with respect to environmental taxation or the ecological tax reform in Germany that could have a positive and accelerating impact on Swiss legislation.

What can be said about the key players' acceptance of instruments? As mentioned above, the narrowly defined domain of application, namely the construction sector, precludes an extensive generalisation of the outcomes discussed in this contribution. Nevertheless, this discussion reflects some characteristics of the general debate. As we have seen, the best instruments from an ecological point of view are not the best from an economic or political perspective, i.e., instruments rated the most ecologically efficient are not the most accepted ones. Nevertheless, the choice between different types of instruments is not a zero-sum game, and **the combination of instruments** is desired by many actors. When asked about their favourite instrument(s), about a third of the actors selected two, and another third more than two – and up to six instruments (Kriesi and Jegen, 2000). With regard to their relatively high political acceptance, collaborative agreements certainly figure among this combination. In order to approximate ecological goals, environmental taxes, which allow for the internalisation of externalities, seem to be indispensable after all, and may, finally, stand a real chance of being accepted in the coming decade. As mentioned above, incentive taxes enjoy larger support among the green-left parties and the environmentalists. Compared to the 1980s, acceptance is also increasing among other central actors of energy efficiency, such as the CVP. But the issue of taxation is highly complex, and the details could not be tapped by our standardised questionnaires. Even if the principle of a tax were accepted by a large majority, its modalities – rate, redistribution – would remain very controversial. Drawing conclusions from these ambiguities and with respect to a sustainable future, we conjecture that external pressure on Swiss politics, coming in particular from European legislation, may be beneficial for an accelerated acceptance of ecologically efficient instruments, such as environmental taxes and the introduction of an ecological tax reform.

Table 5: Abbreviations of the mentioned organisations and institutions.

Public actors	
BFE	Federal Office of Energy
Ct Energy Directors	Cantonal actors at the executive level
Ct Energy Offices	Cantonal actors at the administrative level
BUWAL	Federal Office for Environment, Forest and Landscape
UVEK	Federal Department of Environment, Transport, Energy and Communications
Political Parties	
CVP	Christian-Democrats
SPS	Social-Democrats
FDP	Radicals
GPS	Green Party
LdU	Independents
Electric Power Industry	
Other utilities	Diverse cantonal and municipal utilities
BKW	Utilities of Bern
CKW	Utilities of the central part Switzerland
NOK	Utilities of north-eastern part of Switzerland
EOS	Utilities of the western part of Switzerland
EGL	Electricity company of Laufenburg
VSE	Swiss Union of Electricity Companies
Business organisations	
Energy Forum	Association representing the interests of the electricity industry
Vorort	Swiss Federation of Commerce and Industry
SIA	Swiss Association of Engineers and Architects
Business promotion	Association for the Promotion of Swiss Business and Trade
SGV	Swiss Trade Association
SAB	Working Group of the Alpine Regions
Environmental NGOs	
Solar 91	Working Group Solar 91 (promotion of solar energy)
Greenpeace	Greenpeace
WWF	World Wide Fund For Nature
SES	Swiss Energy Foundation
Greinastiftung	Association founded against the installation of barrages/for the protection of nature
KSU	Co-ordination of environmental organisations
Anti-AKW-Organisations	CAN, GAK, NWA, UeBA, etc.

References

Baumgartner, F.R. and Leech, B.L. (1998) Basic interests. The importance of groups in politics and in political science. Princeton University Press, Princeton.

Borgatti, S.P., Everett, M. and Freeman L. (1999) UCINETV. Analytic technologies. Natick.

Bundesrat (1987) Botschaft über einen Energieartikel in der Bundesverfassung. EVED, Bern.

Freeman, L.C. (1979) Centrality in social networks. Conceptual clarification. *Social Networks, 1*: 215-239.

Grasser, C. and Kiefer, B. (1999) Produktezeichen für Ökostrom und erneuerbare Elektrizität. Kiefer and Partners AG, Zürich.

Handelszeitung (1999) Monopol soll Wettbewerb ermöglichen. Der Bundesrat will die Schaffung einer nationalen Netzgesellschaft notfalls durch Enteignung durchsetzen. 13.1.1999.

Kriesi, H. (1980) Entscheidungsstrukturen und Entscheidungsprozesse in der Schweizer Politik. Campus, Frankfurt.

Kriesi, H. and Jegen, M. (2000) Decision-making in the Swiss energy policy elite. *The Journal of Public Policy, 20/1*: 21-53.

Laumann, E.O. and Knoke, D. (1987) The organisational state. Social choice in national policy domains. The University of Wisconsin Press, Madison.

Le Monde (1999) Le marché européen de l'électricité s'ouvre à la concurrence. 17.2.1999.

Mach, A. (1998) Quelles réponses politiques face à la globalisation et à la construction européenne? Illustration à partir de la révision de la loi sur les cartels. *Swiss Political Science Review, 4/2*: 25-49.

Mach, A. (1999) Globalisation, néo-libéralisme et politiques publiques dans la Suisse des années 1990. Seismo, Zürich.

Mayntz, R. and Scharpf, F. (1995) Gesellschaftliche Selbstregelung und politische Steuerung. Campus, Frankfurt.

Meadows, D. and Meadows, D. (1972) The limits to growth: A report for the Club of Rome's project on the predicament of mankind. Earth Island, London.

NZZ/Neue Zürcher Zeitung (1999a) Allianz Westschweizer Stromproduzenten. Schulterschluss zwischen der Waadt, Freiburg und Neuenburg. 22.6.1999.

NZZ/Neue Zürcher Zeitung (1999b) Verstärkte Wirkung – dennoch ungenügend. Das Aktionsprogramm Energie2000 im 9. Jahr. 14.9.1999.

Sabatier, P.A. (1993) Advocacy-Koalitionen, Policy-Wandel und Policy-Lernen: Eine Alternative zur Phasenheuristik. *Politische Vierteljahresschrift, Sonderausgabe 1993*: 116-148.

Sabatier, P.A. (1998) The advocacy coalition framework: revisions and relevance for Europe. *Journal of Eurpean Public Policy, 5*: 98-130.

Scharpf, F. (1997) Games real actors play. Actor-centered institutionalism in policy research. Westview Press, Boulder, Oxford.

Schlegelmilch, K. (1998) Energy taxation in the EU and some member states: Looking for opportunities ahead. Wuppertal Institute for Climate, Environment and Energy, Wuppertal.

Sciarini, P. (1995) La Suisse face à la Communauté européenne et au GATT: le cas test de la politique agricole. Georg, Genève.

Scott, J. (1991) Social network analysis. A handbook. Sage, London, Newburry Park, New Delhi.

Wasserman, S. and Faust, K. (1998) Social network analysis. Methods and applications. Cambridge University Press, Cambridge.

Changing Things – Moving People.
Strategies for Promoting Sustainable
Development at the Local Level
R. Kaufmann-Hayoz/H. Gutscher (eds.)

Solar and Alternative Power Supply: An Instrument towards Ecologically Sound Power Consumption?

Sonja Gehrig and Nicole North

Solar power and power produced by other renewable energies and supplied by power utilities are means towards replacing fossil fuels and nuclear plants for power production. Promoting these energy sources is therefore an instrument towards more environmentally friendly electricity production and consumption. We classify solar and alternative power supply primarily as a service and infrastructure instrument according to the typology of policy instruments presented in this volume. Solar power supply shares many features with other instruments – depending on the actor's perspective – such as economic instruments (subsidies), communication instruments (stimulating self-commitment), or even collaborative agreements, as in the introduction of labels for "green" power products under changing framework conditions, i.e., the liberalisation of the electricity market.

Increasingly, solar power supply in Switzerland is being promoted by privately or publicly owned power plants and power supply companies – most successfully through a kind of a stock exchange system. Although partly subsidised by the government, the price of solar power for the end consumer – volunteering to pay more for more ecologically produced electricity – remains as much as ten times higher than for conventional power provision. For this reason, the amount of consumed power generated by solar and other alternative technologies is still very small.

The guiding questions of this article are whether the solar power market can succeed under the conditions of a liberalising electricity market and whether it can fully realise and even expand the market potential among end consumers in order to achieve a measurable ecological impact.

1. Introduction

1.1. Aim of this contribution

This chapter analyses various aspects of the solar power market in Switzerland at the end of the 1990s under the conditions of a liberalising electricity market and under the assumption that solar power is a more environmentally friendly power source than power generated by fossil and nuclear sources. It gives an overview of the present solar power supply market in Switzerland and assesses the ecological impact of solar power in section 1. Section 2 discusses interactions among key market actors and describes customer profiles in the developing solar power market. In sections 3 and 4 we examine success factors in and barriers to increased implementation of solar power supply and demand from the suppliers' and the consumers' perspectives. Interactions of solar power supply with other energy policy instruments that aim at more ecologically sound power production and consumption are assessed in section 5. Section 6 predicts further developments in the solar power market with a view to the liberalisation of the electricity market. Finally, conclusions in section 7 are drawn with regard to two aspects: Can the end-consumer market potential for solar power be expanded in order to achieve a measurable ecological impact, and can business actors in the solar power market succeed under the conditions of liberalisation of the electricity market?

The analysis is based on two surveys: semi-structured interviews of 10 representatives of power utilities that offer solar power to their customers (INFRAS, 1999) and an extensive literature and information survey. The utilities interviewed are representative for Switzerland regarding the size of the utility, the products supplied, the area to be distributed, the different solar power supply models, experience (in years), and success.

1.2. Overview of solar power supply in Switzerland

1.2.1. Development and status at the end of 1999

Today 61% of the electricity used in Switzerland is produced from hydroelectric power, 35% from nuclear power, and almost 4% from fossil fuels. New renewable or alternative energies, such as photovoltaic[1] or wind power[2] and electricity produced from biomass and fuel cells, account for 0.16% of the total power supply (BFE, 2000). A new peak in total power consumption in Switzerland was faced in 1999, when the overall final consumption amounted to 51.2 billion[3] kWh

[1] also PV; see glossary → Photovoltaic, → Solar power.

[2] In Switzerland, 7 net-coupled wind systems with over 300 kW are in use (Frischknecht et al., 1996).

[3] 1 billion = one thousand million = 10^9.

(or 51.2 TWh), while power production increased by 9.4% to 66.7 billion kWh per year. This resulted in an increasing export surplus of 10.2 billion kWh (BFE, 2000).

The first solar power plant was erected in 1982, when a 15 kW PV power system was put into operation on the roof of Lugano's technical college, and at the time, it was the largest PV power system in Europe. Seven years later, a 100 kW PV system was installed along national highway N 13 close to Chur, and in 1992 the ambitious 500 kW solar power plant "Mont-Soleil" began production (VSE). At that time, solar power supply was not yet a big issue. It gained greater importance in the early 1990s, when national amendments and decisions were passed, such as:

- the federal government's Decree on Energy Use *[Energienutzungsbeschluss]* (1990)[4] and the respective Ordinance on Energy Use *[Energienutzungsverordnung]* (1992),
- a ten-year moratorium on the building of new nuclear power plants, an initiative that was passed by a popular vote in 1990,
- the launching of an extensive energy action programme "Energy 2000" (E2000) in 1991, with a follow-up programme into the new millennium ("Energy Switzerland"), and probably also
- the introduction of a general VAT, which increased the price of electricity for end consumers by about 5-6% (VSE).

The increasing interest in and demand for solar power has resulted in the installation of more and more PV (roof) systems totalling 13.3 MW_P[5] up to 1999, corresponding to an average consumption of about 1.9 W_P per capita. The E2000 programme has set its ambitious target for the end of the year 2000 at 50 MW_P (starting from initial 3 MW_P) installed PV power, corresponding to a consumption of 7.2 W_P/cap. Also, 0.5% of electricity and 3% of heat consumption were to be covered by new renewables in the year 2000. However, by the end of 1999, only one-fourth of the target for the year 2000 was reached.[6]

Of a total of approximately 900 Swiss electricity suppliers, about 100 offer solar power or other alternative electricity (VSE). 1.8 million, or more than half of all households (and other customers), have the opportunity to subscribe to a solar power distributor. Up to the end of 1999, 21,000 customers (0.5% of all households and 1% of the households with potential solar power supply) had made use of this offer or had their own share in alternative electricity production by consuming a total of 8.3 GWh of solar power per year (BFE, 1999). This is 0.014% of electricity consumed (VSE)[7] and corresponds to the average annual electricity consumption of 1,160 persons or the use of a colour television for more than 12 hours a year for all Swiss inhabitants.[8]

---

4 Legal measure propagating financial assistance for solar constructions, as well as for information, consultation, training, research and development, pilot and demonstration units, use of waste heat, and renewable energies.

5 Swissolar; whereof 2.4 MW_P stem from isolated PV systems (BFE, 1999).

6 For an evaluation of a part of this E2000 programme, refer to BÄTTIG AND BALTHASAR.

7 The total final electricity consumption in Switzerland was 51,213 GWh in 1999 (BFE, 2000).

8 The average electricity consumption per capita for 1999 is 7,148 kWh (estimation: BFE, 2000). With 1 kWh, a colour television can be used for about 11 hours (VSE).

One third of the area of Switzerland is potentially provided with solar power by at least one of the 80 solar power suppliers. However, there are large areas that are not at all supplied publicly with solar power, particularly the French- and Italian-speaking parts of the country, some regions in central and eastern Switzerland, and the sparsely populated alpine regions. Some of these areas produce a large part of their electricity from hydropower plants in the Alps. On the other hand, people living in remote alpine regions are not connected to the public power supply network and may have their own isolated PV systems.

1.2.2. Models of alternative and solar power supply

The power packages that suppliers offer to the end consumers vary to a great extent. Summarised, the following packages can be distinguished (the first is the most commonly offered and the last the least frequently offered "standard" package):

Power Package	**Description**
1. Conventional power package	Represents the country's average mix; mainly nuclear and hydropower, with no – or only few – renewables such as solar power/PV.
2. Solar power package	Supply of solar power/PV as a single product (not in a product mix), covering all the electricity demand or only part of it, as a separate addition to the conventional package, with a high surcharge added to the price of the conventional package.
3. Hydro and solar power package	Supply of hydro and solar power in a mixed product, covering all the electricity demand or only part of it, as a separate addition to the conventional package, with a moderate surcharge added to the price of the conventional package.
4. Wind and solar power package	Supply of wind and solar power in a mixed product, covering all the electricity demand or only part of it, as a separate addition to the conventional package, with a moderate surcharge added to the price of the conventional package.
5. Mixed alternative power package, incl. PV	Supply of wind, hydro, solar power/PV, and in some cases also of electricity produced from biomass in a product blend, covering all the electricity demand or only part of it, as a separate addition to the conventional package.
6. Mixed alternative power package	The same as 5., but without solar power (PV) in the mixed package, at a favourable price.

As additions to the conventional power package (1), solar and alternative power packages (2-6) are offered to customers. From the suppliers' perspective, five basic models can be distinguished with regard to various possibilities of solar/alternative power **production and distribution** (Linder Kommunikation, 1998).

Solar power supply models	Description
A) Pooling	Bilateral or multilateral exchange (purchase and sale) of solar power among power utilities.
B) Allocation procedure	Acquisition of solar power by power utilities from a third party and cost transfer to the customers.
C) "Stock exchange"	Purchase of solar power by power utilities from a third party to supply it to customers.
D) Self-building	Purchase of systems by power utilities to generate solar power and supply it to customers.
E) Participation	Customers buy their own shares in the system to help finance solar power systems.

The most common basic models in Switzerland are the "stock exchange" (C) and "self-building" (D) models, which are often combined. In Germany, the participation model (E) also exists. From the consumers' point of view, there are three main models – with various calculation and contract bases – for financing their solar power demand (see Figure 1).

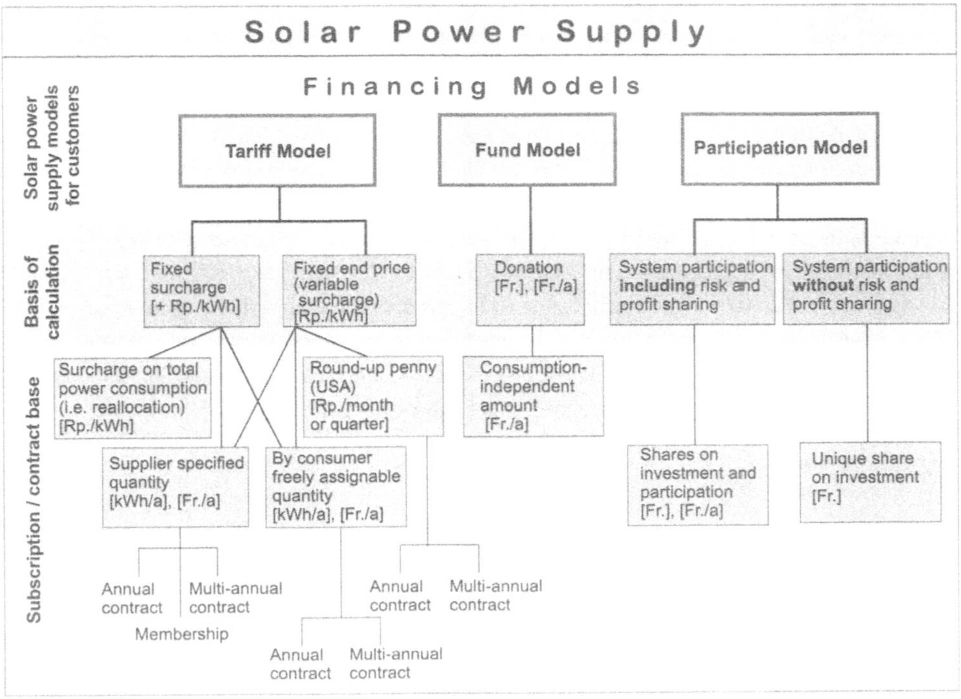

Figure 1: Alternative electricity model types for production and distribution from the consumer's perspective. Source: Öko-Institut (1998) and INFRAS (1999). Graphic by INFRAS. Fr. = Swiss francs, Rp. = Swiss cents.

The **tariff model** is the most widely spread in Switzerland. The surplus price increase for solar power compared to the conventional power package price is charged either as a surcharge or as part of an increased end price. Usually customers have to decide whether solar power should cover all or only part of their overall electricity consumption. The units are either specified by the power distributor or chosen by the customer, per amount in francs or kWh. The **fund model** and the **participation model** are consumption-independent. Donors or shareholders pay a one-time fee or an annual amount to a community system (the participation model may or may not include risk and profit sharing).

All these models give customers the opportunity to consume solar power even if they have no roofs to install their own systems or they are not willing to invest in their own system.

1.3. Ecological impact of solar power and soundness of power producing systems

How "clean" is solar power in comparison with other energy sources? The preference for solar power out of the entire spectrum of energy sources can be explained in energetic and overall ecological terms. Compared to other energy sources, solar power is inexhaustible, flexibly and peripherally useable, modularly applicable, and has very low emissions. The main resources used for and the emissions generated by PV solar power generation result from the electricity used in producing the solar cells and panels (Frischknecht et al., 1996). The energy input for the production of an average Swiss PV system is paid back after 5 to 6 years of system use. Technological progress will significantly lower the pay-back period in coming years. A Japanese study already claims an energetic pay-back period of only 1.6-2.4 years for a small roof-top PV system with a minimum annual production of 10 MW.[9] Since a PV system can easily have a life span of 25 years, and since there are no considerable emissions during use of such a system, solar power – with special focus on PV systems – is regarded as a clean energy source that reduces overall energy consumption in the power production process. A comparison of the ecological impact of different power generating systems is shown in the following figure:

[9] The same holds for the new production site in Gelsenkirchen, Germany, established by Shell by the end of 1999 (Swissolar; NZZ, 2000).

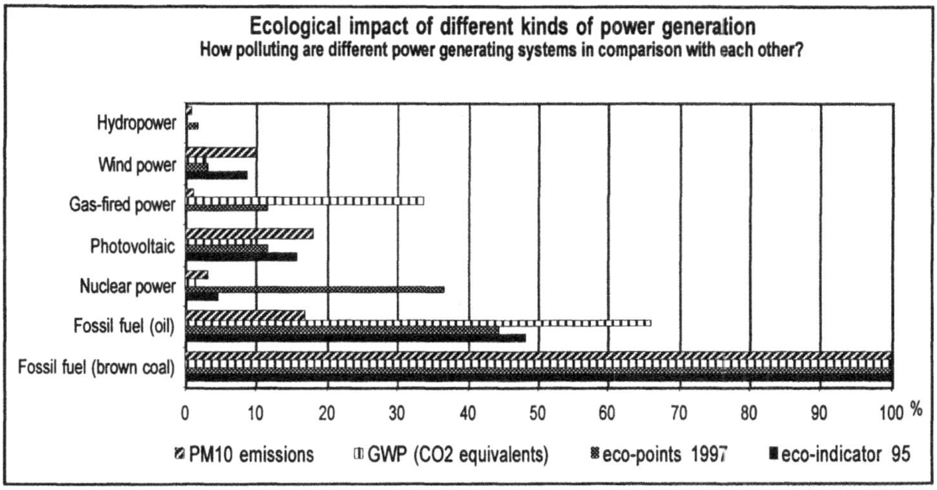

Figure 2: Ecological impact of different kinds of power generation: Emissions of PM10 (particulate matter < 10μm) as an example for air pollution, GWP (global warming potential in CO_2 equivalents), and a comparison of two assessment methods with aggregated ecological impacts, eco-points 1997 and eco-indicator 95, that include pre-combustion and material intensity of the system (according to Econcept, 1999; graph by INFRAS). 100% stands for the most polluting power generation system.

Hydropower – followed by nuclear and wind power – is the least polluting system with regard to **air pollution** criteria (SO_x, NO_x, NMVOC, particles), the **emission of greenhouse gases** (GHG), and **global warming potential** (GWP) per kWh of produced electricity. In terms of **eco-points per kWh,** electricity produced from hydropower – both storage and river-run power plants – is also the best (20 eco-points), followed by wind power, PV, and gas-fired power stations (both 150 points). Therefore, **hydropower** achieves **the best eco-performance** of all the different power generating systems (better than wind power and PV systems). The upcoming and market-gaining **gas-fired power stations** are still worse than PV, wind, and hydropower, especially if external costs (amount of GHG or GWP) are taken into account. Comparison with **nuclear energy** – which produces no direct emissions into the air – mainly depends on calculation of the "risk aversion" and nuclear waste criteria (Econcept, 1999).

Alternative power (water, wind, PV) is ecologically sound as compared to nuclear power and fossil energies. Yet, it must be considered that greater production of solar power does not necessarily have a positive ecological impact if it does not – at the same time – compensate for, and thus save, nuclear and fossil energy. Still rising power consumption in Switzerland (NZZ, 1999a) gives evidence that hardly any fossil fuels are being replaced by solar power[10] and that solar power

[10] On the contrary, power production in conventional fossil plants increased in 1999 by 12% to 2.55 bn kWh (BFE, 2000).

(especially PV) could be regarded as additionally available electricity. Furthermore, if the new systems are built in rural rather than in city areas, negative impacts on nature and landscape are too strong. On the other hand, one may argue that solar power's contribution to ecology consists in preventing, in an expensive manner, the additional import of less ecologically produced electricity from other countries to cover the growing power demand.

2. Overview of key actors in the solar power market

2.1. *Important market players*

Figure 3 gives an overview of the main actors in the solar power market and their interactions. The following actor categories can be distinguished:

① **"Promoters"** (on the legislation and regulation side): In Switzerland, programmes fostering renewable energies have been implemented since the early 1990s under federal, cantonal, and municipal authority. Besides setting the overall legal framework conditions, the federal government pushed the development of renewable energy and rational energy use by launching the Energy 2000 Action Programme (E2000) in 1992 (see section 1.2.1. for programme targets and realisation). Within the federally co-ordinated E2000 programme, an increase in voluntary measures is supported by promotional measures, information and consulting activities, and by the establishment of related actor networks. Another national programme promotes thermal and PV solar energy.[11] The 26 cantons execute the federal legislation and may support the E2000 programme by topping up the E2000 subsidies or creating new incentives in favour of renewable energy sources or solar power.

② **Power producers and suppliers** (supply side): There are a few large-scale power producers that supply electricity to the *power distributors,* which may be small-scale power producers themselves. Section 2.2. gives three examples on different scales (utility co-operation on small, medium and large scales) to illustrate the possible development of network intensification on the production and supply sides.

[11] *Förderprogramm thermische and photovoltaische Sonnenenergie* (Promotion of Thermal and Photovoltaic Solar Energy): In 1997, 4 million CHF had been spent within this programme.

③ **Customers and consumers** (demand side): Today most solar power consumers are households. Solar power distributors supply hardly any institutional or industrial customers. The market potential in this customer sector would therefore be high. Nevertheless, it is difficult to increase this market share due to the non-competitive market prices of solar power in comparison with conventional power (see section 6).

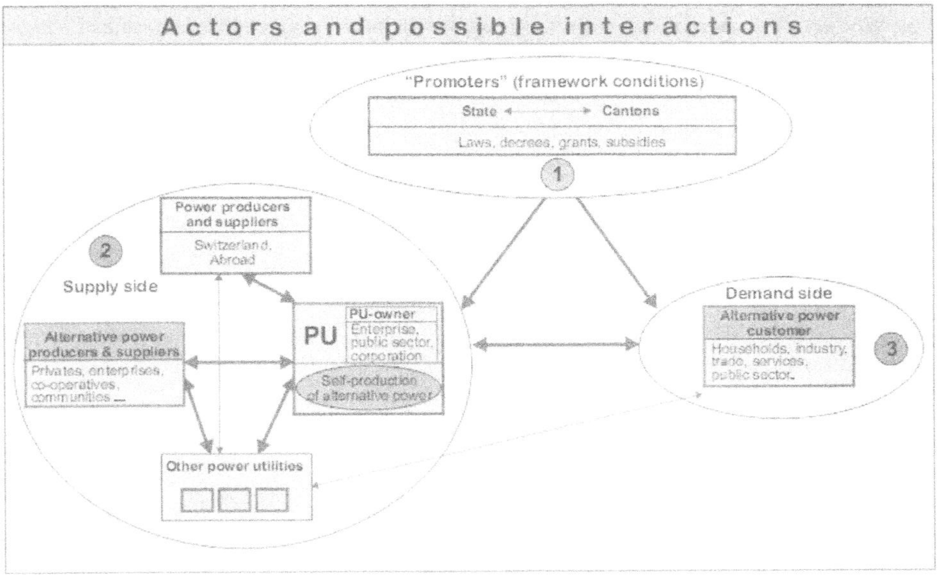

Figure 3: Key actors and their interactions in the promotion, production, supply, and consumption of solar power; PU = power utilities (graphic by INFRAS, 1999).

2.2. *Interactions, networks, and strategies of producers and suppliers*

Interactions between actors in the (solar) power network are manifold. The reactions of power utilities to changing framework conditions due to national policy will sooner or later be transferred to the power demanders.[12] On the production and supply side, three examples of interactions are illustrated below:

[12] The policy instruments on the national level defining the overall framework conditions (see section 5) influence power production and distribution through changing price relations (with changing legislation, energy taxation incl. reimbursement and redistribution strategies, subsidies, national energy programmes, liberalisation policies, etc.).

- **Small-scale distributors:** The network among the small-scale distributors is not yet very dense, but it is growing. Some *solar power pools* have been created recently.[13] They pursue different strategies, but the majority of the power utilities involved in "pools" merely exchange solar power from a surplus to an insufficient supply for reasons of better supply-demand adjustment.
- **Medium-scale distributors:** Seven utilities[14] have been interacting through co-acting. In 1988, the utilities formed the *IGSS network* to strengthen their market position and to gain solar power market shares of large (institutional) consumers (having branches spread all over Switzerland). In view of the step-wise electricity market liberalisation by 2008, the IGSS network has become an "energy corporation" that expanded considerably in December 1999[15] and now holds a 20% share of annual Swiss electricity sales (approximately 9,000 GWh).
- **Large-scale distributors:** The co-operation and unification of power utilities into large-scale companies will likely intensify as they face the market liberalisation. In the spring of 2000, the market activities of the Axpo Corporation began (NZZ, 1999b). Axpo is the result of the union of north-eastern Swiss power producers (NOK) and five (inter)cantonal utilities[16] in eastern Switzerland in a joint commercial and sales company. Axpo's declared target is to become a holding in order to strengthen the market position within Switzerland and to slowly expand to other European countries in an opened market.

2.3. Characterisation of customers

German surveys (ESH, 1998; Öko-Institut, 1998) characterise customers with the highest potential for buying solar power as "responsible and conscious," "actively involved in many causes," and "active value-pluralists." Further, typical solar power consumers are described as environmentally committed, critically minded people in their mid-forties. Market surveys conducted in Switzerland confirm this characterisation (EWZ, 1999; INFRAS, 1999). The 2,000 solar power consumers replying to a questionnaire show an average age of 46 years and 40% have children. The respondents' average income (CHF 6,800.-/month)[17] is an indication of a higher education level. It

---

13 E.g., the *Glarner Solarstrombörse* of the canton of Glarus (incl. 15 out of 18 utilities), the *Solarstrom-Pool Thurgau* (incl. the 4 utilities Weinfelden, Kreuzlingen, Amriswil, Arbon), the bilateral acceptance contracts between EKZ and CKW (two cantons), EWZ and IBC, or EW Heiden and EW Rheineck (two city administrations).

14 The largest municipal utilities: Zürich, Basel, Bern, Luzern, Winterthur, Schaffhausen/Neuhausen, and St. Gallen.

15 Municipal utilities in Aarau, Biel, Interlaken, Chur, Davos, EWZ Mittelbünden, Frauenfeld, Kreuzlingen, Weinfelden, and Zug.

16 These five cantonal utilities comprise the cantons of Aargau, Zürich, Schaffhausen, Thurgau, and St.Gallen together with Appenzell (AI+AR).

17 6,800 CHF correspond to 4,420 Euro. See glossary for approximate exchange rates from CHF to Euro and USD.

is interesting that income is quite inelastic in relation to the amount paid for solar power. An increase in income of CHF 1,000.- results on average in an increase in solar power demand equivalent to only CHF 1.- (≈ 0.65 Euro). Age, gender, and occupation, however, are not significant indicators for characterising the "typical" solar power consumer (EWZ, 1999). But the findings show that the willingness to pay and the share of customers participating in solar power programmes is likely to be higher in urban than in rural areas. This is because rural areas are often supplied by larger utilities, and for many customers the distance to the supplier is greater than is the case in city areas. Also, in rural areas with large suppliers, the PV systems are larger and tend to be centralised at few locations. This again does not help to bring solar power closer to customers' minds, since the installed systems are hardly in plain sight. Even if customers do take note, they are sometimes not willing to support large-scale power systems, because of the fact that this may not lead to the installation of new, decentralised solar panel systems.

Typical solar power consumers in Switzerland are private households, independent of age, gender, occupation, or political inclination. Some power utilities estimate, however, that families in particular (mostly on the initiative of women) show higher participation in solar power programmes than single-person households.

3. Success factors for increasing the market potential for solar power

3.1. Suppliers' views

Suppliers in Switzerland assess the following factors as important measures to increase the market potential of solar power successfully:[18]

- Reliability of the supplier
- Price decrease for renewable energies
- Integration of hydropower into the solar power package
- Regionalisation of the supply, decentralisation
- Development of ecological criteria (labelling) for various energy sources (see section 6.5.).

German studies (ESH, 1996) and experts in Switzerland (INFRAS, 1999) acknowledge that one of the most important incentives for people to participate in a "green pricing" programme is the **reliability of the supplier** (Figure 4). Reliability can be achieved through providing adequate information to customers, through communication and PR, and by guaranteeing cost and price transparency in the products. Voluntary/additional suppliers' contributions or third party financial administration are not (yet) regarded as important contributions to improving the utility's reliability (INFRAS, 1999).

[18] Survey carried out by R. Wüstenhagen (IWÖ-HSG, University St. Gallen) in 1998.

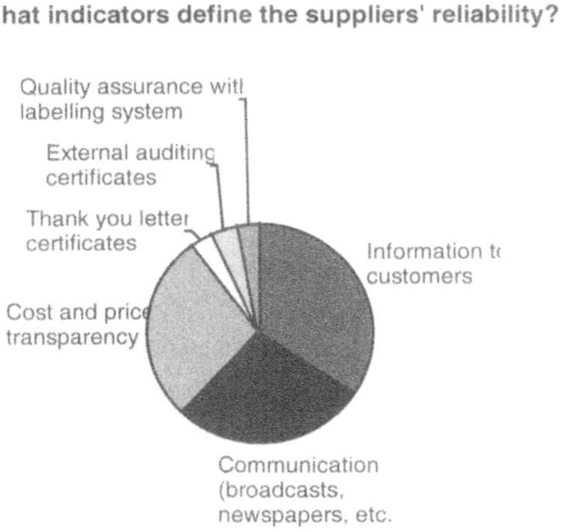

Figure 4: The experts' opinions on the indicators that count as "reliability factors" of a solar power utility. Expressed as percentage of all denominations (=100%).

In addition to the success factors above, effective **marketing** seems crucially important for an increase in solar power demand. The most efficient marketing measures are direct mailings or the mailing of (personally addressed) flyers to customers with their electricity bills. Public relations efforts – conducted most effectively through local/regional broadcasts and newspaper articles – should accompany the personal mailings. Event marketing, such as "open doors", or target group oriented marketing (e.g., promotion through a quiz) at special events and a centrally located customers' centre are also regarded as success factors. Non-addressed flyers (e.g. solar mail), information posters at exhibits, and non-targeted marketing or decentralised customers' centres are found to be rather ineffective and therefore inefficient. Internet web sites may sustain marketing, but they are not regarded as effective in gaining new solar power customers (INFRAS, 1999; Linder Kommunikation, 1999).

3.2. Consumers' views

On the side of the consumer, in addition to the above-mentioned "transparency about the means used" or "regular information", the following factors (ESH, 1996; INFRAS, 1999) are important: Customers would like to know that the surcharge they pay has a **positive impact** on the environment and is helping to construct **new systems** producing alternative/solar power. People are less willing to subscribe to solar power if their money is used to amortise already existing old PV systems instead of fostering the building of new systems. The "stock exchange" model (section 1.2.2.), where the solar power distributor sells electricity from a third-party system provider (the utility does not own the system itself but compensates third parties at a certain price/kWh for the PV power generation), is also estimated to be slightly more accepted among customers than the self-production of solar power. **Convincing solar energy projects** and additional offers of special energy-saving measures further improve customers' confidence in the supplier.

Also essential from the customers' point of view is **flexibility** in choice of the surcharge to pay. It is not very important whether the surcharge is variable (additional amount per kWh) or a fixed part of a higher end price. The following is recommended to the suppliers (INFRAS, 1999):

- **Fixed vs. variable end price**: The integration of the surcharge into a fixed end price is to be preferred by single utilities (a stable end price is easier to communicate). In the face of the opening market and in the case where some utilities are co-operating or have/are retailers, a fixed surcharge resulting in a variable end price (varying according to the price fluctuations of the basic power package) is to be preferred.
- **Basis of calculation**: It does not matter very much whether the solar power surcharge is calculated on the basis of kWh used or as a fixed annual amount. The first gives a stronger relation to the product; the latter is easier to communicate.
- **One single bill**: It is more transparent and more customer-friendly if customers receive only one electricity bill that includes all subscribed power packages (no separate bill for solar power).
- **Integration of municipal administration**: It is helpful to include municipal authorities in the solar power programme or to at least keep them informed, not least because they themselves constitute a possible target group.
- **Institutions and companies** have to be addressed differently than households. Face-to-face consulting appears to be the most effective, even more so if a package including consulting in energy-saving measures is offered.
- **Customer proximity**: PV power is more likely to be demanded if power generation is spread out over inhabited areas and is highly visible to customers (successful are roofs of school houses and other public buildings).
- **System size**: Several small or medium sized PV systems increase the demand for solar power as opposed to 1 to 2 large systems (in remote areas).
- **Self-participation**: Power utilities should be ready to guarantee an investment advance to third parties for the erection of solar power systems.

4. Barriers to the increased implementation of solar power programmes

4.1. Barriers on the supply side

Sometimes arguments for protection of the landscape are brought against the installation of solar panels on the rooftops of buildings. Also, the process for receiving a permit for the installation of solar panels, and the restrictions involved – depending on cantonal legislation and enforcement – is an important factor that has so far prevented an increased spreading of PV systems. In future these restrictions are likely to be loosened.

Further difficulties for increased implementation of solar power supply result – mostly at start-up and during the consolidation phase – from the power utilities' incapability to provide the requested kWh of solar power to their customers. Demand often exceeds the possible supply, above all in urban areas. In contrast, for smaller and rural-based power utilities it is often difficult to find enough customers willing to pay for solar power. As a consequence of this supply-demand imbalance – and also for reasons of institutional strengthening – several power utilities have joined together and created solar power pools in the last few years (refer to section 2.2.).

For some power utilities with solar power supply, an internal barrier is the lack of acceptance of "green" products by their own staff. Further, some rather small power utilities argue that a major barrier is the absence of a clear solar power marketing strategy due to budget restrictions. Finally, to launch solar power products successfully, it is crucial that management is highly engaged and convinced (INFRAS, 1999).

4.2. Barriers on the consumer's side

The most important barrier preventing people from demanding solar power is the continuing large difference in price between solar power and the "conventional" power mix. For electricity produced by PV systems, there is still an average surcharge of between CHF 1.- and 1.20 per kWh to be considered.[19] Even though newly erected PV systems could generate electricity at a slightly lower price, it would still be considerably higher than the price of other power products. High minimum amounts for annual subscriptions (tariff model, see Figure 1) or system participation at an unusually high one-time fee (participation model) are clear barriers to customer participation in solar power programmes. The customer should also not be obligated to subscribe for a period longer than a year.

In one study, 70% of the customers surveyed within a solar power supply area did not know that it was possible to subscribe to solar power in their community. Above all, a large part of

---

[19] Approx. 0.65-0.8 Euro, corresponding to a price 6 to 8 times higher than for conventional power provision.

customers stated that they had never been personally contacted about subscribing to solar power (Linder Kommunikation, 1999). Furthermore, there are lots of psychological barriers on the demand side: It seems hard for people to accept that persons already high in ecological awareness should pay more for using clean energies only to see the high polluters in society getting away with cheaper electricity (INFRAS, 1999).

Active and reliable marketing could weaken or break down some of these barriers. Examples in Switzerland show that the most successful solar power suppliers have invested quite heavily in marketing (section 3). On the other hand, for some people it is very important that the higher price of solar power is justified due to market conditions and that surplus money is not used for marketing measures. They would prefer to see the full revenue reinvested in new solar power facilities.

5. Interlinkages between the solar power market and energy policy instruments

Government's energy policy instruments and measures at different levels influence the promotion and dissemination of solar and other renewable energy production.[20] What is the contribution of these **top-down policy instruments** to the spreading of solar power and – over all – towards more ecological power production? And what is, on the other hand, the ecological contribution of political or individual **bottom-up initiatives** of "demanders" (consumers)?

5.1. Top-down policy instruments

The top-down policy instruments at the national level that affect solar power supply are mainly economic and command and control instruments according to the typology. Such instruments include (federal or cantonal) subsidies for installing solar power systems, energy taxation programmes (such as the introduction of an ecological tax reform or any kind of energy tax or levy), the CO_2 reduction law[21], or newly introduced or changed laws and regulations[22] (see also section 2.1.). The government-run E2000 Action Programme is based on voluntary participation and

---

20 For a discussion on the acceptance of energy policy instruments, see the chapter by JEGEN.

21 The CO_2 reduction law introduces "voluntary" agreements for industries to reduce CO_2 emissions. If the target of a 10% CO_2 reduction by 2010 (compared to 1990 levels) does not seem to be achievable by 2004, the law foresees the introduction of a carbon tax.

22 The electricity market legislation will be a progressive energy bill. It aims at lowering energy prices for industries, which will improve their international competitiveness. There is still some disagreement (e.g., about hydropower discharge) focussing either on Swiss competitiveness or ecology.

implements economic as well as communication and diffusion instruments (see chapter by BÄTTIG AND BALTHASAR).

- E2000 promotes the (internationally early) development of "green pricing programmes" in Switzerland. As early as in 1991, E2000 supported some pioneer solar power suppliers. But despite this fostering of renewable energy sources and the providing of wide support in PV capacity building – which could increase people's acceptance of solar power programmes launched by the utilities – large actor groups are not reached, especially with regard to the training programmes. Additional flanking measures and communication programmes are regarded as important (INFRAS, 1999).
- To increase ecological power production, regulative federal instruments such as taxes, levies, laws, and regulations are important framework conditions that vary in their effectiveness. People respond better to price incentives than to restrictions by legal national framework conditions, e.g., when an excise tax is levied on non-renewable energy sources. But the price differences between conventional power and renewables, such as PV, are still too large. An incentive tax placed on conventional energies would therefore have a remarkable positive impact. Energy is still too cheap to have any real impact on changing lifestyles, and this prevents the successful application of energy-saving measures. Negative impacts are sometimes linked to measures that are dictated by the government and are thought to hinder voluntary bottom-up measures.

5.2. *Political bottom-up initiatives*

Political initiatives emerging from activities on the demand (and supply) side are provoking changes in policy strategies or political framework conditions through polls, mainly at the national level. As mentioned in section 1.2.1., the nuclear power moratorium initiative in 1990 has had an impact on the Swiss electricity market policy. Some recent Swiss initiatives influenced energy taxation activities at the federal level, resulting in federal counterproposals to the initiatives.[23] With the defeat of the initiatives and federal counterproposals in September 2000, these environmental policy instruments with potential ecological steering effects – prior to the votes being regarded as a compatible free-market option – have lost political priority. The power utilities, however, assess as minimal the overall influence of such measures as ecological tax reform, energy levy, or bottom-up initiatives with national implications for the expansion of the solar power supply (INFRAS, 1999).

---

[23] The "Energy Environment Initiative" provoked a federal counterproposal that aimed at setting a milestone for an ecologically oriented tax reform. This led the initiators of the Energy Environment Initiative to withdraw the same. The "Solar Initiative" also gave rise to a federal counterproposal (*Förderabgabe-Gesetz*). All three energy proposals were voted on in September 2000, but all three of them were put down by the democratic majority. This means that the reduction of CO_2 emissions according to the Kyoto agreement has to be realised with the CO_2 reduction law through voluntary agreements or through a carbon tax starting from the year 2004.

5.3. *Individual bottom-up initiatives*

Individual bottom-up initiatives, i.e., instruments initiated by power consumers, consist in voluntary energy-saving measures, such as insulation of buildings, replacement of old machines, switching off stand-by modes, and – above all – modification of energy-excessive lifestyles. These measures and lifestyle changes, mostly linked to communication and diffusion instruments according to the typology of instruments, clearly have a higher potential to have a positive ecological impact than the generation of solar power alone, but they are not directly linked to the latter. Energy-saving measures are also triggered through intensive energy saving campaigns by power utilities. Utilities nowadays promote quite credible energy-saving measures such as free consulting services, customer centres, and so on. All power utilities acknowledge that there is a high ecological potential, not yet exploited, in rational energy use. At the same time, the utilities will probably phase out these consulting services, particularly in view of market liberalisation, or restrict such services to large-scale consumers only (INFRAS, 1999).

6. Outlook in view of the liberalisation of the electricity market

6.1. *Market conditions in the liberalised electricity market*

Market liberalisation in Switzerland will presumably begin in the year 2001 – according to the Law on the Liberalisation of the Electricity Market *[Elektrizitätsmarktgesetz]*. The stepwise implementation will at first allow about 100 large consumers free access to the market. In a second step – presumably in 2004 – medium-sized power consumers will follow and finally – in 2007 at the earliest – all companies and households will have free access to all power suppliers. Parallel to the gradual liberalisation, power-supplying utilities can freely choose the origin of a specified amount of the electricity they supply (according to the power demand of large consumers, etc.). The details of the framework for the stepwise market liberalisation have still to be discussed in Parliament and are, in part, closely related to the energy initiatives and the federal counterproposals (see section 5.2.).

Within the changing legal framework conditions, the traditional value chains – production, transport/transfer, distribution – will be reorganised, and new or reorganised market players will enter into the competitive market. This process has already begun, as evidenced by the formation of a big corporation (Axpo), the power companies merging and co-operating in the different cantons, the enlargement of the IGSS network of medium-scale power distributors, and slow privatisation.

The competitive market conditions will transform the energy market from a supply to a demand focus. Pessimists estimate that two thirds of today's power suppliers will not survive under liberalised market conditions. Experiences in the British energy market show that with customers a kind of "passive satisfaction" with the previous power supplier prevails.[24] The urge to make changes on the demand side is therefore hardly evident in the first phase of liberalisation, as long as no shortages in supply occur (NZZ, 1999c). On the other hand, current market observations indicate that in the long run, there are no reasons for customers to favour loyal behaviour towards their suppliers – as can be observed in today's telecommunications market. Also, the majority of the 100 large consumers who can profit first from the electricity market liberalisation has already signed a medium-term contract (of about five years) with a supplier at favourable price conditions.

6.2. Development of conventional power prices

A liberalised market will – at least temporarily – lower the price of conventional electricity. The price development in Europe will be of increasing importance for the electricity sector in Switzerland. A price comparison among European countries reveals rather high prices for household electricity in Switzerland and even top end pre-tax prices for electricity for Swiss industries (INFRAS, 1998; 2000). This is the case even without the energy, carbon, sulphur, or other taxes up to now – besides the still low VAT – that other countries already have. As a result of the rather high electricity prices, Swiss power suppliers have already started to respond, for instance by building up new alliances (NZZ, 1999c).

A survey of energy customers in the British liberalised market show that 18% look to the price, 25% to the supplier's image, 18% to the supplied products, and less than 11% to services, loyalty rebates, and support on the phone.[25] Attractive prices have thus only a small influence on the customer's readiness to change suppliers. While the liberalised market is undoubtedly influenced by competitive prices, there is at this time no evidence that price alone will dominate the decisions made (NZZ, 1999c).

6.3. Development of solar power prices

The existing price differences between solar power and other energy sources will gradually diminish in the long run and within an overall system analysis considering external costs. The price for PV power generation decreased annually by 22-27% during recent years, while at the same time, the production of solar cells increased by 25-30% per annum (NZZ, 2000).

Figure 5 shows experts' opinions on how lower prices of renewable energy – with a focus on solar power – could be reached (INFRAS, 1999). A decrease in solar power prices can best be achieved through the supply of an electricity package including, but not consisting only of, solar

24 According to a study carried out in 1998 by Pricewaterhouse-Coopers (NZZ, 1999c).
25 According to a study carried out by Pricewaterhouse-Coopers (NZZ, 1999c).

power/PV and through an increase in technological system efficiency (panel efficiency, technology leaps, R&D). The product cost can be further gradually lowered through standardised and expanded processing.

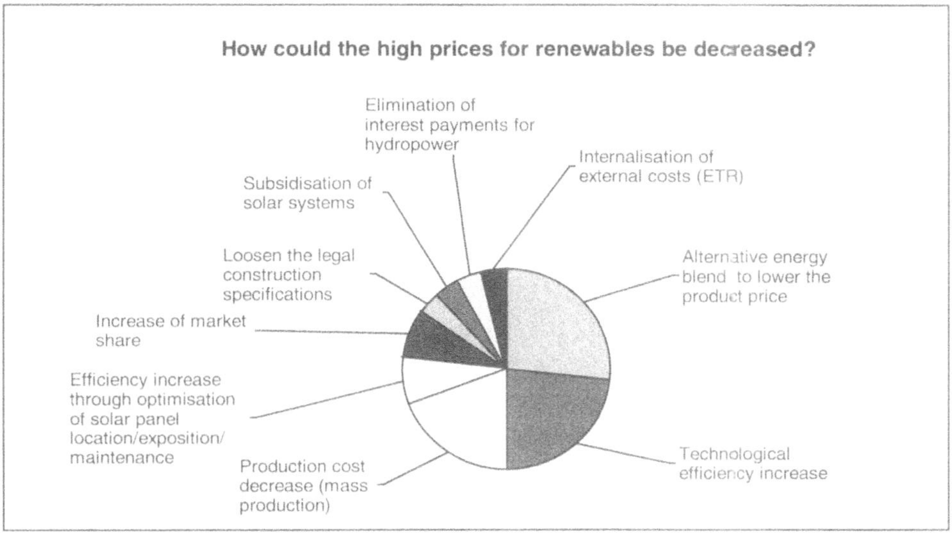

Figure 5: The experts' opinions on how today's high prices in Switzerland for renewables, especially for solar power, could be lowered. Expressed as percentage of all denominations (=100%).

6.4. *Market potential for solar power*

The PV power potential for Switzerland was estimated in 1991 (Frischknecht et al., 1996). The technical potential of 168 million m^2 in built-up areas and 186 million m^2 in alpine regions is reduced by legal, economic, and social restrictions to 18 million m^2 and 4 million m^2, respectively. This area could be used for a solar power production of 300 GWh per year, equivalent to about 4.5-5.5% of Switzerland's current total electricity generation (66.7 billion kWh/yr). Whether or not a real break-through and diffusion of PV systems will take place in the near future depends largely on the price differential to other power supplies – and thus on the prices of silicon extraction and panel production. On average, Swiss PV power is now produced at prices 6 to 8 times higher than conventional electricity. Looking at other new renewable energy sources, fuel cells, although regarded as having a great potential, will probably reach a breakthrough only in about a decade's time or even later.

Market liberalisation will probably improve the chances for solar power marketing. Experiences in the United States indicate that the green power's market potential is increasing with the decreasing market regulation under more competitive framework conditions. This is due to increased product-specific marketing and the sharing of responsibilities. In a fully liberalised electricity market, the share of customers participating in green power programmes could rise to 10-30% of all customers, according to experts' estimations. Today, the suppliers EWZ (Zürich), EWB (Bern), and AEW (Aargau) have the largest share of subscribers to solar power in Switzerland, with 3-4% of all customers. Based on the market liberalisation experience in the United States, it is likely that the market potential in Switzerland will be much higher in future. To fully realise this potential, lower prices for at least some alternative power packages, comparable to conventional packages, would be important (see section 1.2.2.). This could result from more favourable framework conditions, a higher diversification of alternative power products (mixed packages, not only PV), communication campaigns that are informative, progressive, and more appeal-oriented, and intensive marketing (Wüstenhagen, 1999).

6.5. *Solar power labelling*

The introduction of a label for solar power or for an alternative power mix was mentioned above as a success factor (section 3). As an instrument, product information labels come under the category of collaborative agreements in the typology of instruments. They aim at an increase in the acceptance, reliability, market share, and quality of solar power. Quality standards for ecologically sound power (production) exist so far only abroad (Niederberger, 1999). In **Germany**, five organisations (e.g., the national environment protection office) offer labels. In the already deregulated power market of the **United States**, private companies in various states (such as California and Pennsylvania) are promoting the supply of "green power" (Wüstenhagen, 1998; 1999). Labels also exist in **the Netherlands, Denmark, Sweden, Canada,** and **Australia**.

For some years the introduction of a national "ecological power" label has been discussed in Switzerland (Kiefer and Partners, 1999). A recently founded business association consisting of the major market players (including non-governmental organisations, nature protection organisations, power utilities, research institutions, and others) proposes two labels. All electricity produced from renewables would be labelled "nature-made basic", whereas power produced from PV, wind, and biomass would be labelled "nature-made star". The latter label for ecological leaders could also be applied to small and particularly sound hydropower utilities, if they complied with stringent criteria for residual water masses; re-naturation of creeks, and the like according to a reorganisation catalogue.[26]

[26] Criteria for hydropower qualification should be available by the end of 2000. As an example, a minimum of 0.5% of the total power sales have to come from renewable energy sources.

Whether or not hydropower should be regarded as just as ecological as other alternative power systems is important with regard to the promotion of alternative energy sources because of existing price differences. The conditions under which the integration of hydropower in an "alternative power package" should be considered are shown in Figure 6 (INFRAS, 1999): The readiness to consider all hydropower as ecologically sound power comparable to other alternative energies is high. On the other hand, there is wide acceptance of a differentiated labelling system integrating a certification of hydropower plants according to prior defined criteria.

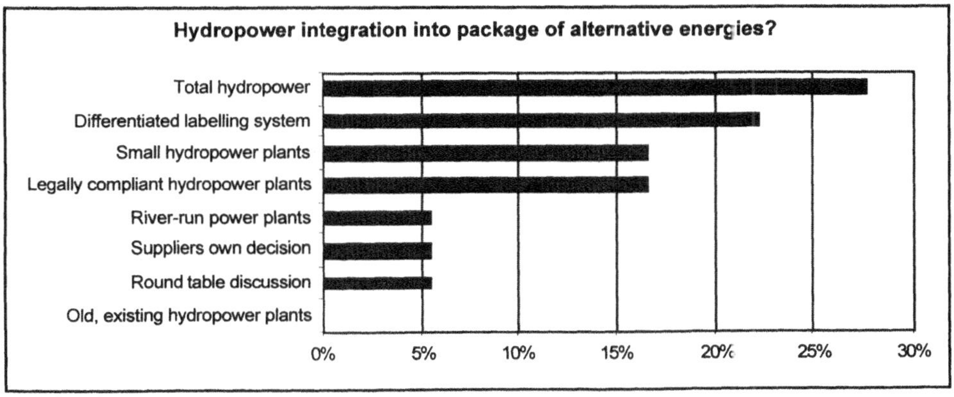

Figure 6: Experts' opinions of criteria for integrating hydropower into a "green power" package. Expressed as percentage of all denominations (=100%).

Actually, there are no reasons why hydropower should be declared less "green" than PV generated electricity (see section 1.3.). Labels should therefore integrate hydropower unless the power systems have or have had strong negative impacts on landscape and biodiversity, and presupposing that the system is legally compliant (residual water mass, etc.). On the other hand, neither PV nor wind systems built in open landscapes should be given labels since such systems have a strong negative impact on the landscape. With the benefit that there is a possibility to steer the quality of the selectively labelled products, labels could play an important role in a shift towards more environmentally friendly power production. After the first phase of liberalisation, when the fight for market shares on the supply side will to a large extent be price-guided, the product name/label will play an even more important role in a second phase in a consolidated actor network (NZZ, 1999c). Strategies to create or strengthen a supplier's own national product name/label (brand) will thus be decisive for long-term survival in an opened market.

7. Summary and conclusions

By analysing the solar power market in Switzerland we have sought answers to the following main questions:

- Is solar power supply a means towards more ecologically sound power consumption, and has it the potential to replace power generated by fossil fuels and nuclear plants?
- What factors determine the success of solar power supply in increasing the market potential for its demand?
- How are the business strategies of solar power suppliers evolving in view of the liberalisation of the electricity market in Switzerland?
- By what governmental policies and bottom-up initiatives can solar and other alternative power be promoted and supported in order to increase more environmentally sound power consumption?

The ecological impact of solar power is still minimal: Solar power is ecologically sound compared to nuclear power and fossil energies, but as long as total power consumption continues to increase each year, greater production of solar power does not necessarily mean that any fossil energies or nuclear power will be replaced. Energy-saving measures (such as insulation of buildings, replacement of old machines, switching off stand-by modes) and above all, modification of energy-excessive lifestyles still have a higher potential to reduce overall power consumption than the generation of solar power. The ecological contribution of solar power goes far beyond the replacing of additional imports of (less sound) electricity from other countries to cover the growing power demand. The ecological impact of solar and other alternative power can be increased only if the market potential can be further realised, and if – at the same time – consumers are prepared to pay a higher price for ecologically sound products.

Success factors to increase the market potential for solar power: The technical potential for solar power production in Switzerland is estimated at 5.5% of the total power consumption, whereas the theoretical customer potential is estimated at 10-30% of all power consumers. The activities of solar power suppliers today relate to this potential increase in the demand for solar power. Besides reliability of the supplier and decentralisation of the supply, a lowering of the prices for solar and alternative power products is an essential success factor for increasing the market potential in the short term. Effective marketing will be of growing relevance for increasing the solar power market share in future. In the longer term, additional success factors could include the introduction of an "alternative power" mix product (including cheaper renewables such as wind and hydropower) that could lower the prices and the introduction of labels, i.e., defining the quality standard, for ecologically sound electricity. This could increase the demand especially among environmentally conscious customers, as the product name will play an important role on the liberalised electricity market.

Business strategies of power utilities in view of market liberalisation: The power market in Switzerland has long been dull, monopolistic, and supply-driven rather than demand-oriented. As a consequence, there were no real incentives for power utilities to introduce new, alternative power products. With the announcement of the electricity market liberalisation, a diversification of the power utilities' portfolio becomes interesting even economically. Power utilities may either begin to offer solar power supply as a proper or a mix product, or some utilities may diversify. Monopolistic suppliers may become unified multiple products and services enterprises that also provide natural gas, district heating, fresh water, telecommunications, and financial services. Such synergies may lower administration costs and may help to lower the price of electricity. A diversification of the product portfolio reaching far beyond power supply may also help to bind customers (households, and especially the industry sector) in a liberalised market. Co-operation and union among power utilities will intensify. As a result, there is already now a gradual reduction of structures that had been settled for decades in a typical Swiss federalist coexistence. Some important market players, largely dependent on fossil fuel production and trade today, are taking the opportunity to become forerunners in research and development on renewables (such as the large petroleum suppliers BP and Shell). A reorientation in the energy sector towards less polluting and more ecological energy sources has begun. It only needs to increase the pace in future.

The optimal mix of instruments towards more ecologically sound power production and consumption: National framework conditions, such as the Decree and Ordinance on Energy Use, the launching of an extensive energy programme, and the introduction of a general VAT, have been decisive in the implementation of solar power products on the market. Both the forthcoming liberalisation of the electricity market and the possibility of the introduction of economic instruments (launched by popular initiatives and federal counterproposals) that aim to decrease the price differences between conventional fossil fuels and alternative energies can contribute to a further increase in solar power production/demand. Even though the impact has hardly been big enough to create a radical, measurable change in suppliers' and consumers' actions towards a more ecological power generation/consumption, there has been a moderate start in the right direction. What is needed in addition are further bottom-up efforts and technical innovations. These could grow and develop on the supplier's side by further developing and promoting alternative energies. On the consumer's side, more attention could be paid to ecological criteria in power consumption, that is, consumers could reduce personal power use, consume larger amounts of renewable energies, or launch ideas to be publicly discussed in a political process.

Glossary

AEW	Aargauisches Elektrizitätswerk (power utility of the canton Aargau).
Alternative electricity:	Non-fossil and non-nuclear electricity production. For Switzerland that means photovoltaic, wind, biomass, fuel cell, and – depending on the plant and/or definition – hydropower.
Axpo	Axpo Corporation is the result of the unification of the north-eastern Swiss power producers (NOK) and five (inter)cantonal utilities (AG, ZH, SH, TG, SG/AI/AR) from eastern Switzerland to a joint commercial and sales company. Axpo Corp. began operations in the spring of 2000. Later it will merge to a holding.
CHF	Swiss Franc; 1 CHF = 0.65 Euro = 0.6 USD (currency exchange rate of 31 July 00).
EAWAG	Eidgenössische Anstalt für Wasserversorgung, Abwasserreinigung und Gewässerschutz (Swiss Federal Institute for Environmental Science and Technology).
Eco-points	Negative environmental impact of a process or system, calculated within a life cycle assessment (including pre-combustion, use, and disposal) using the method of ecological scarceness (ecologically sound systems have fewer eco-points).
EWB	Elektrizitätswerk der Stadt Bern (power utility of the city of Bern).
EWZ	Elektrizitätswerk der Stadt Zürich (power utility of the city of Zürich).
GHG	Greenhouse gases (e.g. CO_2, CH_4, N_2O, HFCs, PFCs, SF_6) causing global warming.
GWP	Global warming potential, calculated in CO_2 equivalents / kWh.
IGSS	"Interessengemeinschaft Schweizer Stadtwerke", an association originally consisting of the seven largest communal utilities Zürich, Basel, Bern, Luzern, Winterthur, Schaffhausen/Neuhausen, and St. Gallen. In December 1999 the network expanded to an "energy corporation" including the utilities Aarau, Biel, Interlaken, Chur, Davos, EWZ Mittelbünden, Frauenfeld, Kreuzlingen, Weinfelden, and Zug.
MW_p / W_p	(Mega)Watt peak (1 MW=1 million Watt): Unit for the performance of a photovoltaic system during maximum sun exposure.
New renewable energy sources (renewables):	Solar, hydro and wind power, energy produced from biomass, wave, tides systems, running water energy, terrestrial heat, fuel cells.
Photovoltaic	A photovoltaic cell (solar cell) converts sunlight directly into electricity (direct current). The solar cell can convert only one part of the received radiation into electricity; the remainder is converted into heat. Most photovoltaic (PV) systems are connected to the electricity supply grid. In remote areas, isolated photovoltaic systems (not connected to the grid) are also used.
PV	Photovoltaic.
Solar energy	The term "solar energy" includes solar power and other systems using solar energy, e.g., to produce heat.
Solar power	**Direct** solar power, active systems: PV, fuel cells (passive and hybrid systems are not listed). **Indirect** solar power: water, wind, biomass, tides, wave energy, terrestrial heat, etc. (solar energy for heat generation is not listed). Electricity from PV systems is often referred to as "solar power." In a broader sense, the term solar power also covers electricity generated from hydropower, wind, biomass, fuel cells, tides, wave power, etc. (= alternative power). In this paper, solar power refers to direct solar power, i.e., photovoltaic, and excludes indirect solar power, i.e., wind and hydropower, unless indicated otherwise.
VAT	Value added tax. The general VAT level in Switzerland is presently 7.5%.

References

BFE/Bundesamt für Energie (Swiss Federal Office of Energy) (1999) Schweizerische Gesamtenergiestatistik 1998. Bern.

BFE/Bundesamt für Energie (Swiss Federal Office of Energy) (2000) Elektrizität 1999: Rekorde in Produktion und Verbrauch. Pressemitteilung und Statistiken. Bern.

Econcept (1999) Ökologische Qualiäten der Wasserkraft und anderer Stromproduktionsarten – eine Übersicht über den Stand des Wissens. SWV/Schweizerischer Wasserwirtschaftsverband (Swiss Association of Water Management), Zürich.

ESH/Energiestiftung Schleswig-Holstein (1996) Grüner Tarif. Klimaschutz durch freiwillige Kundenbeiträge zum Stromtarif. Akzeptanz, Umsetzung, Verwendung. Energiestiftung Schleswig-Holstein, Kiel.

EWZ/Elektrizitätswerk der Stadt Zürich (electric power utility of the city of Zürich) (1999) Analysis of solar power consumers' choice and optional behaviour: Result of the questionnaire. Zürich.

Frischknecht et al. (1996) Ökoinventare von Energiesystemen – Grundlagen für den ökologischen Vergleich von Energiesystemen und den Einbezug von Energiesystemen in Ökobilanzen für die Schweiz. 3. edition. Bundesamt für Energiewirtschaft (ed.), ETH Zürich, PSI Villigen, ENET Bern.

INFRAS (1998) Aussenhandelsverträgliche Ausgestaltung der Energiebesteuerung: Übersicht über die Regelungen in ausgewählten Ländern Europas. Commissioned by the Federal Office for Foreign Economic Affairs (BAWI), Zürich.

INFRAS (1999) The solar power market in Switzerland. A survey among 10 power utilities. Internal working paper. Zürich; *unpublished.*

INFRAS (2000) Übersicht über die Besteuerung von Energieträgern in Europa. Commissioned by the Federal Office for Energy (BFE), Zürich.

Kiefer and Partners (1999) Businessplan: Qualitätszeichen für Elektrizität. Incl. additional information of 2 June 1999: Produktezeichen für Ökostrom und erneuerbare Elektrizität. Zürich.

Linder Kommunikation (1998) Steckbriefe der Grundmodelle. Overview; *unpublished.*

Linder Kommunikation (1999) Erfolgsrezepte für das Solarstrom-Marketing. Linder Kommunikation AG and Ipso Sozial, Marketing und Personalforschung, BFE (Swiss Federal Office of Energy), (ed.), Zürich; *Vorabdruck.*

Niederberger, W. (1999) Ökolabel für Schweizer Strom. *Tages Anzeiger, 11. 11. 1999.*

NZZ/Neue Zürcher Zeitung (1999a) Schweizer Stromverbrauch auf Rekordstand: Hoher Exportüberschuss. 19.11.1999: 23.

NZZ/Neue Zürcher Zeitung (1999b) Grundstein zu einem Schweizer Stromkonzern, NOK und Kantonswerke auf dem Weg zum Zusammenschluss. 25.11.1999: 21.

NZZ/Neue Zürcher Zeitung (1999c) Und plötzlich gibt es Kunden am Strommarkt: Notwendiger Kulturwandel in der Energiebranche. 10.12.1999: 23.

NZZ/Neue Zürcher Zeitung (2000) Abschiednehmen von den fossilen Brennstoffen: Gute Aussichten für den Durchbruch erneuerbarer Energieträger. 16.02.2000: 69.

Öko-Institut (1998) Green Pricing: Welchen Beitrag können freiwillige Zahlungen von Stromkunden zur Förderung regenerativer Energien leisten? Freiburg (D).

Swissolar: Online in the WWW: <http://www.swissolar.ch>

VSE/Verband Schweizerischer Elektrizitätsunternehmen: Online in the WWW: <http://www.strom.ch>

Wüstenhagen (1998) Green electricity in Switzerland: A mere eco-niche or the first step towards a sustainable energy market at large? *In:* International Association for Energy Economics/Gesellschaft für Energiewirtschaft und Energiepolitik: Energy markets: What's New? *Proceedings of the 4th European IAEE/GEE Conference, 9./10. 09. 1998*, Berlin.

Wüstenhagen (1999) Ökostrom-Marketing zwischen Nische und Massenmarkt. Vergleichende Analyse von Marketingstrategien und Marktentwickung in den USA und in der Schweiz. *Bulletin SEV/VSE, 22/99*: 27-32.

Changing Things – Moving People.
Strategies for Promoting Sustainable
Development at the Local Level
R. Kaufmann-Hayoz/H. Gutscher (eds.)

Strategic Networking and Implementation of Communication and Diffusion Instruments to Develop Local Energy Policies

Christoph Bättig and Andreas Balthasar

This contribution focuses on the network management strategy adopted by the Swiss federal government in connection with the programme "Energy 2000 for Municipalities". This approach emphasises the importance of optimising co-operation among all parties involved in policy implementation. In doing so, it is strongly focused on communication and diffusion instruments used at an operational as well as strategic level. Strategically oriented services at a higher level should prepare the ground for an active energy policy in general by, for example, strengthening the networking among municipalities and other energy policy players. Action-oriented services, on the other hand, aim directly at the planning and implementation of measures in specific areas.

The first section summarises the context, approach, and major measures and instruments of the "Energy 2000 for Municipalities" programme. Section 2 analyses and evaluates the effects of the programme. On this basis, section 3 seeks to evaluate the interaction of communication and diffusion instruments, in particular with regard to the strategic and operational levels. Finally section 3 briefly discusses the combination of network strategy with economic instruments, as well as with command and control instruments.

1. Initial situation

1.1. The programme "Energy 2000 for Municipalities"

In September 1990, the Swiss electorate voted in favour of the Decree on Energy Use *[Energienutzungsbeschluss]* and accepted a ten-year moratorium on the expansion of nuclear energy. In order to exploit the energy savings potential during this period, the federal government complemented the statutory provisions of this Decree with voluntary measures. In this way the Energy 2000 Action Programme evolved through the co-operation of private, business, and government organisations. The Programme is financed by the federal government granting mandates to private organisations for the introduction of voluntary measures. The programme currently consists of seven departments, each focusing its activities on a specific target group (hospitals, fuels, housing, industry, trade and services, renewable energies, and public services). The present contribution concentrates on the efforts of the public services department, which seeks to motivate municipalities to promote a more active lenergy policy. The programme is called "Energy 2000 for Municipalities" (Balthasar, 2000).

Prior to the launch of the Energy 2000 Action Programme, the federal government had made little effort to motivate the municipalities towards an active energy policy. As in other policy areas, the cantons are primarily responsible for the implementation of national energy policy objectives. Within the framework of this executive federalism, it is the responsibility of the cantons to involve the municipalities in the execution of energy policy. In view of this legislative background, the Programme had no choice but to limit itself to "voluntary" measures. Therefore, with regard to the typology introduced in this volume, only collaborative agreements and communication and diffusion instruments may be considered.

Voluntary measures for municipalities did indeed exist before the launch of Energy 2000 for Municipalities. In this context, two points in particular are worth mentioning with regard to the activities of Energy 2000 for Municipalities:

- Firstly, prior to 1990 several cantons had already tried to support the energy policies of their municipalities through communication measures (communication and diffusion instruments). Besides these measures, the co-ordinated efforts of the cantons within the INFO Energy framework should be mentioned in particular. This model, co-financed by the federal government, developed a network of energy consulting offices providing free-of-charge consulting on implementing procedures for municipalities and private organisations.[1]
- As early as the 1980s, the Swiss Energy Foundation *[SES – Schweizerische Energiestiftung]*, as well as WWF Switzerland, had made efforts to gain the support of selected municipalities for an active energy policy. In this regard, they offered long-term consultancy services to municipalities with active energy policies on fully exploiting their scope of action within their energy policies. Municipalities responding positively were granted the honorary title of "Energy Town" (see Wortmann and Rieder, 1994).

---

[1] Four Competence Centres were financed in order to give expert support to the variously organised cantonal networks of energy consultancy offices. Consulting services are and were very different in the various cantons. Probably the most developed is this system in the Canton of Bern, with about a dozen regional energy consultancy offices (see also Beltz and Honegger, 1995).

Services offered to municipalities by "Energy 2000 for Municipalities"

Strategic level

- **ERFA-Association:** The Programme organises regional and national events for the exchange of experience ***(ERFahrungsAustausch)***, where exemplary solutions already implemented in Switzerland are presented and discussed. The Programme tries to recruit municipalities as ERFA members in order to bring continuity into the discussion process and to establish a distribution channel (the equivalent of network creation and maintenance).
- The **"Energy Town" label** is granted to municipalities that have already decided upon or carried out measures selected from a label catalogue. The label granting procedure also includes the evaluation of the institutional establishment of an energy policy (e.g. employment of an energy policy officer, establishment of an energy policy committee). The label is to represent proof of a continuous, consistent, and result-oriented energy policy. Following the granting of the label, the implementation of this energy policy is audited at regular intervals. Until recently, granting of the label Energy Town was not tied to adherence to energy-policy execution duties by municipalities. Nowadays the label is granted only with the consent of the competent cantonal authority for energy policy execution, which means that the label is the equivalent of a legal compliance certificate.

Action-oriented level

- **Energy Accountancy** is offered as an instrument for energy-optimal building management. The energy consumption of municipality-owned buildings is recorded and analysed. The key measurements obtained provide information on operating and technical faults. This service is the one that the municipalities have bought the most, and it is the product that was expected to make one of greatest contributions to achieving the goal in the short term.
- **Property Management Courses** are one-day seminars on how to record the energy consumption of buildings, to determine weak points, and to take effective action (maintenance, modernisation). Property Management Courses are a classic setting for applying communication instruments (presenting and persuading about facts, options, appraisals, goals, and norms).
- During **Energy-Policy Weeks in schools,** pupils analyse energy consumption and determine opportunities and measures for energy savings in order to improve the school building's energy consumption in the long term. The municipality makes a commitment to implement one of the best proposals for improvement and to involve the pupils in the financial results. Active communication means (exhibitions, newspaper articles) convey information on the results beyond the school. During Energy Policy Weeks several different communication instruments (giving feedback, sending appeals, providing experience of reality) are used and combined with incentives as parts of action campaigns.
- The **Transport Action Proposal** offers municipalities tailored consulting services for parking space management, speed reduction, and the improvement of facilities for pedestrians and cyclists. These do not necessarily, or even primarily, have to be technical solutions. The standard should rather be a two-stage approach. The design of an initial action plan involving politicians and affected groups (trades in particular) would be followed by the actual implementation at the second stage. During the first stage of action planning (namely participative planning and problem solving procedures) communication and diffusion instruments play a central role and aim at the involvement of local trades in policy development. During the implementation phase (second stage), communication and diffusion instruments are also applied (e.g., feedback on compliance with 30 km speed limit) and often combined with service and infrastructure instruments (e.g., creation of bicycle parking spaces).
- During **Energy Saving Weeks at Municipal Authorities**, all the employees should be involved, including property managers, the energy policy officer, town council members, or office equipment purchasers. Regularly communicated measurements demonstrate the energy consumed in administration, as well as any success achieved through energy-saving actions. The main communication instruments are feedback, prompts, and appeals.

1.2. The concept of the programme "Energy 2000 for Municipalities"

The starting point for Energy 2000 for Municipalities is the finding – confirmed by international comparative studies (Burns, 1985; Gaskell et al., 1986) – that municipal authorities are the most suitable vehicles for the implementation of energy-saving measures.[2] However, Swiss municipalities are a long way from exploiting all their potential for energy savings (see Bierter et al., 1985). Energy 2000 for Municipalities pursues a double approach in order to realise this neglected potential. On the one hand, municipal authorities should become more active within their own scope of action and take direct measures, for instance by making their own administration work more energy-efficient, or by modernising their office buildings. On the other hand, municipal authorities should be motivated to take measures within the framework of an active community policy to encourage actors outside their administration (households, trade, and industry) to save energy.

Energy 2000 for Municipalities is a typical experimental programme, in that it underwent several instances of conceptual change during its initial years. However, a consolidated strategy began to emerge in the mid 1990s. The Causality-Effects Model[3] assumed by the Programme views the municipalities as policy addressees. Service offers ("outputs") are produced on their behalf, consisting mainly of the **communication and diffusion** type of instruments (see box below). These offers include consulting services and support for the formulation and implementation of energy policy measures at the municipal level. We need to distinguish between the action-oriented offers and rather more strategy-oriented offers. The latter prepare the ground in general for an active energy policy by promoting networking between municipalities and other players within this energy policy. Action-oriented offers, on the other hand, are aimed directly at the planning and implementation of measures in specific areas.

In this context it is essential that the services offered should optimise the implementation process of municipal energy policies by means of communication and diffusion instruments. Energy accounting, for example, provides an instrument for the capture of energy consumption data of community-owned buildings, including supportive consulting for capturing and analysing such data. The Programme's intended impact is to cause a municipality to take action based on the accounting results and to seek improved energy-efficiency management of buildings and their modernisation. With regard to transport policies the municipalities are also supported through a process including the use of communication instruments (e.g., feedback in the form of recorded speed measurements, round-table talks with local business, and so on) to reach a consensus on policy measures. All programme offers generally function along these lines; they should contribute to

---

[2] BRUPPACHER AND ULLI-BEER also use the example of the Global Action Plan to point out the important role to be played by the municipalities with regard to the implementation of environmental policy measures.

[3] The effect model described here builds on the understanding of evaluation research distinguishing between services provided by the Programme players (outputs), the changes (impacts) caused with regard to the behaviour of target groups (policy addressees), and the effects with regard to the (environmental) problem to be solved (outcomes; see also section 2.6).

expanding the scope of energy-policy action by the municipalities and enable them to take appropriate action. The Programme's outcomes can be seen first in a change in public behaviour and second in energy savings.

It should be noted that Energy 2000 for Municipalities finances "only" the development costs for the instruments applied, the necessary training for energy consultants, and lump sum payments for the consultants' first talks with the municipalities. This means that it is not the change of behaviour of the policy addressees that is subsidised, as would be the case with economic instruments (in this context this would be the equivalent of subsidising energy-policy measures of municipalities).[4] Therefore, the Programme relies strongly on the self-motivation of municipalities and assumes that they would be more proactive in the development of energy policies if they only knew how, or if they were motivated by action proposals by competent energy consultants.

2. Effects of Energy 2000 for Municipalities

How should the Programme's effects be evaluated? In a three-step evaluation, a first step examines the distribution of programme products in Swiss municipalities (output distribution). The second step concerns the question as to whether the application of Energy 2000 products actually results in a more active energy policy in the municipalities involved (impact). Finally, the third step tries to evaluate the changes in public behaviour and the energy savings achieved (outcome).

This evaluation is based on standardised surveys in municipalities in German and Italian-speaking Switzerland carried out in the spring of 1999. Analyses of the results are then complemented by findings from in-depth case studies in nine municipalities[5] carried out concurrently with the surveys (see Knoepfel et al., 1999).

2.1. Distribution of output

This section seeks to evaluate whether Energy 2000 was actually able to achieve the planned output, i.e., whether the municipalities utilised the services the Programme offers.

First it has to be noted that the activities of Energy 2000 for Municipalities have acquired a very high recognition level. More than 90% of surveyed municipalities indicated that they had already

4 This waiver of economic incentives is in keeping with the philosophy of Energy 2000 for Municipalities. In other programmes, on the other hand, economic incentives (subsidies in particular) play a very prominent role (see contribution by GARBELY AND MÜLLER in the present volume).

5 All municipalities in German and Italian-speaking Switzerland with more than 1,000 inhabitants were contacted. Of a total of 972 municipalities, 315 returned the questionnaire (response rate of 32.4%). The in-depth case studies were carried out in the municipalities of Adliswil, Arbon, Münsingen, Schlieren, Gubiasco, Sursee, Wittenbach, Solothurn, and Zofingen.

heard of specific programme offers for municipalities.[6] Almost one fifth of surveyed municipalities stated that they had already utilised service offers from the Programme (outputs, see Table 1). The Programme's own statistics also indicate a similarly wide distribution (see Table 1).[7] Conservative estimates would assume that the Programme achieved outputs in approximately 15% of Swiss municipalities.

Table 1: Participation according to Programme statistics and community surveys.

Offer	Programme Statistics (June 1998)		Survey (April 1999)	
	No. of Municipalities	Percent**	No. of Municipalities	Percent***
Property management courses	-	-	44	14
Energy accountancy	137	14	37	12
ERFA membership	92	9	30	10
"Energy Town" label	46*	5*	26	8
Participation in transport activities	-	-	21	7
Energy-Policy Weeks in schools	-	-	13	4
Energy-Saving Weeks in municipal buildings	-	-	8	3
Total	-	-	62	20

* June 1999; of a total of 46 municipalities participating in the label procedures, 17 had been awarded the label by June 1999.
** According to the Programme statistics, the proportion of all the 972 Swiss municipalities with more than 1,000 inhabitants (German and Italian-speaking Switzerland; source: *Gemeindestatistik des Bundesamtes für Statistik*).
*** The proportion of municipalities stating that they had utilised the services mentioned (N=315; municipalities above 1,000 inhabitants only).

"Property Management Courses" and "Energy Accountancy" were the most frequently utilised service offers. These can be called "classical", because some cantons had already offered them prior to the launch of the Programme, and because the application of this type of communication instrument (organising education and training) are frequently in general use. Energy Accountancy is also a service offer with strong technical features.[8]

6 The extensive base marketing of the overall Programme Energy 2000 probably contributed strongly to this result.

7 The slightly higher estimate of Programme participation indicated by the questionnaires results from the fact that the response rate from participating municipalities was higher than from municipalities not participating (see Knoepfel et al., 1999).

8 Although the Programme comprises both the sale of products and comprehensive accompanying guidance and consulting by energy consultants for the evaluation and implementation of accountancy results, the Programme evaluation results (see Knoepfel et al., 1999) indicate that too often the sale of products is at the centre, and guidance and consulting of participants were rather neglected.

The Programme managed to place products primarily in municipalities employing an energy officer (working at least part-time, one day a week). Almost half of these municipalities participate actively in Energy 2000 for Municipalities, whereas this is true of less than one sixth of the remaining municipalities. This points to the – more or less obvious – fact that the functioning of the impact model assumed by the Programme depends on the existence of a minimum of self-motivation with municipal authorities, or a "cell" at this level with the appropriate responsibility (protection of the environment, energy savings). This notion is also supported by the reasons given by non-participating municipalities for refusing the Programme offers. The most frequently stated reasons by far for refusal were not a lack of knowledge of these services or their wrong orientation, but a lack of money and time (see Knoepfel et al., 1999).

2.2. *Effect of the programme on the design of municipal energy policies (impact)*

The following sections are concerned with the issue of the effects of the Programme's products on the design of municipal energy policies. We distinguish between effects with regard to the networking of the players involved in energy policies, effects with regard to the design of the implementation process, and effects with regard to the implementation of practical measures.

Table 2: Improvements in the networks of participating municipalities.

Contacts to...	New Contacts	Positive Influence on Contacts	Total
Energy consultants	38.7%	25.8%	64.5%
Cantonal energy and other offices	11.3%	53.2%	64.5%
Other municipalities	35.5%	9.7%	45.2%
Engineering firms	27.4%	17.7%	45.1%
Federal Office of Energy	17.7%	16.1%	33.8%
Other departments at community level	3.2%	19.5%	22.7%

The proportion of participating municipalities (n=62) for which Programme participation resulted in new contacts, or positive influences on existing contacts, to the departments listed.
Question in the questionnaire: "Has participation in 'Energy 2000' had a significant influence on your contacts concerning your municipal energy policy? Please state whether your participation in 'Energy 2000' created new contacts or whether it has had a positive or negative influence on existing contacts."

2.2.1. *Networking of municipal energy policies*

The networking among municipalities and other partners with authority for energy policies can be seen as a first indicator for the existence of an active energy policy. Energy 2000 for Municipalities strives to strengthen these networks (see 1.2.). This objective was achieved very successfully in municipalities participating in the Programme (see Table 2). More than half of these municipalities was able to establish new contacts with energy consultants and cantonal offices or strengthen existing contacts. Networking among municipalities was also improved: over one-third of the participants stated that the Programme helped them to establish new contacts with other municipalities. It is surprising that participation in the Programme also improved networking within municipal administrations. Whereas the improvement of contacts with other municipalities is based almost exclusively on ERFA events, the improvement of internal administration networking took place mainly in connection with the planning and implementation of measures.

2.2.2. *Improved implementation of municipal energy policies*

As already stated in 1.2, the Programme's logic assumes that the application of communication and diffusion instruments (in the form of networking in particular) can significantly improve implementation of measures. Has the implementation of measures really been improved by Energy 2000? The following part seeks to identify impacts by comparing Programme participants with non-participants. The survey addressed participants as well as non-participants with regard to their (subjective) impressions of their own energy policy's impact. A positive effect of the Programme would be indicated by a better assessment of the participating municipalities' own energy policies.[9] Table 3 shows that participating municipalities indeed judge their energy policies as better than non-participating municipalities judge theirs. Interesting to note are the advantages of participating municipalities with regard to the activation of support for policy decision-makers. In this area the communication instruments applied by the Programme have apparently proved useful.

Table 3: Evaluation of participating and non-participating municipalities' energy policies.

	Assessment "good"	
	Participating (n=57)	Not Participating (n=217)
Implementation of measures	33.3%	24.0%
Establishment of support for decision-makers	35.1%	18.9%

Proportion of municipalities answering the question "How do you assess the effectiveness of your own energy policy with regard to...?" with "good" (the answer choices were "good", "quite good", "not so good", and "bad").

[9] We base this on the very plausible assumption that participating as well as non-participating municipalities generally assess their own energy policies by similar parameters. In other words: the psychological effect of effort justification can be seen as of little influence and is therefore ignored. The results of the in-depth case studies even lead in the opposite direction: participating municipalities assess the impact of their energy policies rather more critically than non-participating municipalities.

2.2.3. *Additional and more diversified measures*

The survey asked municipalities how they assessed the influence of Programme participation on the design of appropriate measures. Almost two-thirds of the municipalities utilising Programme offers saw positive effects from these offers on practical measures. Almost one-quarter (24.2%) of responding participants even implemented measures that had not been planned initially. Four out of ten municipalities implemented measures earlier, in different ways, or with a much larger scope. A good third (35.5%), on the other hand, could not detect any influence of the Programme support on the design or implementation of measures.

Therefore, the majority of municipalities detected an impact of the Programme on the design of their energy policies. This leads to the question as to whether the Programme has had an impact on the general design of energy policies, or whether the profile of municipal energy policies has been extended due to Energy 2000. A comparison between major measures in participating and non-participating municipalities allows some conclusions to be drawn.

Figure 1 compares the distribution of implemented "major measures"[10] to different issue areas in participating and non-participating municipalities. This results in obvious differences. In particular, measures in the areas of public relations, organisation, and transport are of higher importance in participating than in non-participating municipalities. This result seems to indicate the Energy 2000 for Municipalities has been successful in diversifying the bundle of energy-policy measures, or in establishing a broader understanding of municipal energy policy extending beyond the planning and construction areas.[11] This impact – as deduced from the survey results – can be supported by the findings of in-depth case studies. All of the participating municipalities investigated by these in-depth case studies stated that the Programme provided significant momentum for the expansion of the energy-policy horizon beyond the traditional scope of action.

---

10 The municipalities were asked in an open-ended question to name the **three most important** measures of the previous five years regarding energy policies.

11 The analysis of data shows that the stated difference does not result from the (hidden) influence of other variables such as type or size of the municipality (see Knoepfel et al., 1999).

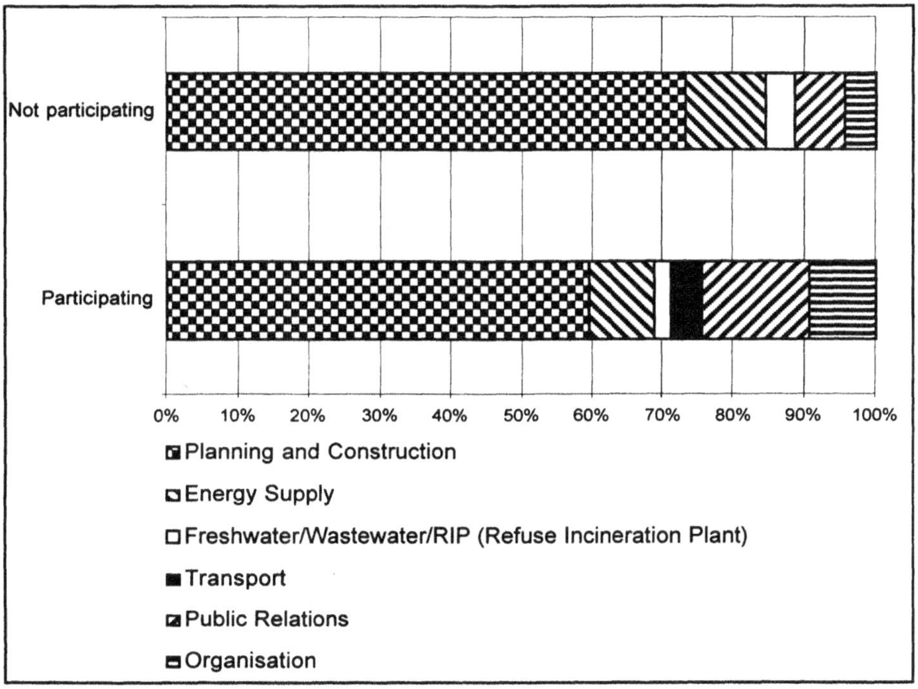

Figure 1: Major measures according to different issue areas (participating vs. non-participating municipalities). Participation was defined as a municipality's utilisation of at least one Programme offer. The 62 participating municipalities recorded a total of 148 important measures (equivalent to 100% in the lower bar of Figure 1), and the 201 non-participating municipalities recorded 427 important measures (equivalent to 100% in the upper bar of Figure 1).

2.3. *Effect on changes in public behaviour and energy consumption*

Finally the question is whether the measures caused or improved by the utilisation of Programme services had any impact on public awareness and on energy consumption as the target parameters proper (outcomes) of energy policies. Again a comparison between participating and non-participating municipalities allows some conclusions with regard to this question. As shown in Table 4, the respective outcomes cannot be demonstrated: participating municipalities evaluate the impacts of their energy policies with regard to outcome hardly any better than do non-participating municipalities. Even when municipalities are asked directly how they would evaluate their Programme participation with regard to these outcome dimensions, the reply is similarly sobering.[12]

---

[12] However, in-depth case studies (see Knoepfel et al., 1999) established that the municipalities lack reliable data to quantify the impact of their measures.

Table 4: Outcome evaluation of energy policy by participating and non-participating municipalities.[13]

	Assessment "good"	
	Participating (n=57)	Not Participating (n=217)
Public awareness	18.0%	15.1%
Savings and substitution effects in private households	7.0%	6.9%

Proportion of municipalities answering the question "How do you assess the effectiveness of your own energy policy with regard to...?" with "good" (the answer choices were "good", "quite good", "not so good", and "bad").

There are two main reasons responsible for the municipalities' low marks at the level of outcomes:

- On the one hand, the measures have been implemented too recently to have influenced outcomes;
- On the other hand, the changes in behaviour and energy savings achievable with these measures are of minor importance in comparison with overall energy consumption.

It can be concluded from both explanations that, at the municipal level, effects on the overall outcome level can only be expected when various measures are integrated within a long-term strategy. This conclusion supports the Programme's approach aiming mainly at strong support from businesses and the public for consistent municipal energy policies.

3. Combination of measures

The previous sections concentrated on a global assessment of the Programme's impacts and outcomes from the application of various communication and diffusion instruments. The following section adds three remarks on the interaction of these instruments among themselves as well as in interaction with economic command and control instruments.

[13] The data result from the municipalities' self-evaluation on the questionnaires. Case studies attempted to demonstrate energy savings effects based on existing data from the municipalities investigated. This revealed that no municipality has "hard" data suitable for such conclusions (see also Knoepfel, 1999). Therefore it has to be assumed that the survey responses are not based on "hard" data either (see also the contribution by MAUCH ET AL.).

3.1. *ERFA membership as a strategic instrument*

Within the interaction among the more strategic and rather more measure-oriented Programme offers, ERFA events and the permanent establishment of municipalities within an ERFA association are primarily used to increase a community's willingness to utilise measure-oriented Programme services. The results of our survey now confirm that ERFA membership is indeed associated with a more extensive utilisation of Programme offers (see Table 5).

Table 5: Number of Programme services utilised by ERFA members and non-members.

	ERFA Member (N=30)	Non-member (N=32)
No offer utilised	2	2
1 to 2 offers utilised	10	27
More than 2 offers utilised	18	3

Of a total of 315 replying municipalities, 62 participated in the Programme, with 30 of them as members of the ERFA association and 32 of them as non-members of ERFA.

Table 6 shows that the said correlation between ERFA membership and the utilisation of outputs is even more apparent when it comes to innovative instruments (communication instruments such as giving feedback, presenting prompts, and so on). A total of 24 of the 30 ERFA member municipalities surveyed also utilised these newer communication instruments. On the other hand, rather more traditional offers, such as Energy Accountancy and Property Management Courses, are well accepted by municipalities that are not ERFA members, whereas more innovative instruments meet with problems in these municipalities. Therefore, the establishment of ERFA as an institutionalised association seems to have had a positive effect on the sale of Programme products (outputs).

One could be critical in saying that the discussion so far does not support the assumption of causal relations between ERFA membership and higher utilisation of Programme services. Nevertheless, interviews with participants in the case studies contradict this criticism. The municipalities were unanimously adamant that the exchange of experience encouraged them to implement new measures, even in areas alien to traditional energy policies.

From the Programme's perspective, ERFA events as a "meta-measure" should be evaluated as useful for the promotion of measure-oriented offers. However, it should be pointed out that ERFA events in comparison with other offers are seen as of moderate value by participating municipalities. But the case studies clearly indicate that the municipalities criticise ERFA not so much as an institution or in terms of its events, but with respect to the contents it focuses on. Established "pioneer" and "new" participants differ considerably in their requirements, as the former ask for more innovative contents and nation-wide contacts, and the latter for tried and tested solution approaches and more regional contacts. Programme management will have to face the challenge of giving both interest groups equal attention.

Table 6: Number of ERFA members and non-members using Programme services.

Offer	ERFA Member (N=30)	Non-member (N=32)
Property Management Courses	21	23
Energy Accountancy	20	17
Energy-Policy Weeks in Schools	9	4
Energy-Saving Weeks in Community Centres	7	1
Transport Measures	14	7
"Energy Town" Label	20	6

Of a total of 315 replying municipalities, 62 participated in the Programme, with 30 of them as members of the ERFA association and 32 of them as non-members of ERFA.

3.2. *Combination with economic instruments: subsidies (grants)*

The concept of Energy 2000 for Municipalities – as discussed in section 1.2 – is strongly based on a municipality's own interest in participating in the Programme, and subsidies for the implementation of measures are generally not granted. Nevertheless, the Programme's actual implementation plan contains some subsidy incentives. On the one hand, some cantons subsidise the participation of municipalities; and on the other hand, the Programme (contradicting its own basic philosophy) takes over a part of the costs for the planning and implementation of measures for transport policies. All in all, over half the municipalities were recorded as having benefited from cantonal or federal subsidies.

The analyses show that services offered by Energy 2000 trigger measures not planned initially only if they are coupled with financial contributions (see Table 7). Out of the 15 municipalities implementing initially unscheduled measures, 12 received financial support from federal and cantonal sources. Only three municipalities started new measures without such subsidies.

3.3. *Combination with command and control instruments*

Although the Programme itself does not apply any command or control instruments, it is part of a regulative environment that is not the same as that of the respective canton. In-depth case studies of nine municipalities and interviews with cantonal energy policy officers have shown that complementary as well as competitive relationships between the Programme and cantonal regulations can develop. Ideally the Programme services can help municipalities to improve the implementation of command and control instruments imposed by the cantons. Conversely some municipalities try to avoid the implementation of – from their viewpoint nonsensical – cantonal regulations by pointing out equivalent measures to be implemented with the help of Energy 2000. In extreme cases, such competition scenarios can – as some cantonal officials fear – result in resources and/or legislative powers for the implementation of regulative measures being revoked by Cantonal Parliaments, with the argument that "voluntary" measures initiated by the federal government are quite sufficient. With regard to Energy 2000 for Municipalities this competition scenario has in part successfully been avoided by coupling the Programme explicitly with energy policy legislation. The "Energy Town" label, for example, is only awarded with the consent of the cantonal department(s) responsible for implementing energy policies.[14]

Table 7: Influence of participation in relation to subsidised participation.

	Subsidies				**Total (N=33)**	
	Yes (N=29)		No (N=33)			
	Number	Percent	Number	Percent	Number	Percent
Measures triggered	12	41.4	3	9.1	15	24.2
Measures altered or carried out earlier	13	44.8	12	36.4	25	40.3
No influence	4	13.8	18	13.8	22	35.5

4. Summary and conclusions

The results presented here demonstrate that the network approach of Energy 2000 for Municipalities has contributed to the improved establishment of energy policies in the participating municipalities. First, the networking of energy-policy players (municipalities, cantonal authorities, energy consultants) has been strengthened. Second, the Programme has had positive effects on the implementation process within the municipalities and especially promoted the development of energy-policy support for decision-makers. Third, the Programme has led to new or expanded measures in many municipalities and helped them to build a more varied range of instruments for their energy policies. The Programme particularly promoted the application of more innovative forms of communication and diffusion instruments (organising participatory planning and problem solving processes, giving feedback, using prompts, and the like).[15]

14 This procedural regulation intends to ensure legal compliance with the "Energy Town" label (see the contribution by KAUFMANN-HAYOZ ET AL., section 3.4., in the Theoretical Framework, on descriptions of the instruments).

15 For a more detailled discussion of the network approach and its use for environmental policy, see Bättig, 2001.

However it remains problematic that participating municipalities have come – at least for the time being – to a critical judgement about the measurable effects of the Programme on changes in public behaviour and energy consumption. Nevertheless, the overall conclusion is a positive one: The development and maintenance of networks, in combination with the consulting services offered for practical and specific measures, represents a promising way to motivate municipalities to develop more active energy policies without enforcing regulations. The research results indicate in particular that the integration of municipalities within networks is an important condition for the activation of municipal energy policies. One can assume that the maintenance of these networks will contribute significantly to the continuity of such activities in municipal policies that will have the associated ecological effects in the long term.

The study indicates that a strategy strongly oriented to the application of communication and diffusion instruments could become more powerful by the addition of financial incentives (subsidies, grants). In particular, new and previously unscheduled measures depend strongly on such grants. Finally, optimum co-ordination of a strategy based on voluntary participation with command and control measures of other state institutions is required.

Generally, the strategy initiated by Energy 2000 for Municipalities appears to be suitable as a model for expanded sustainability policies at the municipal level. What remains to be demonstrated is whether the established activation of energy policies results in measurable energy savings.

References

Bättig, C. (2001) Netzwerkmanagement als neues Instrument der Umweltpolitik; *forthcoming*.

Balthasar, A. (2000) Energie 2000: Programmwirkungen und Folgerungen aus der Evaluation. Rüegger, Zürich.

Beltz, G. and Honegger, J. (1995) Evaluation der Energieberatung: Synthesebericht. Bundesamt für Energie, Bern.

Bierter, W., Binder, H.-M. and Rüegg, E. (1985) Energiepolitische Handlungsmöglichkeiten auf kommunaler und kantonaler Ebene. Nationales Forschungsprogramm 44, Zürich.

Burns, T.R. (1985) The Social Logic of Energy Conservation in Industrial Society. IIG Discussion Papers 85/24. Wissenschaftszentrum Berlin, Berlin.

Gaskell, G., Joerges, B., Ester, P., Midden, C., Monier, E. and Mueller, H. (1986) A Multinational Study of Local Energy Conservation Schemes. *In*: Consumer Behavior and Energy Policy. An International Perspective, Monier, E., Gaskell, G., Joerges, B., Ester, P., Midden, C., LiPillonne, B. and Puiseux, L. (eds.), Praeger Publishers, New York.

Knoepfel, P., Bättig, C., Varone, F., Balthasar, A., Mauch, C. and Eggerschwiler, C. (1999) Evaluation des Programmes Energie 2000 für Gemeinden. Bundesamt für Energie, Bern.

Wortmann, K. and Rieder, S. (1994) Evaluation des schweizerischen Projekts "Energiestadt" im Vergleich zum deutsch-dänischen Projekt "Brundtlandstadt". Bundesamt für Energie, Bern.

Changing Things – Moving People.
Strategies for Promoting Sustainable
Development at the Local Level
R. Kaufmann-Hayoz/H. Gutscher (eds.)

Economic Instruments for Wastewater Disposal

André Müller and David Kramer

This case study evaluates two economic instruments of water pollution control in Switzerland: wastewater taxes and wastewater charges, as cantonal and municipal instruments respectively. We investigate the issues in relation to how these economic instruments should be designed; how their interaction with other instruments (command and control, subsidies, and communication and diffusion) should be organised; what obstacles might be faced in their implementation; what ecological impact and related counter-effects have to be dealt with; and how these can be overcome. The evaluation of wastewater charges has shown that the municipal task of making an ecology-related wastewater charge should be based on the "polluter-pays" principle and limited to the financing of disposal services. The cantons and the federal authorities should introduce cantonal and federal wastewater taxes in order to provide the necessary prerequisites that would enable municipalities to contribute to a sustainable development of water pollution control. Wastewater taxes would be a sensible complement to the current range of instruments based on command and control. Wastewater taxes can have a positive effect in relation to the discharge of pollutants into rivers and lakes. This will create an incentive for improving processes at sewage treatment plants. Communication and diffusion instruments should be employed in order to overcome the obstacles both to implementation and to the success of economic instruments.

1. New challenges for the financing of wastewater disposal

The financing of wastewater disposal in Switzerland is undergoing major change. In the 1960s, 1970s, and 1980s, wastewater disposal was developed by means of massive public subsidies. During those years, federal and cantonal authorities carried an average of 37% of all wastewater disposal costs.

No other environmental protection area, with the exception of transport, received subsidies of this order. The success rate is impressive: most households are linked to a wastewater treatment plant, and the quality of rivers and lakes has been greatly improved over the last 40 years.

However, the situation changed fundamentally during the 1990s. The reduction or abolition of subsidies and the introduction of wastewater taxes were considered and partially implemented at the federal and cantonal levels in particular, for the following reasons:

- Lack of money at federal and cantonal level: public money was scarce and further tax increases would not have been accepted.
- Enforcement of the polluter-pays principle: subsidies lower wastewater disposal costs and are inconsistent with the polluter-pays principle.
- Re-orientation of financial balancing and sharing of responsibilities between federal and cantonal authorities.

At the municipal level, the financing of wastewater disposal became a major issue for the following reasons:

- Municipalities were confronted with additional costs caused by the high demand for the renewal and reconstruction of plants.
- Future construction and reconstruction works were not subsidised to the same extent as previously.
- The revenue from connection charges decreased dramatically because of reduced construction activities.
- An increase of municipal wastewater charges became inevitable. In view of the fact that water charges were not really orientated to the polluter-pays principle, the question of optimum financing had to be redefined.

The guidelines for the concept of wastewater taxes and wastewater charges are introduced in section 2. below. This is followed by separate analyses of wastewater taxes and wastewater charges:

- How to structure wastewater taxes/charges?
- How to introduce wastewater taxes and wastewater charges, and in the face of what kind of obstacles?
- What kind of ecological effects and counter effects are to be dealt with?
- What contribution can be expected from communication and diffusion instruments aimed at overcoming these obstacles and counter effects?

2. The polluter-pays principle as a conceptual guideline

The discussion of the structure of wastewater taxes and municipal wastewater charges as economic instruments must be based on a transparent and, in itself, consistent concept. This study is based on the polluter-pays principle (PPP), which is largely supported by the public and already has a legislative foundation.[1] The three-stage concept details the costs to be covered for the consistent enforcement of the polluter-pays principle (see Figure 1):

Enforcement of primary PPP: Coverage of the direct costs of wastewater disposal. These construction and operating costs are to be paid by households and businesses in the form of wastewater charges based on the PPP.

Enforcement of secondary PPP: Coverage of the costs incurred by the state. These include subsidies, expenditure for the development of treatment plants and for compensating extreme regional or social burdens. Subsidies lower wastewater disposal costs and are inconsistent with PPP. Any further subsidies for support or compensation should be financed through PPP-oriented taxes. In addition, cantonal services not covered by administrative charges (for tasks such as monitoring, planning, co-ordination, enforcement, consulting, and public information) are also to be financed by wastewater taxes.

Enforcement of tertiary PPP: Coverage of consequential costs for environmental protection (external costs) caused by actual water pollution. An example of external costs in relation to water pollution control would be the reduction of leisure and recreation value through polluted water resources. In this case, the people seeking recreation pay, instead of the actual polluter. As the polluters do not pay the overall costs of water pollution, rivers and lakes are overburdened. From an economic perspective this has to be seen as a market failure. In order to remedy this, external costs should be covered by wastewater taxes to be paid by the polluters. Anybody disposing of wastewater into rivers and lakes should be charged accordingly. Wastewater taxes are levied at the wastewater treatment plants (WTP) according to the volumes of wastewater disposed of into rivers and lakes. WTP would invest in optimising the plant until the costs for additional treatment equal the wastewater taxes paid by the plant. Therefore, these taxes are an incentive to decrease pollution below the limits.

[1] See the lawyer's comment by FRIEDERICH on some legal considerations regarding the polluter-pays principle.

3. Wastewater taxes as a controlling and financing instrument

3.1. The structure of wastewater taxes

The federal and cantonal authorities are responsible for the enforcement of the secondary and tertiary PPP. The following brief explanation of the structure for wastewater taxes is based on a theoretical concept. For a more comprehensive description see ECOPLAN (1993).

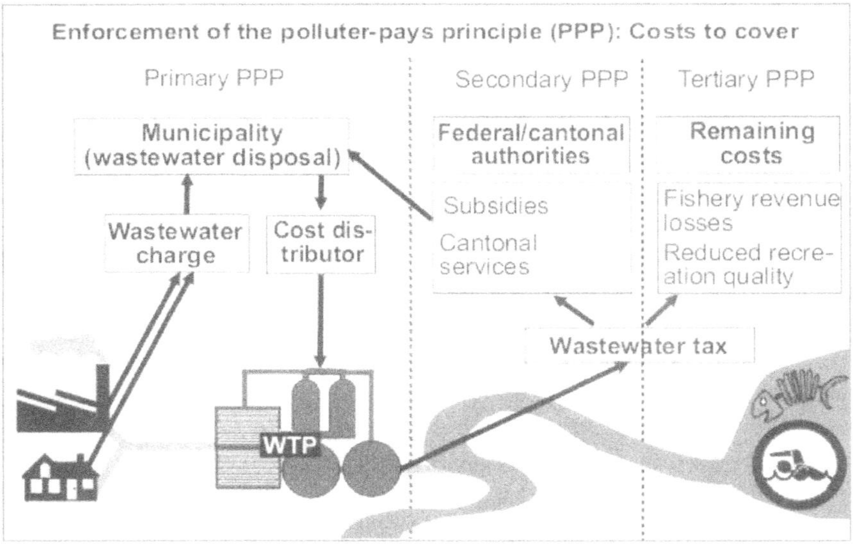

Figure 1: PPP-orientated financing from an economic perspective. Businesses and households cover the costs for wastewater disposal via wastewater charges and cost distributors. Wastewater taxes are levied at the wastewater treatment plants (WTP) in order to cover the costs carried by federal and cantonal bodies and the remaining consequential costs. The WTP then apportions the wastewater taxes via cost distributors and wastewater charges to households and industries as the primary polluters.

Who in Switzerland is entitled to apply the instrument of wastewater taxes?

Based on the existing legal scope of action, wastewater taxes can be imposed by federal or cantonal authorities. However, other arguments also favour their application at the federal and cantonal level: wastewater taxes are aimed at the achievement of regional and supra-regional objectives that are to be put into action by higher institutional levels, i.e., federal or cantonal authorities, even if the subsidiarity principle is taken into consideration.

What constitutes a case for wastewater taxes and who is to pay?

Anyone disposing of wastewater into water resources should pay for damage. The taxes are therefore levied in accordance with the volume of wastewater disposed of. As the various pollution sources (agricultural alluvial, for example)[2] cannot be quantified, the taxes are levied on the so-called direct polluters (in Switzerland, primarily the wastewater treatment plants).[3]

What are the objectives of wastewater taxes?

The wastewater tax is aimed directly or indirectly at covering financing, controlling, and compensation costs.

- **Financing**: According to the concept introduced in the previous section, wastewater taxes are used to finance federal and cantonal subsidies and public services. In addition, they are used to cover the internalisation of external costs.
- **Controlling**: Wastewater taxes set the "correct" prices for wastewater disposal and should have an impact on the behaviour of direct and indirect polluters. Wastewater treatment plants (as direct polluters) can avoid wastewater taxes by improving their treatment performance. Treatment is improved until treatment costs equal the wastewater taxes to be avoided. Apportioning the wastewater taxes by means of wastewater charges to households and businesses (indirect polluters) also should achieve a reduction of wastewater volumes from these sources.
- **Compensation**: Without wastewater taxes, those plants deploying expensive treatment measures suffer from higher costs. In order to achieve regional and supra-regional objectives (reducing nitrogen volumes, for instance) it would make sense to demand an improvement only from those treatment plants that are able to perform additional treatment in the most cost-efficient way (generally the large, municipal plants). However, treatment plants providing such "extra services" in order to achieve regional and supra-regional objectives are "punished" with additional costs. This can be remedied by means of wastewater taxes: "extra services" reduce pollution and, therefore, the tax burden. If the taxes are sufficiently high,[4] a complete balance of the costs for "extra services" can be achieved.

---

2 Different instruments are needed to stem water resource pollution in agriculture, such as, for example, regulations and restrictions about the number of cattle units per hectare, or economic instruments, such as a charge per cattle unit (see ECOPLAN, 1996).

3 Most industrial plants dispose of their wastewater – after internal pre-treatment – to a municipal treatment plant. In Switzerland, unlike Germany, direct polluters are rather rare.

4 The balancing effects of wastewater taxes can be increased by subsidies. Taxes coupled with subsidies can be kept lower in order to achieve the same effect.

What are the problems in structuring wastewater taxes?

From an economic perspective, the actual theoretical concept of wastewater taxes is undisputed. With regard to the practical structure, on the other hand, problems can arise. These, and their consequences, are discussed below.[5]

- **Unknown external costs:** Apart from the financial requirement for state contributions and administrative services, the level of wastewater taxes – in accordance with the theoretical concept – should be based on external costs. However, the costs caused by environmental consequences are not known. One can only state that such external costs are incurred, and that they are carried by the general public instead of the polluters.[6] This is clearly a research gap, with the level of wastewater taxes consequently having to be based not on external costs, but on environmental objectives. Therefore, the level of wastewater taxes has to be guided by impact thresholds or emission targets. Wastewater taxes based on impact targets take into account the varying capacities for self-cleaning of different water resources, and accordingly their different absorption capabilities. Therefore, the taxes should be differentiated by region. Wastewater taxes based consistently on impact thresholds would require different charges for each direct polluter/treatment plant, and should depend on the water quality of rivers and lakes and the avoidance costs incurred by the direct polluter. Such a tax system – apart from possible acceptance problems – would fail due to the information deficit alone. For these reasons a practicable wastewater tax system has to be based on emission targets. The level of taxation is set so that the necessary prevention or treatment measures become profitable investments. With regionally differing emission targets, regional differentiation should not be excluded as an option.[7]
- **Feasibility limits – restricted controlling effects on treatment plants (direct polluters):** Treatment plants cannot retain all pollutants. End-of-pipe technologies in wastewater disposal are capable of treating degradable, biological, and oxygen-consuming substances (at the mechanical/biological stage), phosphorus (by phosphatisation and subsequent flocculent filtering), ammonium (by nitrification), and all nitrogen (by de-nitrification). Although other substances – such as heavy metals – can be partially kept from water resources, they are retained in the sludge, which therefore cannot be utilised for agricultural purposes. Other

5 Obstacles to implementation and effects are discussed in the following two sections.

6 The only comprehensive estimate known to us of external costs from water resource pollution, has been provided by Wicke (1986) who calculated DEM 17.6 billion per annum for Germany, the equivalent of CHF 250 per capita per annum. The most important external cost factor of DEM 7 billion per annum is "reduced leisure and recreation quality", followed by DEM 6 billion per annum of "additional costs for drinking water treatment". With regard to these figures, it has to be taken into account that since the 1960s and 1970s (the period from which these figures were taken) water quality has greatly improved and that Switzerland cannot be compared directly with Germany.

7 See ECOPLAN (1993): The proposal for wastewater charges at a federal level distinguished between regions north and south of the central lakes. The northern lakes were suffering mainly from phosphorus problems, and the southern lakes mainly from nitrogen problems.

problem substances – in particular the range of chemicals with low degradability and endocrinal substances, which are seen as real problems for the future – pass through the treatment plants more or less unhindered. This means that taxes should be levied on substances controllable by direct polluters (treatment plants) in order to influence the behaviour of direct polluters. Such substances are COD[8] chemicals (incentive for the optimisation or construction of a mechanical/biological stage); phosphorus (incentive for the optimisation or installation of equipment for phosphatisation or additional phosphorus elimination; ammonium (incentive for the optimisation or building of nitrification); and nitrogen[9] (incentive for the optimisation or construction of a de-nitrification stage). Taxes on wastewater are levied in order to keep so-called foreign water[10] away from the treatment plant and to improve treatment performance.

- **Feasibility limits – restricted controlling effects on households and businesses (indirect pollutants):** The controlling effects of wastewater taxes depend on the structure of the "apportioning mechanism" for distributing wastewater taxes to the relevant households and businesses. The wastewater taxes are apportioned via distributors to the municipalities, and then in form of wastewater charges to the primary wastewater sources. One has to differentiate between small wastewater sources (households and businesses generating low quantities of wastewater) and large wastewater sources (businesses generating large quantities of wastewater). Measuring the wastewater pollution generated by individual households and businesses generating low quantities of wastewater would be far too expensive. For this reason, municipal wastewater charges take only wastewater quantities into account, and not the pollution quantities (see section 4.3.). Therefore, wastewater taxes tend to influence water consumption only by households and businesses generating small quantities of wastewater. Businesses generating large quantities of wastewater usually pay pollution surcharges. Generally the wastewater taxes could be apportioned via these pollution surcharges, resulting in a controlling effect on pollution volumes. Of particular interest in this context are those substances that cannot be eliminated by the treatment plants. In this case, only the emitters have the means of control. The problem remains, because in reality the necessary measuring devices are too cost-intensive. However, the latest measuring technologies enable more and more cost-efficient capturing of pollutant data. Further progress can be expected in the foreseeable future, which means that an extension of wastewater taxes to problem substances generated by wastewater-intensive businesses will have to be considered.

---

8 COD – Chemical Oxygen Demand: this is an aggregate parameter for biologically degradable substances. Other aggregate parameters would also be suitable: TOC – Total Organic Carbon; DOC – Dissolved Organic Carbon in combination with TUS – Total Undissolved Substances.

9 Nitrogen total, i.e., ammonium, nitrate, and nitrite, including residual nitrogen components.

10 Clean water (such as well water, drainages, etc.) fed into treatment plants.

Wastewater taxes as a substitutive, subsidiary, or complementary instrument?

The following section analyses the relationship between wastewater taxes and command and control instruments. Interaction with communication and diffusion instruments is discussed in section 3.4. Because of their undisputed ecological impact, relatively high targeting regulations cannot be avoided altogether. Three models can be employed to demonstrate the interaction between wastewater taxes as an economic instrument on the one hand, and command and control on the other hand:

- **Wastewater taxes as a substitutive instrument:** Here the taxes replace the majority of regulations. This concept would be irresponsible, as it would be impossible to levy charges on all ecologically relevant pollutants, and because comparison among pollutants from different sources is problematic. With regard to water pollution control in particular, minimum requirements should not be abolished.
- **Wastewater taxes as a subsidiary instrument:** The main objective of this concept would be to enforce regulations. Taxes are levied only if regulations are not complied with. Residual pollution below the upper limit is not taxed. Therefore, the taxes function as a "penalty tax". This concept continues to use regulations as the dominant instrument to achieve water pollution control targets, with the option of narrowing pollution limits in order to further decrease pollution.
- **Wastewater taxes as a complementary instrument:** The regulations are complemented by separate taxes levied irrespective of regulation compliance. This means that taxes are levied on the entire volume of pollution emissions (including residual pollution below the upper limit). Parameters on which taxes are levied are not to be narrowed any further. This means that existing limits achieve the function of "minimum requirements". Further pollution reduction should be achieved by levying taxes (possibly in combination with subsidies) instead of more stringent regulations. For the parameter, without taxes being levied, the regulations remain the dominant instrument.

The decision to introduce wastewater taxes, either as a subsidiary instrument or as a complementary instrument to regulations, depends on the question as to whether and to what extent residual pollution (after compliance with regulations) should be made taxable or not. As the advantages of taxing residual pollution outweigh the disadvantages (see Table 1), wastewater taxes should be levied as a complementary instrument to regulations and controls.

Table 1: Disadvantages and advantages of wastewater taxes on residual pollution – meaning the pollution remaining after complying with regulatory limits.

Advantages of a residual pollution charge	Disadvantages of a residual pollution charge
Charges create incentives for further reductions of pollution after complying with regulations.	The charges would have to be higher in order to achieve the same economic incentive.
Charges on residual pollution create incentives for cost-saving technological progress.	Even with the best technologies and optimum operation, some residual pollution could not be eliminated and would incur charges that cannot be reduced any further by additional investment. This is deemed unfair. Political pressure increasingly demands the abolition of wastewater taxes.
Charges on residual pollution avoid sending out wrong signals with regard to the impact scenario: if residual pollution were not to be charged for, a direct polluter complying with the regulations would not have to pay any charges for discharging pollutants into an already heavily polluted water resource. On the other hand, a polluter not complying with the regulations and discharging into a water resource with little pollution would have to pay charges, although the ecological burden is smaller.	

3.2. *Implementation of wastewater taxes*

Before discussing the individual phases of the implementation process, Switzerland's sad history of wastewater taxes is briefly explained.

3.2.1. *Switzerland's sad history of wastewater taxes*

Some seven years ago, a wastewater tax system was evaluated at the federal level (see ECOPLAN, 1993). The revenue generated by these taxes was to be used for financing federal contributions. The investigation was triggered by the example of foreign models (Germany in particular). The wastewater taxes proposed by the study were based – in contrast to the German model – on the true pollution volumes discharged into water resources, i.e. including residual pollution.

After consulting the cantons and interested parties, the federal government abandoned the introduction of wastewater taxes. The instrument of "environmental taxes" was to be implemented in other, more urgent, areas (volatile organic compounds – VOC, CO_2, etc.). Three cantons, however, decided to introduce wastewater taxes (see Table 2):

Canton of Appenzell Ausserrhoden: Wastewater taxes were introduced on 1st January 1995. The Appenzell AR canton was therefore the first Swiss canton to levy wastewater taxes on residual pollution discharged into water resources. Treatment plants measure the volumes and submit the information to the cantonal authorities (principle of self-declaration). The revenue from the wastewater taxes is used for the financing of subsidies. The taxes are adjusted in accordance with the annual financial requirements.

Canton of Bern: From 1st January 1995, and as a transitional solution, this canton has levied a wastewater tax of CHF 25 per citizen. Since 1999, the canton has also measured pollutant volumes discharged into water resources, and since the beginning of 2000 the wastewater taxes are based on the volume of residual pollutants discharged into water resources. The measuring concept was developed within the framework of a project across cantons (see Künzler and Partner AG and ECOPLAN, 1995). The revenue from the wastewater taxes was tied to the financing of state contributions and for part of the cantonal administration costs incurred in water pollution control.

Canton of Solothurn: In the referendum of September 1998, the majority of citizens in Solothurn voted in favour of wastewater taxes. The introduction is scheduled for January 2001. The revenue from wastewater taxes will be utilised for the financing of state contributions.

Table 2: Cantonal wastewater taxes.

Canton	Charge parameter	Charge level	Controlling and incentive objectives
Appenzell AR (charge levels of 1997)	Phosphorus	11.3 CHF/kg	Optimisation of phosphatisation at WTP
	Ammonium	16.7 CHF/kg	Incentive for WTP nitrification development
	COD	1.2 CHF/kg	Optimisation of WTP operation
	TUS	2.8 CHF/kg	Optimisation of WTP operation
Bern	Phosphorus	30.0 CHF/kg	Optimisation/development of phosphatisation at WTP
	Ammonium	4.0 CHF/kg	Incentive for WTP nitrification development
	Nitrate	1.0 CHF/kg	Optimisation of de-nitrification at WTP
	COD	0.7 CHF/kg	Optimisation of WTP operation
	Dry weather effluent	0.05 CHF/m^3	Foreign water elimination
Solothurn (provisional charge levels)	Phosphorus	25.0 CHF/kg	Optimisation of phosphatisation at WTP
	Nitrogen total	4.0 CHF/kg	Optimisation, poss. (partial) de-nitrification
	COD	2.0 CHF/kg	Optimisation of WTP operation
	Dry weather effluent	0.05 CHF/m^3	Foreign water elimination

WTP = Wastewater Treatment Plant
TUS = Total Undissolved Substances
COD = Chemical Oxygen Demand

3.2.2. The implementation of cantonal wastewater taxes

For analytical purposes, three phases of the implementation of wastewater taxes were distinguished in the study: a) initiation/internal administrative phase of policy formulation; b) parliamentary phase of policy formulation, consultation procedure; and c) the direct democratic phase.

Each of these three phases was approached using different methods of analysis. Based on interviews in six cantons, possible obstacles to the introduction of wastewater taxes were analysed, and solutions were sought to overcome these obstacles. The analysis of the parliamentary phase was complemented by checking the minutes of the *Grossrat* (Upper House) and the Commission, as well as the available results of the consultation procedures of the cantons investigated (see ECOPLAN, 2000a). The analysis of the direct democratic phase was based on telephone interviews with 1021 citizens in the Canton of Solothurn and on various econometric comparison estimates of voting patterns and structural variables (see ECOPLAN, 1999).

a) Initiation / internal administrative phase of policy formulation: The central driving force for initiation and implementation was the competent cantonal authority responsible for the implementation of the water resource protection policy. Wastewater taxes are a very complex instrument with multiple effects, requiring extensive consideration.

The motive for introducing wastewater taxes was financial rather than environmental. The discussion about cantonal wastewater taxes was characterised by the political request to reduce subsidies financed through general public means. Those responsible for water resource protection in the various cantons viewed wastewater taxes as an instrument for financing subsidies based on the polluter-pays principle (PPP). In Bern and Solothurn, wastewater taxes were incorporated into a package for the general stabilisation of state budgets.

Wastewater taxes – coupled with subsidies – always result in redistribution. In one canton, the wastewater taxes were abolished, as redistribution was too much of a disadvantage for one region.

The most frequent argument against the introduction of wastewater taxes was the assessment of the cost-benefit ratio, in particular with regard to implementation costs (measuring and monitoring expenditure) and because of low expectations of controlling effects.

b) Parliamentary phase of policy formulation, consultation procedure: Administrative and financial aspects triggered the most controversy. It is important to keep in mind that in all cantonal concepts, the revenue from wastewater taxes is tied to the financing of subsidies.

It was generally agreed that financing subsidies by means of wastewater taxes, rather than from the general public, was altogether a fairer solution. The argument against wastewater taxes claimed that general tax revenue saved in this way would not be passed on to households and businesses through lower taxation, and for this reason wastewater taxes would not be neutral with regard to the state quota. Objections were also made to the principle of tying revenue from wastewater taxes to specific purposes.

However, in this context it has to be said that a proposal that was not purpose-related would have had no chance of being implemented. Representatives of municipalities and associations viewed the introduction of wastewater taxes as cantonal interference with municipal responsibilities, and municipalities felt that their autonomy was being restricted.

From a socio-political viewpoint it was argued that wastewater taxes would disadvantage poorer households. In most cases, these arguments could be countered by pointing out that these additional burdens would be very small.

With regard to economic policy, it was argued that wastewater-intensive industries in particular (paper and food manufacturers) would be hit hard by wastewater taxes. A compromise was achieved by negotiating exemption rules.

Some environment policy arguments did not have a major influence on the decision for or against the introduction of wastewater taxes.

Table 3 summarises the major obstacles to be overcome in the introduction of wastewater taxes and their respective importance.

c) Direct democratic phase: Solothurn was the only canton where the introduction of wastewater taxes was decided by referendum. Only 51.1% of the electorate voted in favour.
The issue of state finance dominated discussions about wastewater taxes. Initially, financial policy arguments favoured wastewater charges, although quite different motives led to the referendum result: 32 percent voted in favour for environmental reasons (water resource protection, water saving, etc.). For 21 percent, the polluter-pays principle was the most important issue. Just 6 percent voted in favour of wastewater taxes for financial policy reasons. Important arguments against the proposal were that wastewater taxes would threaten the survival of companies and manufacturers and would be socially unacceptable as the burden on poorer households would be too high. This last argument against was also supported by the fact that most households with low incomes voted against wastewater taxes.

Table 3: Major obstacles to the implementation of cantonal wastewater taxes.

Obstacles in the Initiation / internal administrative phase	**Importance**
Complexity of wastewater taxes	+++
Little pressure promoting abolition of cantonal subsidies for water resource protection	+++
Redistribution effects of wastewater taxes	+
Insufficient advantages from wastewater taxes	+++
Obstacles in the parliamentary phase	**Importance**
Structure not state-quota neutral (additional revenue)	+++
General rejection of "fund economy" or purpose-binding	++
Wastewater taxes as "interference" with municipal responsibilities	+++
Socially controversial	+
Too much burden on industry and trade	+++

\+ = low importance
++ = medium importance
+++ = high importance

It is interesting to note that during the parliamentary phase, the frequently repeated argument against wastewater taxes (cantonal "interference" in municipal responsibilities) was of no importance for those voting against the proposal, and for those voting in favour it was actually an argument *for* wastewater taxes. Apparently, cantonal interference in municipal responsibilities for water resource protection is seen as a positive measure.

Conclusion: Findings for the successful initiation and preparation of cantonal wastewater taxes:

- The Water Pollution Control Authority needs to become motivated as the central activist for the "Wastewater Taxes" project.
- Apart from the cost, it is particularly important to identify and quantify the benefits of wastewater taxes.
- Targeted marketing of the proposals is required, by distributing existing fundamental information (studies, expert reports, press conferences, simplified literature, etc.) and direct contacts with the main activists in municipalities and associations.
- Ecologically positive effects of wastewater taxes need to be emphasised more strongly, even if the arguments in favour of the proposal are based on financial policy considerations.
- Distribution of responsibilities between cantons and municipalities needs to be clarified.
- Consequences for businesses and manufacturing plants need to be shown, and general exemption rules should be introduced.
- Social distribution effects need to be shown, and general "buffering measures" for poorer households should be introduced.
- The proposal's complexity needs to be reduced. This contradicts the notion of exemption rules and buffering measures.

3.3. *Effect of cantonal wastewater taxes*

Analyses of the pollutant-related wastewater taxes, as already in force in the Cantons of Appenzell/A.R. and Bern, have shown that wastewater taxes contribute to a sustainable development of water resource protection, mainly by promoting improved operation of treatment plants and resulting in a reduced burden on water resources. The effects of wastewater taxes can be summarised as follows (see ECOPLAN, 2000b):

- WTP operators are currently often "punished" by higher costs (e.g., for phosphorus elimination). Optimisation does not pay. Wastewater taxes remedy this situation and reward the well-operated plants. This results in optimised operation and maintenance.
- This optimisation is an important aspect with regard to the future development of wastewater treatment. With the technologies available today, levels below the legal limits can be achieved. Without wastewater taxes these avoidance opportunities would not be exploited. In the case of pollutant parameters where the achievement of levels below the legal limits would incur addi-

tional costs (e.g., phosphorus), the following situation would result: Without wastewater taxes, the new technology would cause an increase of water resource pollution, as it enables the plant to operate closer to the limit without exceeding it! This results in higher pollution on average, as compared with a less controllable plant. Only pollutant-related wastewater taxes can exploit the avoidance opportunities which, in some cases – phosphorus, for instance –, exceeds 20%.

- A clear incentive effect on the development of treatment plants was detected in the Canton of Appenzell Ausserrhoden. In view of relatively high taxes on ammonium, investment in nitrification was greatly accelerated.
- Wastewater taxes led to increased employment of professionals at wastewater treatment plants.
- The extent to which wastewater taxes can trigger technological innovations for wastewater treatment (dynamic innovation effects of wastewater taxes) remains to be seen.

3.4. Communication instruments to overcome obstacles against the impact and implementation of wastewater taxes

Three areas in particular would be suitable for the use of communication instruments:

- **Removal of obstacles to the implementation of wastewater taxes:** It is doubtful whether other cantons, apart from Appenzell A.R., Bern, and Solothurn, will introduce wastewater taxes. Active networking between cantonal water pollution control experts could increase the motivation of the competent authorities to introduce wastewater taxes. Wastewater taxes that have already been introduced could be utilised as pilot and model projects in order to demonstrate the costs and benefits of wastewater taxes.
- **Removal of obstacles to the effects of wastewater taxes:** Operators of wastewater treatment plants are often ignorant of feasible optimisation measures, or they lack the necessary knowledge and skills. Consulting services and education can remedy this situation. On the other hand, it should be carefully considered as to whether consulting might not prevent more practical outsourcing of operations.
- Communication instruments can also be used to **expand the scope of instruments** in areas where commands, controls, and wastewater taxes are insufficient. This applies to pollutants that are not, or are insufficiently, chemically degradable and to endocrine substances.

4. Municipal wastewater charges as a controlling and financing instrument

4.1. The structure of municipal wastewater charges

The municipalities are responsible for the enforcement of the polluter-pays principle. The following explains what has to be considered with regard to the structuring of municipal wastewater charges:

Who in Switzerland is entitled to apply the instrument of wastewater charges?
Wastewater charges are a form of remuneration for wastewater costs and are paid by those generating wastewater (households and businesses) to the service provider. In Switzerland this is generally the municipality ensuring wastewater disposal through municipal operations, with special financing.

How are charges collected, who pays the charges?
The charges are collected from households and businesses in accordance with the scope of service. The municipality is responsible for the collection and treatment of wastewater and for the drainage of precipitation, unless it is allowed to soak into the ground.

What are the objectives of wastewater charges?
Wastewater charges have to be economically efficient in order to ensure the financing of wastewater disposal based on the polluter-pays principle. The objectives can be summarised as follows:

- **Economic efficiency:** In our economy (even in markets with low competition, such as waste disposal) prices are the main controlling instruments for influencing consumers and manufacturers. At what levels should prices be set in order to send the "correct" signals to consumers of waste disposal services? Efficient pricing should be orientated at the marginal cost level. Therefore, the price paid for wastewater disposal should be in line with the costs for the disposal of an additional pollution unit.
- **Fairness/equality (same service – same price):** Sometimes it is claimed that approximately the same amount of pollutants is generated per capita and that everybody should pay the same price. This argument cannot be supported from an economic point of view. The critical factor is not the treatment service, but the costs incurred. One should state more precisely: "Same costs incurred – same charge to pay", with the objective of preventing cross-subsidisation between regions, communities, and households (e.g., between poorer and richer households) or between households and businesses.
- **Cost coverage:** Prices or charges should not only ensure efficiency, but also generate revenue to cover the costs for wastewater disposal.
- **Administrative costs:** With regard to wastewater disposal, one has to weigh up administrative costs and efficient pricing. The exact measuring of pollutant quantities at all wastewater

generating locations would not make sense, as the additional administrative costs (measuring, monitoring, etc.) would be far out of proportion to the additional benefits gained. A more complex charging system or additional administrative effort would only make sense if the increased accuracy of charges were to result in appropriate efficiency gains (lower costs in the long run, for example).

- **Transparency and Acceptance:** The pricing or charge system has to be transparent and comprehensible to the generators of wastewater. It has to be "fair" in order to meet with the required acceptance.

Where are the obstacles against the "ecological structuring" of wastewater charges?

The above discussion of objectives shows that up to now, wastewater charges have been employed as a municipal instrument with financial rather than environmental objectives. An "ecologically" structured wastewater charge system that aims to contribute to a more sustainable development in the field of water pollution control faces many obstacles:

- Municipal wastewater charges are not suitable for the internalisation of external costs, because external costs affect other municipalities in the form of spill-over. In this case, organisations at higher levels are required (obstacle: municipality as the wrong institutional level, or lack of engagement at higher levels, respectively).
- Comprehensive internalisation of external costs demands that wastewater charges exceed the treatment cost and that the assessments be based on water resource protection targets. However, the legal scope of action for this is limited, and comprehensive internalisation at a municipal level is not possible (obstacle: legal scope of action).
- Additional controlling effects aiming for a sustainable water pollution control policy can be expected from differentiated wastewater charges. However, the benefits of such additional controlling effects should not be overestimated, as the scope of action for polluters is already very limited by the federal regulations for indirect polluters (obstacle: existing regulation environment).
- These limited benefits would also incur high administrative costs for the employment of very expensive measuring technologies at individual locations (obstacle: costs of instruments out of proportion with potential benefits).
- It would be a mistake to demand a wider legal scope of action with regard to the internalisation of external costs at municipal levels. Instead, the cantons and the federal government should be requested to provide the necessary conditions by means of cantonal and federal wastewater taxes, in order to enable communities to contribute to the sustainable development of water pollution control.

The municipal task of introducing "ecological" wastewater charges is therefore limited to financing the disposal services and apportioning the costs to the direct polluters, both based on the polluter-pays principle.

What are the consequences of financing through wastewater charges based on the polluter-pays principle?

The following focuses on a secondary aspect of municipal charges policies that is currently receiving little attention in Switzerland and is also neglected in recommendations from associations and cantons to the municipalities:[11] The question is how fixed and variable wastewater disposal costs should be apportioned to direct polluters. Our analysis sought to determine what costs should be covered by volumetric pricing, and what costs via the flat fee (fixed charge). Frequently, the following argument is heard:

- Charges have to be based on today's average costs.
- Revenue from flat fees should more or less cover short-term "fixed" costs (capital costs in particular), whereas revenue from volumetric pricing should more or less cover "variable" costs (operation, energy, and routine maintenance in particular).

Such a charging structure would be economically inefficient and, in the long term, would not comply with the polluter-pays principle. Volumetric pricing in accordance with short-term "variable" costs would result in very low volumetric prices. As a consequence, high volumes of freshwater and pollutants would be fed into the treatment plant. Treatment plants undergoing renewal or development would have to be designed for this "higher" capacity, although this would not have been necessary if direct polluters had known in advance that high wastewater volumes would incur higher costs at a later stage. If direct polluters had been told in advance – by means of a volumetric price oriented to long-term cost levels – the "correct" price for wastewater disposal costs for the WTP renewal would have been lower. "Correct" pricing can save costs. Volumetric prices and flat fees should be set as follows:

- Volumetric pricing should be based on long-term cost limits and used mainly as a controlling instrument. Long-term cost levels depend on wastewater and pollutant volumes and should include the costs of the treatment plant (construction and operation), the costs of the pumping energy, and part of the costs for sewer network maintenance. The costs for sewer network construction and maintenance depend mainly on precipitation volumes. Wastewater volumes have little effect on the costs incurred by the sewer network.
- The flat fee serves the main purpose of cost coverage. It is therefore foremost a financing instrument, and of course there are options for the introduction of some controlling functions (to promote the ground absorption of precipitation, for instance).

[11] For a more detailed description of financing by means of wastewater charges based on the polluter-pays principle, see Müller (1999) and Gawel (1995).

4.2. Implementation of municipal wastewater charges

We have shown that municipal wastewater charges are to ensure financing according to the polluter-pays principle. Even this narrow task has not been accomplished up to now:

- In many cases the level of wastewater charges is not based on long-term cost limits. Apportioning effects of wastewater charges are frequently neglected or underrated (see the following section).
- Wastewater charges are more strongly oriented to historically developed structures. Although the disadvantages of charge changes are apparent, the advantages are difficult to communicate. The voice of potential losers is always louder than the voice of potential winners.

On the one hand, the lack of knowledge about the interaction between wastewater charges based on the polluter-pays principle needs to be remedied and, on the other hand, communications between authorities and direct polluters needs to be improved. This would increase both the acceptance of charge system changes and public awareness of the charges, thus creating the preconditions for appropriate action.

4.3. Effects of municipal wastewater charges

It is frequently doubted whether municipal wastewater charges have a controlling potential at all, with the argument that direct polluters have no alternative option and that the costs always remain the same regardless of charge levels. We must disagree with this:

- High water prices signal that disposal at certain locations is expensive and ties up many resources[12] and that for a company or a private investor a different location would be more attractive (allocation efficiency). For disposal companies it indicates that other options, such as decentralised wastewater treatment, should be considered. This optimisation with regard to direct polluters results in lower wastewater disposal costs for treatment providers as well as for the national economy.
- Municipal charges have another important controlling function with regard to capacity planning. If charges are too low, water consumption increases, and pollution can also rise. As shown in section 4.1., low-level charges lead to over-sized, and therefore costly, renewal or capacity development of treatment plants.

An econometric analysis (see ECOPLAN, 2000c) proves that direct polluters respond to higher charges by consuming less water and lower the risk of over-sized treatment plants undergoing

[12] For example: widespread settlements or regions with highly polluted water resources and, therefore, high demands for quality treatment.

renewal or capacity development. A cross section analysis of 140 municipalities in the Bern region established a demand flexibility of minus 1.5%. This means that if volumetric pricing is raised by 10%, then water consumption will fall by 1.5%. Even if we had underestimated the demand response (because of causality and data problems), it can still be seen that direct polluters respond to increased volumetric pricing with lower water consumption.

4.4. Communication and diffusion instruments to overcome obstacles to effects and implementation of wastewater charges

Communication and diffusion instruments should be employed mainly in two areas:

- **Removal of obstacles to the implementation of municipal wastewater charges:** The implementation of changes regarding the wastewater charge system has not been hampered by a lack of motivation. The problem lies rather in not knowing what the "right" thing to do is. This means that the main objective must be to remedy this lack of knowledge by providing facts, consulting services, and so on.
- **Removal of obstacles to the effects of municipal wastewater rates:** Here the municipalities face a dilemma: if wastewater volumes can be reduced by means of communication and diffusion instruments, then they have to raise charges, as the overall costs of wastewater disposal will hardly be affected in the short term. The most sensible measure would be to implement these instruments prior to renewal or capacity development of a treatment plant. Targeted campaigns (such as action campaigns promoting water-saving fittings) could be used in order to achieve real influence on the cost of development.

5. Summary and conclusions

We propose a three-phase concept for structuring economic instruments for wastewater taxes and municipal wastewater charges based consistently on the polluter-pays principle (PPP). In the first phase, the direct costs for wastewater disposal are covered by PPP-based wastewater charges (causal rates). In the second phase the costs carried by the state (contributions and administrative expenses) are covered mainly by wastewater taxes. Finally, the remaining consequential environmental costs (external costs) for unavoidable burdens on water resources (diminished leisure and recreation quality) are covered by wastewater taxes.

Wastewater charges as a municipal instrument are still employed as a financing instrument rather than as a means to protect the environment. Many obstacles prevent the ecological structuring of wastewater charges with the objective of contributing to sustainable development of water pollution control (limited controlling effects of wastewater charges; municipalities are the wrong institutional level for the internalisation of external costs). The evaluation of wastewater charges has shown that municipal responsibilities for ecologically structured wastewater charges should be limited to the PPP-based financing of disposal services and to the, also PPP-based, apportioning of cantonal levies imposed on direct polluters. Even this narrowly defined task is far from being fulfilled: municipal wastewater charges in particular are not based on long-term cost levels.

It can be concluded that in view of today's environment, the cantons and the federal government in particular should be requested to introduce cantonal or federal wastewater taxes in order to provide the conditions that would enable communities to contribute to sustainable development of water pollution control.

Wastewater taxes would be a sensible complement to today's range of instruments based on regulations. Wastewater taxes have positive effects with regard to the pollutants reaching our water resources. They create incentives for improved treatment plant operation. This is of particular importance in the future as new treatment plant technologies enable plants to operate closer to the upper limits without exceeding them. The result is higher pollution on average in comparison with a less controllable plant. Only wastewater taxes based on pollutant volumes can counterbalance this avoidance potential, which is more than 20% for some pollutants.

References

ECOPLAN (1993) Abwasserabgabe für die Schweiz. *In*: Schriftenreihe Umwelt Nr. 203, commissioned by SAEFL/Swiss Agency for the Environment, Forests and Landscape.

ECOPLAN (1996) Zukünftige Finanzierung der Gewässerschutzmassnahmen im Kanton Thurgau. Several working papers, commissioned by the Amt für Umweltschutz und Wasserwirtschaft of the canton of Thurgau.

ECOPLAN (1999) Abwasserabgabe im Kanton Solothurn: Analyse der Umfrage zur kantonalen Abstimmung vom 27. September 1998. Working paper in the framework of the Priority Programme Environment of the Swiss National Science Foundation.

ECOPLAN (2000a) Einführung kantonaler Abwasserabgaben: Chancen und Hindernisse. Working paper in the framework of the Priority Programme Environment of the Swiss National Science Foundation.

ECOPLAN (2000b) Die Wirkungen schmutzfrachtabhängiger Abwasserabgaben. Working paper in the framework of the Priority Programme Environment of the Swiss National Science Foundation.

ECOPLAN (2000c) Abwassermengenpreis und die Wirkung auf den Wasserverbrauch. Working paper in the framework of the Priority Programme Environment of the Swiss National Science Foundation.

Gawel, E. (1995) Ökologisch orientierte Entsorgungsgebühren. Ökonomische Analyse von Abfall- und Abwassergebühren als Mittel kommunaler Umweltpolitik. Erich Schmidt, Berlin.

Künzler and Partner AG and ECOPLAN (1995) Die Erfassung von Schmutzfrachten hinsichtlich der Abwasserabgabe nach dem Frachtmodell. Projekt der Kantone Solothurn, Bern, Zürich und St. Gallen.

Müller, A. (1999) Finanzierung der kommunalen Abwasserentsorgung aus ökonomischer Sicht. *Umweltrecht in der Praxis, Vereinigung für Umweltrecht, Vol. 13/6*: 509–538.

Wicke, L. (1986) Die ökologischen Milliarden. Kösel, München.

A Lawyer's Comment on Müller and Kramer

Ueli Friederich

The following brief comments discuss some issues from a legal perspective that arise in connection with the study by MÜLLER AND KRAMER. This short, legal overview was prepared only after the study was conducted. For that reason, it restricts itself to sketching out the most important problem areas and associated problem-solving approaches. It does not intend to serve as a detailed treatment of the study.

The polluter-pays principle as a conceptional guideline in the elaboration of sewage taxes and communal sewage rates is intrinsically oriented towards self-sufficient, autonomous individuals who behave in a rational and economically responsible manner. Instruments that are based on the notion of "causation equity", therefore, are qualified as economic instruments – a designation that also applies in the present contribution.

Financing the implementation of state activities on the basis of other equity *topoi* is conceivable. Thus, the basic principle underlying funding by tax revenue is not causation equity, but solidarity: all members of a community should collectively bear certain costs – as a rule, proportionate to their individual earning capacity – irrespective of whether or to what extent they may benefit from state facilities and services. The baseline in this context is not *homo oeconomicus*, but rather those individuals who, owing to their lack of means, are reliant on their fellow men and on the welfare of the community. Whether a specific state activity should be financed by imposing "causation-equitable" rates or, alternatively, by levying taxes based on a "unity of interests" is not determined simply as a matter of course, nor is this necessarily within the province of legal discretion, but is a political issue in each case.

Apart from essentially being an instrument for the implementation of a certain notion of fairness, a control function may also be attributed to the polluter-pays principle, in that (impending) financial consequences are intended to deter potential polluters from behaviour that is detrimental to the environment.

1. Purport and legal basis of the polluter-pays principle

In a broad sense, the polluter-pays principle responds to the question of who should defray the costs for certain measures. It constitutes a "cost distribution principle" and presupposes that specific state expenditures should be borne by a quantifiable group of individuals rather than by the general public as such. Environment law stipulates that the costs in connection with the prevention, elimination, or toleration of adverse impacts on the environment should not be borne by the affected parties or by the general public, but by the polluters. The polluter-pays principle is incorporated as a general principle in Article 2 of the Federal Law on Environment Protection of 7 October 1983, and has recently also been embodied in Article 3a of the Federal Law on Water Pollution Control of 24 January 1991.

As previously mentioned, the polluter-pays principle may also serve to deter certain types of behaviour; as opposed to injunctions and directives, however, it is not stated conclusively what measures are to be adopted or what types of behaviour are to be prevented.

2. The polluter-pays principle and the covering of costs

As a "cost distribution principle", the polluter-pays principle primarily addresses the issue of who is obliged to assume certain costs and, in this connection, also refers to the magnitude of the costs for which a predefined group of individuals are liable. Total cost recovery is set down as a standard; in other words, the comprehensive coverage of the costs related to the measure in question. Therefore, if revenue insufficiently covers public spending, a violation of the polluter-pays principle is incurred, in that the onus of liability is indiscriminately shifted to the general public irrespective of its behaviour.

The so-called "cost covering principle" also pertains to the scope of cost liability for attributable services. In its legally decisive form as a "total cost recovery principle", it implies that tax revenue that is earmarked for a particular purpose may not exceed the total operating costs of the relevant administrative department. Accordingly, the cost covering principle only narrowly regulates the basis of taxation. First, it only defines a tax ceiling and does not necessarily foresee total cost recovery; second, interdepartmental subsidisations within the radius of government operations are possible. The cost covering principle applies, for example, to so-called administrative fees, but not implicitly to other forms of revenue. Thus, an electric power plant may make a profit on power consumption rates.

From a logical viewpoint, the applicability of the polluter-pays principle is always a precondition for the cost covering principle, in that the latter foresees that specific costs are to be assumed by those groups of individuals that have directly generated the need for the cost-related activities of the administrative department in question.

3. Attributable costs

The polluter-pays principle is based solely on specific, tangible costs defined by law, i.e., costs incurred "for measures pursuant to this law" (Article 2 of the Law on Environment Protection; Article 3a of the Law on Water Pollution Control), or – more concretely in connection with (communal) sewage rates – on "the costs for the construction, operation, maintenance, renovation and replacement of sewage installations which serve a public purpose" (Article 60a Paragraph 1 of the Law on Water Pollution Control). Therefore, it does not allow for the internalisation of so-called external costs, in any case not to an undetermined extent. Moreover, the applicable law does not provide a basis for the global enforcement of the "polluter-pays principle in a broader sense" as referred to in the present contribution. The diminished recreational value of polluted water as such is only then taken into account if, as a consequence, public expenditure is incurred.

4. The polluter-pays principle as a basis of taxation

In extension to its form and content as a "cost distribution principle", the polluter-pays principle may also serve in some instances as a guideline in establishing a basis of taxation, in that individual polluters are required to share costs in proportion to their actual causation of them. In other words, fees, charges, and other forms of taxation are to be structured and accordingly levied in line with causation equity.

The polluter-pays principle thereby modifies the "softness" of the cost covering principle, which only takes the total revenue (fees, charges) and the total operating cost of an administrative department into account. As an illustration, the polluter-pays principle rules out the financing of waste disposal solely via basic rates that are not scaled to reflect the level of consumption. In practice, the decision has been taken in the past to separately determine the sectors of taxable polluters for a contribution to the construction costs of a particular installation on the one hand, and for differential user rates on the other hand (a procedure that, upon closer inspection, is questionable).

5. The temporal scope of application of the polluter-pays principle

With regard to the significance of the long-term costs involved in sewage disposal as touched upon in the present contribution, it should be noted that the temporal aspect of the polluter-pays principle foresees to the greatest possible extent an evenly distributed polluter liability. This owes to the premiss that all polluters contribute in basically the same manner to eventually necessitating the construction, renovation, or replacement of installations. The temporal contiguousness to the financial investments required for these works is, under certain circumstances, purely coincidental and does not constitute any definitive inequality, which must be borne in mind in light of the precept of equality before the law. Hence, provision is fundamentally made for the equal treatment of current and future polluters.

With regard to the past, however, precisely this precept of equality before the law has been of limited validity. In the event of former "sins of omission", the burden of assuming the costs for "remedying the situation" would – once again, according to the polluter-pays principle – necessarily (also) have to be placed upon the then offending polluters. Apart from the practical obstacles entailed in this approach (e.g., tracing those liable to pay tax or their legal successors), legal provisions would, as a rule, preclude its feasibility, in that the belated taxing of former polluters would be tantamount to inadmissibly enforcing the law with retroactive effect. Accordingly, the only viable option is to impose liability on both current and future polluters.

Since its revision in 1997, the Law on Water Pollution Control has included the binding directive that both prospective and retrospective measures shall be taken into account in the structuring of sewage rates. Moreover, in determining these rates, the necessary write-offs for the preservation of substance, interest, and the projected need for capital expenditure for the maintenance, renovation, and replacement of the installation as well as for internal restructuring to meet legal requirements or operational optimisation shall also be taken into account (Article 60a Paragraph 1). The investment demand is primarily ascertained in line with what investments have been made or have failed to be made in the past. Accordingly, past "sins of omission" are to be rectified by levying commensurate rates. Financing "subsequent improvements" by taxation measures is only permissible as an exception, namely if rates which are structured to cover costs and reflect the level of consumption should jeopardise an ecologically sound disposal of sewage (Article 60a Paragraph 2).

The legal boundary for rectifying "sins of omission" or for financing future projects has been reached in instances where higher taxes or other charges are imposed in order to compensate for previous deficient cover, thus violating the principle of equivalency. In other words, if – from the standpoint of those liable to pay tax – the tax levied was obviously disproportionate to the actual value of the service to be compensated by the tax. In such cases, the conditions pursuant to Article 60a Paragraph 2 of the Law on Water Pollution Control would presumably also be fulfilled.

6. The need to concretise the polluter-pays principle

Although the polluter-pays principle is peremptory for sewage rates, it nevertheless requires concretisation to a large degree. The Law on Water Pollution Control serves to concretise the overriding principle in Article 60a, but does not, however, provide a definitive taxation structuring directive for sewage disposal. Within the scope of this law, the cantons and municipalities are allowed considerable flexibility with regard to structuring charges into basic or flat rates and rates that reflect the actual consumption volume. In simplified terms, it may be said that the polluter-pays principle in its form and content as a cost distribution principle (who pays for what?) is essentially binding.

However, as a basis of taxation for the structuring of rates and other taxes scaled in line with causation equity (who among those liable to pay is obliged to pay for which proportion?) it leaves cantonal and municipal legislators considerable leeway.

Within the framework of concretising the polluter-pays principle, legislators must respect the general constitutional principles relevant to the levying of taxes, such as the exigency of a legal basis, the precept of isonomy, the unlawfulness of arbitrary rule as well as, under certain circumstances, the cost covering principle and the principle of equivalency.

Changing Things – Moving People.
Strategies for Promoting Sustainable
Development at the Local Level
R. Kaufmann-Hayoz/H. Gutscher (eds.)

Mobilising Resources for More Sustainable Lifestyles: Views of Households and Local Authorities

Susanne Bruppacher and Silvia Ulli-Beer

This contribution examines the strategy taken by Global Action Plan (GAP) to promote and support the development of environmentally sustainable lifestyles by combining various communication and diffusion instruments. Potentials and effects of the instruments applied are discussed with regard to co-operation between municipalities and non-governmental organisations (NGOs) like GAP. Two studies combine a bottom up and top down perspective: The system dynamics perspective on the municipality is complemented by a study of the viewpoints of the individual citizen and household. Results indicate that a combination of communication instruments with other policy instruments, such as economic and command and control instruments, is well suited to lead a municipality towards sustainability by improving the preconditions for environmentally sound citizen behaviour. The focus is on the ecological and social dimensions of sustainability. Due to their common relevance for overall life goals and quality of life, our findings further suggest that these two dimensions can sometimes be improved by one and the same measure, e.g., in traffic safety.

After a short overview of the Global Action Plan environmental programme and the actors involved in it (section 2), the methods used in the two studies are briefly described (section 3). In the fourth section, the potential and effectiveness of the GAP strategy is discussed with special reference to both the municipality and the household levels. Finally, the fifth section concludes with special requirements that are to be met when implementing a strategy like GAP, as well as further implications for municipal and national environmental policies.

1. Introduction

Statistics provided by the Swiss Federal Statistics Office (SFSO) and the Swiss Agency for the Environment, Forests and Landscape (SAEFL) (SFSO and SAEFL, 1997) show the importance of private households in energy and water consumption, transport, and waste production. A summary is provided in Table 1.

Table 1: Private households as the consumption leader (adapted from SFSO and SAEFL, 1997).

Energy	Transport	Water/Waste water	Waste
Households are the most important user group with 31%!	About 63% of passenger kilometres are travelled for leisure and shopping purposes!	Households and small businesses consume the largest share of drinking water (58%)!	The largest share of waste is municipal waste (4.3 million tons per year)!

Table 1 illustrates the importance of private households in energy and water consumption, transport, and waste production.

Taking it as a given that participants of the GAP EcoTeam programme are highly motivated to change their behaviour in a sustainable direction, this sample is particularly appropriate for the testing of two questions. First, do certain measures of municipalities really facilitate environmentally responsible behaviour of their citizens? Second, where are there still too many obstacles, and where are barriers so high that even GAP participants do not succeed? (see also Gessner, 1996, and the chapter by VATTER ET AL.).

2. The GAP strategy to promote sustainable lifestyles[1]

2.1. *The environmental action programmes of the non-profit organisation Global Action Plan (GAP) in Switzerland*

GAP, an international environmental NGO, was founded in the United States and is now active in several Western European countries, including Switzerland since 1993. Its objective is to support the development of sustainable lifestyles in as many municipalities as possible. Several action programmes have been developed for municipalities, households, schools, and enterprises. They propose concrete practical steps to preserve the environment and, in many cases, to save money at the same time.

1 See also the study by MAUCH ET AL. The authors discuss the GAP strategy in comparison to other programmes in the mobility sector and with regard to Swiss environmental policy.

Households or individuals who decide to participate in the GAP programme for private households form "EcoTeams" of five to eight households and use a **workbook** with about 90 suggested practical actions in order to adopt more environmentally sustainable behaviour. It takes approximately six months to work through the programme, each month being dedicated to a particular area: garbage, water, energy, transportation, consumption, and handing on[2]. Participants choose the suggestions that they consider reasonable and suitable for implementation in their own households. The teams are offered **counselling** by GAP. A lifestyle questionnaire, filled out before and after the programme, evaluates improvements and makes them transparent to the participants. Moreover, GAP also provides information to its members and active participants in the form of a newsletter, which summarises the Swiss data in international comparison.[3]

According to GAP statistics, more than 1,000 Swiss households have completed the programme. It attracts above all families with school children in suburban areas (Graf, 1997). Känel et al. (1998) found significant improvements in the garbage and consumption domains and – to some extent – in water conservation. However, most of the suggested practical actions concerning transportation and energy conservation were not followed. These two areas seem to be most resistant to behaviour change (see also the chapter by MAUCH ET AL.). Furthermore, it appears that the programme mainly attracts people who have already adopted quite sustainable lifestyles compared to the average population. This was also found in Dutch studies (Harland, in press). These studies indicate that the effects achieved by the GAP programme are lasting, and in some areas, the effects even continue to improve two years after participation.

However, GAP's recruitment strategy of having team initiators personally ask neighbours or friends to join an EcoTeam has proved only moderately effective in Switzerland and in the Netherlands. People do not seem to be willing to act as "missionaries" in order to spread the programme (Känel et al., 1998; Harland, in press). It seems therefore sensible to place the focus on municipalities that are trying to get involved in Local Agenda 21, the United Nations plan for communities. Ideally municipalities provide support for new teams through an employee trained by GAP as team counsellor. Several municipalities in the German-speaking part of Switzerland[4] have succeeded in starting up new teams.

---

2 "Handing on", or "empowerment", aims to enable the team to help others (friends, neighbours) take action by forming new teams. In this way, the sustainable lifestyle practices will spread.

3 Due to problems within the administration of GAP Switzerland, the Swiss newsletter was published only until the fall of 1999.

4 Ittigen b. Bern, Radelfingen-Dettligen, St. Gallen, Therwil.

2.2. *The actors*

The focus of the present contribution lies on four actors at different levels:

- **The private household:** The GAP programme focuses on the household level and on individual behaviour. Even if only one household member actively participates in the GAP team and attends the meetings, this affects the household as a whole.
- The **GAP EcoTeam** consisting of several households and a team initiator: Working through the programme as a team has several advantages (see Tables 2, 3, and 4). The team provides material and inspirational support and includes the element of self-commitment.
- **The municipality:** A municipal partnership with GAP has three important effects: Municipal support makes the programme an official policy and, therefore, residents take it more seriously. Participation of the local authorities in a pilot team sets an example and serves as a model. Finally, the municipal authorities take on the role of multipliers.
- **The national GAP co-ordination** helps municipalities to start up new teams, provides the workbook, publishes the newsletter, trains team initiators, and collects and evaluates statistical team data for feedback purposes.

3. Methods

3.1. *Approach via the individual:Qualitative interviews*

Qualitative interviews were conducted on people's perceptions of their community lives, their social networks, and their personal behaviour, e.g., their investment behaviour.[5] The sample consisted of 44 individuals[6] in 29 households in 5 municipalities[7] in the German-speaking part of Switzerland. One or more members of 18 GAP households were interviewed before and after participation in the programme. A control group of 15 adults from 11 households was interviewed just once. The first interviews took place when the GAP households had just begun participation in the programme (March-July 1999). After completion of the programme, the adult GAP participants (N=24) were interviewed a second time (January-March 2000).

Interviewees were between 7 and 56 years old (mean 39). Their political orientation can be characterised as neither very conservative nor extremely progressive. GAP participants are slightly more socially oriented than the control groups.

The in-depth interviews were analysed by qualitative content analysis (Mayring, 1995).

---

5 Because most of the questions were very personal in nature, a qualitative approach with open questions and face-to-face interviews was chosen.

6 40 adults, 4 adolescents and children.

7 Gipf-Oberfrick AG, Radelfingen-Dettligen BE, St. Gallen SG, Therwil BL, Würenlos ZH.

3.2. *Approach via the experts: Group model building*

To investigate the municipalities we chose the method of group model building (Vennix, 1994). Its backbone is the systems theory concept (Forrester, 1968), which focuses on the causalities of a specific problem and its environment. It assumes that the behaviour of a system is a consequence of structural system characteristics of problems. In our study we focused on the influences of municipal "structures" interacting with the behaviour of inhabitants.

Three group model-building sessions were held to map the problematic causal chains in the municipality. Ten experts from various fields (local government, local administration, consulting firms serving the municipality in the areas of energy, transportation, water/wastewater, waste, and consumption) shared their mental models about problems of environmental policy in order to explain the main impediments to change for citizens. The aim was to create among the experts a shared reality of the causes and consequences of problematic behaviour of the inhabitants. Based on the subjective models, an overall model for responsible environmental behaviour (Model-REB, Ulli-Beer, 2000) was synthesised.

There is evidence that this transdisciplinary model building procedure can be an important component of the policy-making process. The multiple perspectives on the public interest and the political agenda are merged to a consensus. In accordance with post-positivist theory, the researchers establish a participatory relationship with the political agents and the experts in the municipalities. Decision-makers are interested in elucidating and understanding the internal dynamics of overall system processes.

4. Results: potential and actual effects of the applied instruments

4.1. *The GAP strategy as a combination of communication and diffusion instruments*

This sub-section comments on potential effects of the strategy applied by GAP with regard to the typology of policy instruments on the basis of the interview study (Bruppacher, 2000). The strategy of the GAP programme is based on a combination of the three types of communication instruments and – to a lesser extent – service and infrastructure instruments. For an overview of the applied communication instruments see Tables 2, 3, and 4. In the following, examples and specific results of the interview study concerning single instruments or combinations of instruments are presented. Instruments are printed in bold.

Table 2: Communication instruments without direct requests: Potential and actual effects of the GAP strategy with regard to the typology of policy instruments for sustainable development.

Communication instruments without direct requests	Effects of the programme (workbook)	Team effects (monthly meetings)	Potential of municipal support
Presenting facts and options	Workbook provides background information and provides practical action suggestions. Team initiators inform about important news on the market and stress the main points of the monthly meeting.	Teamwork fosters the elaboration of knowledge. Each meeting is led by a different team member who takes extra effort in presenting the topic area.	Counselling, e.g., when the GAP team initiator is a member of the environmental department.
Presenting appraisals, goals and norms	Workbook provides model behavior. Team initiators inform team about experiences of other teams.	Participants are responsible for themselves, but model persons do set norms. Feeling supported by the team can encourage behaviour.	Municipal authorities can set norms by model behaviour. Feeling supported by the municipality encourages behaviour, above all, investment behaviour.
Providing experiences of reality	Encourages new behaviours as suggested in the workbook.	Mutual exchange of experiences; events / demonstrations organised by the team during the meetings; team members show their own equipment, e.g., their solar panels or heating systems.	Promotion of municipal demonstration projects, e.g., installation of environmentally sound heating systems or solar panels on official buildings.
Presenting model behaviour	Workbook describes model behaviour.	People on the GAP team as model persons.	Municipal authorities as model: e.g., installing an alternative heating system in the town hall.
Giving feedback and enabling self-feedback	Feedback through newsletter. Self-feedback through lifestyle check and measuring resource use.	Feedback during discussions at team meetings.	-

Table 2 gives an overview of the communication instruments without direct requests that are applied within the GAP strategy and summarises potential and actual effects of this strategy.

Table 3: Communication instruments with direct requests: Potential and actual effects of the GAP strategy with regard to the typology of policy instruments for sustainable development.

Communication instruments with direct requests	Effects of the programme (workbook)	Team effects (monthly meetings)	Potential of municipal support
Persuading about facts	Workbook provides background information.	Teamwork fosters the elaboration of knowledge.	Municipal support can increase credibility of programme.
Persuading about possibilities	Suggestions provided. GAP offers a course on becoming a team initiator.	Reinforced through discussions. Team members participate in courses together, can give courses if they are specialised in an area.	Municipal support can increase credibility of programme suggestions. Support of people who give courses, e.g., in organic gardening.
Persuading about appraisals, goals, and norms	Knowledge and possible steps of action in their own households confront participants with contradictions between their own behaviour and values.	Reinforced through discussions at the team meetings. Personal contacts among team members may persuade others to join in the activities.	Municipal support can increase credibility of programme.
Sending appeals	Suggestions, no moral appeals.	Appeals can but do not necessarily take place.	-
Presenting prompts and reminders	Suggestions provided.	Encounters with other participants have the effect of a reminder.	Municipal authorities can support the installation of prompts and reminders by granting permission and providing locations.
Stimulating self-commitment	Commitment in front of the team and the organisation.	Public self-commitment is more effective than private.	Commitment extends to the municipality. Offerings and subsidies stimulate commitment and action.

Table 3 gives an overview of the communication instruments with direct request that are applied within the GAP strategy and summarises potential and actual effects of this strategy.

Table 4: Diffusion instruments: Potential and actual effects of the GAP-strategy with regard to the typology of policy instruments for sustainable development.

Diffusion instruments	Effects of the programmme (workbook)	Team effects (monthly meetings)	Potential of municipal support
Establishing direct personal contact	Knowledge transfer to family and friends, "handing on", starting up a team creates networks of households within municipalities.	Personal contacts among team members may persuade others to join in the activities.	GAP teams can initiate participatory processes regarding problems in a municipality. GAP teams that include members of municipal authorities strengthen communication processes between citizens and authorities.
Establishing contact via person to person media	-	The "handing on" procedure of the GAP acts as small scale information campaign.	Municipals can support GAP teams to start an information campaign.
Establishing contact via mass media	Workbook as "container" of several communication instruments.	Some GAP teams publish their experiences in local newspapers.	GAP teams can launch an information campaign.

Table 4 gives an overview of the diffusion instruments that are applied within the GAP strategy and summarises potential and actual effects of this strategy.

In the GAP workbook and the packet of materials for GAP team initiators, presenting facts and options, giving feedback, enabling self-feedback, and stimulating self-commitment are stressed as main strategies. Instead of appealing to people only on moral grounds, the workbook provides simple instructions and practical suggestions that lead to visible results, and members are encouraged to support each other and to share their experiences with the suggested practical actions. **Providing experience of reality** within the GAP strategy also means to create positive associations to the team experiences. GAP stresses the fact that environmentally sound behaviour can be fun and has conceptualised the programme accordingly. Each meeting is organised by a different team member who is invited to present the material in an informative but special and entertaining way. Of course the actual impact of entertainment depends on the team members, but most participants in this sample pointed out that the team meetings were very entertaining and motivated them to take further steps in the programme as well as to keep in contact with other members of the team.

Our results show that also further instruments are applied and that they are of considerable importance when partner municipalities co-operate with the NGO and the participants (Bruppacher, 2000). **Presenting facts and options, persuading about facts and options, presenting model behaviour** and **providing experience of reality** are not only important within the EcoTeam, where the members exchange and copy successful solutions for their own households and demonstrate their own household equipment to each other. Municipal authorities take on an important role as model persons. In one municipality[8] a heat pump was installed in the town hall (a pilot and demonstration project), and it has become very well known among the residents. In addition, people building a house in the municipality are subsidised if they install a heat pump in their own homes. Two households in the sample of this study living in this municipality took advantage of the opportunity and reported that they were very motivated to undertake further steps towards sustainable household management. Demonstration projects and their promotion are an effective way for a municipality to **present** and **persuade about appraisals, goals, and norms. Providing experience of reality** fills technological knowledge gaps and fosters environmentally sound investments.

Persuasion about facts and options can be reinforced through the team. Three of the five teams investigated had one or more team members who have special knowledge in a particular field of action. An energy expert, for example, provided important knowledge that aided the installation of a solar panel on the roof of another team member's home.

In one case, participation produced an important side effect concerning **persuasion about options**. Of the 18 households of the GAP sample, one was about to build a house together with three other parties that had no contact with the programme. The participant reported that he would have realised many of the investments and installations whether or not he participated in the programme, but he probably would not have been motivated to convince the three other parties to install a heat pump as well.

Participants get **feedback** on the achievement of Swiss GAP teams in a world-wide comparison through a newsletter. In addition, they evaluate their own achievements by means of a lifestyle questionnaire (**enabling self-feedback**). Suggestions on how to practice self-management in order to make personal environmentally responsible behaviour more manageable are mentioned in the workbook and discussed at team meetings (such as the use of prompts). The effects of feedback and self-feedback within the GAP programme have been investigated extensively in the Netherlands by Weenig et al. (1994). They found that the feedback as provided in the GAP programme leads "to the notion that many small efforts lead to large effects and give participants an impression of the effectiveness and usefulness of the program" (Weenig et al., 1994: 16).

Particularly when members of a GAP EcoTeam live in close proximity – an important factor according to the participants – **presenting and persuading about norms** in combination with several other instruments takes place. First, a certain commitment towards the other team members to behave in a sustainable way arises quite automatically when members discuss possible steps of actions. Second, people are confronted with contradictions between their values and their behaviour, which enforces **persuasion about appraisals and norms**. Third, encountering other team members when shopping or on the street after the programme has been concluded serves as a **re-**

[8] Gipf-Oberfrick

minder. All three instruments seem to increase people's willingness to take action, and residential closeness fosters a sustainable programme effect.

As far as the application of **diffusion instruments** is concerned, there is certainly room (and need!) for improvement. Participants were not satisfied with either the amount of support from GAP or from the municipality. GAP Switzerland has had too many changes of staff in the recent years. Therefore, there was a great deal of uncertainty in all the municipalities under study due to communication difficulties with the NGO. As a result, the "handing on" or "empowerment" part of the programme did not take place as planned. The municipalities could not execute their role as promoters because of the NGO's lack of professionalism.

Establishing direct personal contact is especially essential at the household level. If practical steps towards sustainability are to be taken, they usually affect the entire household and not only household members participating in the GAP meetings. Participants view it as essential that the whole household is informed about the programme. Actions and solutions are to be discussed to prevent arguments that lead to inaction and dissatisfaction among all household members. But knowledge transfer to friends also occurs. In one municipality[9] participation in the programme resulted in the forming of a municipal traffic group in order to solve problems of traffic safety in the municipality, especially the problem of children's safety on their way to school. In this case, an **infrastructure instrument** has also been applied, in that certain dangerous crossings have been modified. The main reason was a social one, but an environmental effect will also be achieved.

Establishing direct personal contact as means to integrate oneself in a municipality motivates participants to go through the programme, particularly people who are relatively new to the area. In two municipalities, questionnaire respondents mentioned a cleft between people born and raised in the municipality (old-established residents) and "newcomers". Participating in a GAP EcoTeam seems to be a good and effective opportunity to get in contact with neighbours or the municipality on a personal basis, and therefore two goals – environmental protection and social integration – can be achieved at the same time. At the level of the municipalities, networks between both municipal authorities and residents and between municipalities and NGOs can be established, but in the present situation unfortunately, the latter has not been the case. For more information about the effectiveness of network management, see the study by BÄTTIG AND BALTHASAR.

Establishing contact via person to person media is an option for handing on the GAP programme. However, direct personal contact has proven to be more effective (Graf, 1997; Harland, in press; Market Street Research, 1996; Vatter, 2000).

One municipality[10] – instead of handing on the programme to further teams – started an **information campaign via mass media**. Each team member – in this case members of the municipal environment and energy commission – published an article on one of the five areas of action (garbage, water, energy, transportation, consumption) in the local newspaper.

---

9 Gipf-Oberfrick

10 Therwil

The following tables (Tables 2, 3, and 4) give an overview of the communication instruments applied. The column to the very right is dedicated to the potential of the strategy if the municipality supports the GAP programme. Experience shows that an effective way to become a partner municipality is when members of the municipal authorities (such as energy committee members) follow the programme themselves and start up a pilot GAP team in the municipality.

4.2. Effects of the GAP programme on people's behaviour

In the 18 GAP households studied a total of 116 practical action suggestions were carried out. Energy was the area where the most action suggestions were realised (34), followed by water (29) and waste (25). Figure 1 shows the distribution of the suggestions carried out according to action areas:

The action suggestions concern three different types of action:

1. Actions that involve a one-time effort, such as installing something in the home or taking a course to gain know-how. After this one-time effort, everyday life continues without major changes.
2. Actions that become a habit, because they have to be performed on a regular basis, such as turning off the light when leaving a room.
3. Actions that involve changes in the lifestyle of a person or household, e.g., using public transportation in leisure time instead of one's private motor vehicle.

Most of the action suggestions carried out belong to the first two types. While nearly all of the type one suggestions (e.g., installing water-saving devices in taps and showerheads) were carried out because members acquired knowledge on their relevance and technological options, type two actions were performed simply because attention was drawn to an already familiar specific behaviour or habit (such as switching off lights, avoiding wasteful packaging when shopping). For type three actions, it was mostly the motivation and stimulation effect that led to action and rarely the acquired knowledge. Most of the action suggestions carried out in the area of transport belong to type three, while most of the action suggestions in the areas of water and energy conservation are of type one or two.

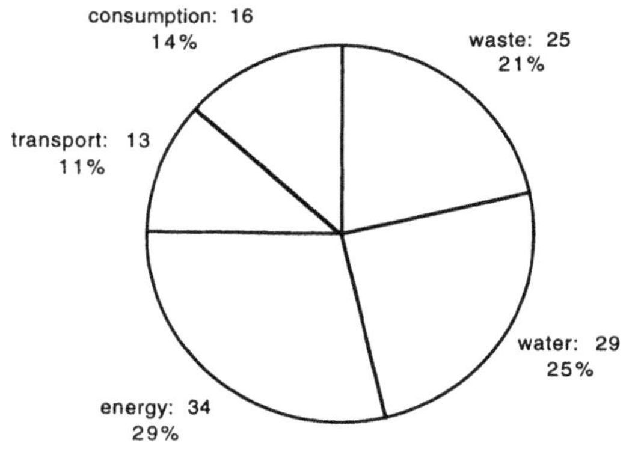

Figure 1: Total number of suggestions for practical action carried out per action area.

Compared to previous Swiss studies (Graf, 1997; Känel et al., 1998), the results of the interview study (Bruppacher, 2000) show some improvements of programme effects in the area of energy conservation. Participants readily implement changes that incur only small additional expenditures of time, money, and personal effort, and if they are one-time investments or efforts. As far as transportation and consumption are concerned, however, the picture has remained the same. Suggestions that most participants failed to carry out still involve one or more of the following problems: rather costly investments and considerable sacrifice of time or comfort, which can affect well-being; impediments because of inadequate or missing infrastructure; difficulties with gaining the co-operation of other people who have little or no ecological interest.

4.3. A system dynamics point of view of the municipality: Key factors for responsible environmental behaviour

Based on the mental models developed by the participating experts and on research findings we have synthesised ten clusters of factors that are main influences on the individuals' intentions to act in more or less environmentally responsible ways (see Table 5). There are five important clusters of external factors that influence the actors' surroundings and their perceptions of the situation, and there are five important clusters of internal factors that influence personal attributes.

Table 5: Internal and external factors influencing individual environmentally responsible behaviour.

A)	Aspects of the surrounding and perceptions of the situation	B)	Aspects of personal attributes
1)	"Social and political integration"	6)	"Understanding of inter-relationships"
2)	"Opportunities to act"	7)	"Personal audit"
3)	"Costs of living"	8)	"Convenience, laziness"
4)	"Capital expenditure"	9)	"Habits"
5)	"Social and legal norms"	10)	"Overall orientations" (moral, values, environmental concern)

Table 5 shows internal and external factors influencing individual environmentally responsible behaviour. The factors are shown in quotation marks, which indicates that they are interpreted as key-variables of the system.

Figure 2 presents the qualitative model that includes the ten important clusters of factors influencing intention and their connections to the two subsystems of household consumption patterns and household metabolism. Perceptions of the situation in the two subsystems (household metabolism and consumption patterns) again have an influence on intentions to act in a more or less environmentally responsible way. The perception of options and constraints influences consumption patterns that in turn influence household metabolism.

The intention to put more environmentally responsible behaviour into practice can be formed or can be reinforced (for a theoretical explanation of ways to induce behaviour change, see the contribution by FLURY-KLEUBLER AND GUTSCHER and Figure 1 in the chapter by KAUFMANN-HAYOZ AND GUTSCHER in the Introduction section, page 23). As we can see in the model presented in Figure 2, the intention to act in a more environmentally responsible way is regarded as a key factor in changing consumption patterns and subsequently decreasing household metabolism.

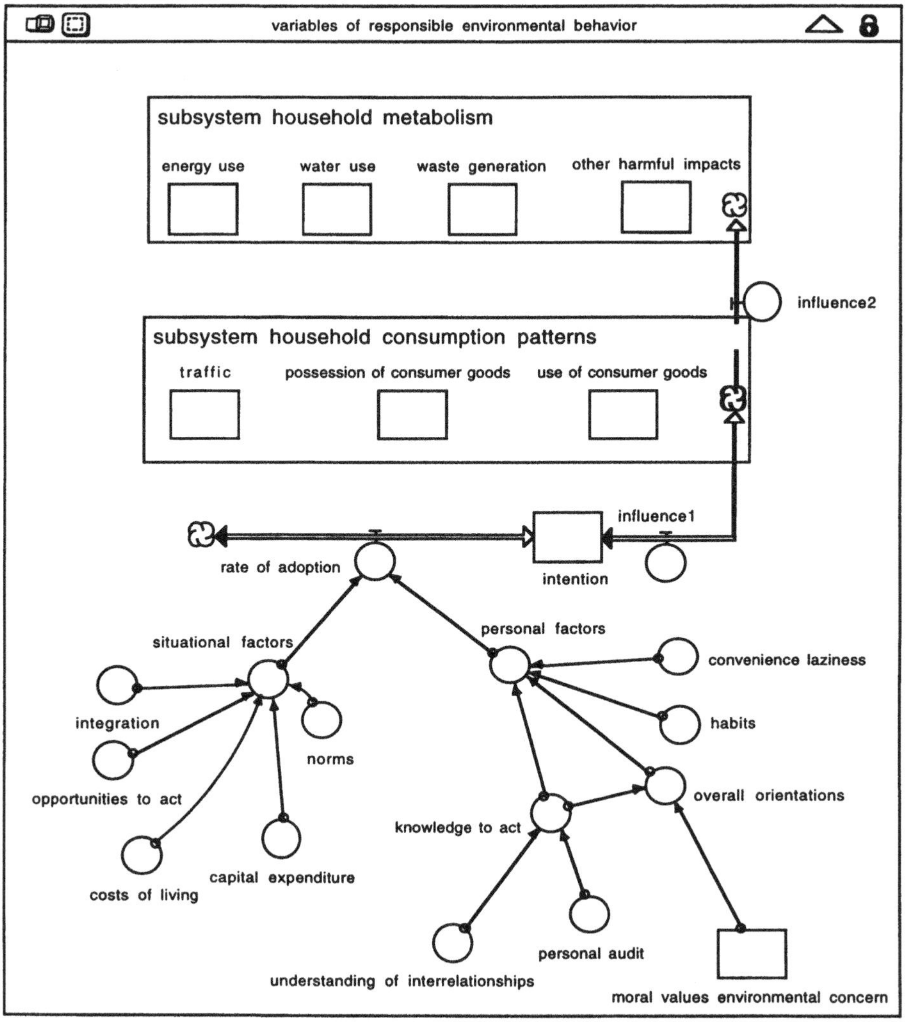

Figure 2: Variables of responsible environmental behaviour: Model REB.

This model can serve to guide different political measures. The clusters of factors can be entry points for political interventions in order to change consumption patterns in a more sustainable direction, as shown in Figure 2. The groups of factors can be interpreted in two ways:

On the one hand, they can be interpreted as ...	On the other hand, they can be interpreted as ...
• groups of variables or factors influencing the everyday management of households, leading to **unsustainable** lifestyles; • groups of barriers to responsible environmental action.	• groups of variables or factors influencing the every day management of households, leading to **more sustainable** lifestyles; • groups of resources, catalysts for responsible environmenta action.

The main challenge to politicians is to develop political strategies and implement measures and instruments that create and modify structures in the municipalities in such a way that the factors will mobilise the resources and act as catalysts for more responsible actions and lifestyles. The GAP campaign in municipalities can help to bridge the gap between environmental concern and environmentally responsible behaviour. Besides the impacts discussed in the previous sections it can help to diminish and solve typical common pool problems as reported by Diekmann and Franzen (1995).

For some of the factors discussed above there are complementary results from the interview study that focused on the point of view of the citizens. We will first discuss the factor of integration and then shed some light on overall orientation at the concrete level of quality of life in a municipality. Finally, we will give an example of creating opportunities to act.

4.3.1. Complementary views from the citizens' point of view: Social support and integration

Previous studies (Graf, 1997; Känel et al., 1998) suggest that social networks influence performance in the GAP programme considerably in both negative and positive ways. Models, support from the social environment, and the team aspect of the programme (co-operation, sharing of experiences, sense of community) are perceived to foster the success of the programme. Negative attitudes toward action suggestions on the part of family members, fear of social control, and loss of prestige by exposing oneself as "green" represent obstacles. These findings are confirmed by the present study. The support of the municipality, i.e., the municipal authorities' activity level on environmental issues is very relevant for GAP-participants. At both interview times the problem of "self-exposure as green" was frequently mentioned spontaneously. Participants do not want to take on an image as environmental missionaries. Feeling supported by the municipality counteracts the fear of becoming a social outcast when standing up for environmental issues, because the programme has received an official label of support by the municipality.

Yet the influence of the social network on the ecological performance is only one dimension of influence. The programme also affects the social network. Participants pointed out the combined effect of the programme on ecology and their social networks. Most of them reported that they had formed new friendships during the programme and intensified social bonds with other team members. People who had just recently moved to the neighbourhood in particular reported that they profited immensely from the opportunity to build a social network. The special circumstance that two of the investigated municipalities are very rural has to be considered when interpreting these results: In such municipalities, people from the outside face a relatively settled community life that makes it difficult for them to integrate.

As far as the question of integration in a municipality is concerned, residents and local authorities face the same problems under different perspectives. While private actors are concerned about their own integration, local authorities are concerned about overall integration. Their comments draw special attention to the integration of people from different cultures. This kind of integration is also mentioned by the interviewees of the qualitative interview study, but they are even more concerned about the integration of Swiss people moving into their municipalities, and, in rural areas, also about the integration of political conservatives and progressively oriented people for better co-operation in municipal matters.

4.3.2. Complementary views from the citizens' point of view: Overall orientation

Where do citizens see a need for municipal intervention in order to improve the quality of life in their municipality?

When asked what measures for improving quality of life[11] they would implement if they were municipal politicians, 34 out of 43 persons came up with one or more proposals. Table 6 lists the measures mentioned in order of frequency:

As can be seen in Table 6, a large share of the measures suggested concern traffic problems. Comparing this result to the GAP participants' achievements as far as transportation behaviour is concerned (see Figure 1), we see that on the one hand, transportation behaviour is difficult to change and, on the other hand, traffic problems are among the most prominent ones affecting quality of life. Residents seem to perceive rather little control over their transportation behaviour (without too many sacrifices) and traffic safety and demand actions from their authorities.

Traffic and social aspects make up two thirds of all mentions (one person is only counted once per area). Traffic concerns were to a large extent directed at safety and health targets. Almost two thirds of them were justified by a concern about the safety of pedestrians and bikers. Parents were particularly concerned about the safety of their children on their way to school.

---

[11] Quality of life is defined subjectively in this study by the interviewees. The content analysis revealed that they see their quality of life as depending on

- how they can fulfill their everyday needs (work, school, shopping, and so on), here called everyday function;
- their social life (social function);
- the quality of recreation (recreational function), and
- the extent to which their immaterial and material values are present in their lives (value function).

Table 6: Suggested measures to improve the quality of life in the five municipalities under study (Bruppacher and Kaufmann-Hayoz, 2000: 5).

Areas of intervention	Measures / specific areas mentioned	Frequency
Traffic	Measures against speeding, emissions, security, zoning	19 (35%)
Social issues	• Social integration and co-operation (6) • Social institutions (5) • Education(5) • (Governmental) social contributions (4)	17[12] (31%)
Zoning	(Building sector)	11 (20%)
Energy supply and consumption		3 (6%)
Waste management		3 (6%)
Conservation of local spring water		1 (2%)

Table 6 shows the areas in which measures were suggested to improve the quality of life in the five municipalities under study.

4.3.3. Complementary views from the citizens' point of view: Opportunities to act – Changes in everyday life as windows of opportunity to influence investment behaviour

When asked about investments like washing machines, dishwashers, and electrical appliances, participants reported that they pay attention to the energy consumption of their purchases. When it comes to heating systems and building insulation, the situation seems to be less clear. Of course, few of the participants in a programme like GAP are about to build a new house or to buy a house and renovate it. Most of the time, heating systems are treated as given. Results from the qualitative interviews indicated that people are not very informed about new and alternative heating technologies and mostly rely on experts, in this case architects. This is not per se an inefficient way to act, but there seems to be a gap between the legitimate interest of actors in the building sector to build houses as cheaply as possible and the environmental sustainability of those houses.

Here a combination of command and control instruments or subsidies (economic instruments) with communication instruments appears to be a solution, because communication instruments foster the acceptance of command and control instruments, and subsidies make the content of communication instruments more likely to be applied. Noteworthy is that the timing of the communication instrument is best when information is delivered while people are facing or planning major changes in their everyday lives, particularly when they are moving and building their own homes. Homeowners are key actors in the field of energy conservation (see also the contribution by GARBELY AND MCFARLANE).

12 The various social measures do not add up to 20, because three people mentioned more than one social measure.

5. Summary and conclusions

Factors in environmentally responsible behaviour are not only diverse but also involve dynamics that are yet to be revealed. Ten key factors could be identified for the areas of waste, wastewater, energy, consumption, and transportation. The communication and diffusion instruments applied by the GAP strategy seem to be suitable to influence the key factors "social and political integration" and "opportunities to act", especially when combined with service and infrastructure instruments, economic instruments, and/or command and control instruments.

Private households constitute an important group of actors. Municipalities can influence the structures and preconditions to facilitate environmentally responsible behaviour by applying an adequate mix of instruments. The GAP strategy can be part of such a mix. The strengths of the GAP programme are certainly that it addresses a variety of behaviours in diverse fields of action and that it can be implemented at low cost for both the municipality and the private household. Effects are greater when GAP teams are formed in neighbourhoods, because some applied communication instruments work better with a certain amount of social control. Also, infrastructure instruments are more likely to be initiated by participants when they all share the benefit.

A yet unsolved problem with the GAP programme remains: As with most other educational programmes, it attracts mainly people who are already well on the way to the desired goal. We see three possible solutions to this problem: First, for easier implementation by municipalities, the GAP programme needs to be institutionalised, e.g., within a national energy action programme such as "Energy 2000" of the Swiss Federal Office of Energy. Institutionalisation would lend the program a more professional touch and probably raise its acceptance on the part of municipal authorities as well as on the part of inhabitants. Second, municipalities should focus on the overall benefits of ecological measures: Improvements to the quality of life can be an important source of people's motivation to engage in municipality projects with common ecological and social aims, e.g., in the area of traffic safety. Third, in order to disseminate programmes like GAP effectively, a careful implementation of diffusion instruments is crucial (see the study by GUTSCHER ET AL.).

Future research should concentrate not only on isolated instruments, but also on the dynamics that result from combinations of instruments and their societal impacts and environmental outcomes.

References

Bruppacher, S. (2000) Umweltverantwortliches Handeln im Privathaushalt. Chancen und Grenzen der Selbstmodifikation. IKAÖ, Universität Bern; *Typoskript.*

Bruppacher, S. and Kaufmann-Hayoz, R. (2000) How to reveal common potentials of environmental, social and economical sustainability. Proceedings of the Third Biennial Conference of the European Society for Ecological Economics, May 3-6, Vienna; *CD-Rom.*

Diekmann, A. and Franzen, A. (eds.) (1995) Kooperatives Umwelthandeln. Rüegger AG, Chur, Zürich.

Forrester, J.W. (1968) Principles of systems. MIT Press, Cambridge.

Gessner, W. (1996) Der lange Arm des Fortschrittes. *In*: Umweltproblem Mensch. Humanwissenschaftliche Zugänge zu umweltverantwortlichem Handeln, Kaufmann-Hayoz, R. and Di Giulio, A. (eds.), Haupt, Bern.

Graf, E.O. (1997) Das Tun der Einen ist das Ziel des Tuns der Anderen. Erfahrungen mit Energiesparaktionen, die sich an Individuen richten. *In*: Bundesamt für Energiewirtschaft: Energiesparaktionen – Individuelles Energiesparen. Tagungsbericht. Workshop vom 25. Juni 1997, Bern.

Harland, P. Striving for sustainability via behavioral modification at the household level: Psychological aspects and effectiveness. *To appear in*: How green is the city? Sustainability assessment at the local level, Devuyst, D., Lens, L. and de Lannoy, W. (eds.), Columbia University Press, New York; *in press*.

Känel, E., Magun, B., Öhri, R. and Sanchez, A. (1998) Umweltverantwortliches Alltagshandeln beim Global Action Plan: die Bedeutung sozialer Netze. IKAÖ, Universität Bern.

Market Street Research (1996) Assessing the market potential for the household Ecoteam Program. Surdna Foundation, New York; *Manuskript*.

Mayring, P. (1995) Qualitative Inhaltsanalyse. Grundlagen und Techniken. Ed. 5. Deutscher Studien Verlag, Weinheim.

SFSO/Swiss Federal Statistics Office and SAEFL/Swiss Agency for the Environment, Forests and Landscape (1997) The environment in Switzerland 1997 – Facts, Figures, Perspectives. SFSO and SAEFL, Bern. Online in the WWW: <http://www.buwal.ch/e/themen/umwelt>.

Ulli-Beer, S. (2000) A transdisciplinary system dynamics approach to "Responsible environmental behavior". Proceedings of the Third Biennial conference of the European Society for Ecological Economics, May 3-6, Vienna; *CD-Rom*.

Vennix, J.A.M. (1994) Group model building. Facilitating team learning using system dynamics. John Wiley and Sons, Chichester, New York, Brisbane, Toronto, Singapore.

Vatter, M.D. (2000) Einstellungen, Handlungserfahrung und Kooperation im Umweltbereich. Unveröffentlichte Lizentiatsarbeit, IKAÖ, Universität Bern.

Weenig, W.H., Gijsman, N. and Langezaal, S. (1994) The Ecoteam Program in the Netherlands. Study 2: A study into the effects of three types of feedback. Centre for Energy and Environmental Research, Leiden University, Leiden.

A Lawyer's Comment on Bruppacher and Ulli-Beer

Jürg Wichtermann

The following brief comments discuss some issues from a legal perspective that arise in connection with the study by BRUPPACHER AND ULLI-BEER. This short, legal overview was prepared only after the study was conducted. For that reason, it restricts itself to sketching out the most important problem areas and associated problem-solving approaches. It does not intend to serve as a detailed treatment of the study.

Modern society has evolved into a communication society. In the wake of this evolution, there has been a change in the perception of how and via what channels the state should promulgate its activities. Today, the generally accepted view is that the polity should also be able to address the people by accessing communication channels that may diverge from the hitherto conventional understanding of governmental intervention. This, however, should not disguise the fact that despite an informal approach, state activities must, without exception, be embedded within a legal framework.

As is shown in the contribution by BRUPPACHER AND ULLI-BEER, the strategies of the Global Action Plan (GAP) have, in the first instance, been devised to effect a change in the environmentally relevant behaviour of the general public or of private households. These strategies are put into action mainly by means of communication and diffusion instruments. Support in the implementation of these campaigns is also sought at a municipal level. However, as opposed to the involvement of private initiators or organisations, communal involvement immediately raises fundamental legal issues once the behaviour of the people is to be modified and controlled. Does the state – in this case, the municipality – have a right to influence the behaviour of the population by means of appeals, endorsements or via other communication channels? And, if so, under what conditions?

It is an achievement of modern constitutional government that the state may, as a matter of principle, only become operative in cases where it is empowered to act. Phrased in simplified terms, the people must first specifically authorise the state or its administrative bodies to perform certain activities or fulfil certain functions. The historical development of constitutional government has also led to an increased formalisation in the activities of public offices. A municipality may not simply pronounce rulings or take arbitrary action, but must undertake to act solely within a predefined and circumscribed procedural order. These restrictions serve to guarantee adequate legal certainty, but have also given rise to the notion that state activities are fundamentally of a formal nature. However, it must not be overlooked that the state has often taken informal action in the past (or has been obliged to do so, in order to fulfil its functions with a reasonable expenditure of funds and resources). Furthermore, the need for such "softer" methods of intervention will presumably increase in future.

What does this mean in the event of a municipality wishing to support GAP activities? Because the scope of such activities – as described in the contribution by BRUPPACHER AND ULLI-BEER – usually ranges outside of conventional government channels, the central issue arises of what legal requirements must be satisfied to legitimate communal participation. Communication and diffusion instruments do not follow unequivocal "cause and effect" rules, and the method of approach precludes sanctions. For example, prompting the public to sort waste prior to disposal, issuing an endorsement of energy-conservation measures, an appeal to use public means of transport, or initiating projects which, owing to their role-model function, stimulate people to modify their behaviour are all strategies having basically no binding force. They do not incorporate injunctions or directives, nor can compliance be enforced among those individuals who choose not to save energy or use public transport.

Notwithstanding the non-binding character of "informal" state activities, such interventions may not be conducted outside the realm of a legally established framework. Indeed, the basic question immediately arises of whether the state is entitled to communicate in the course of fulfilling certain functions. Even sanction-free communication can trigger more or less strong reactions – which is precisely the goal of GAP activities. First and foremost, it lies within the power of the people to decide whether or not municipal authorities should become active in any given project. Within this context, there are three alternatives: the people may charge the authorities to promote a particular project, to hinder it, or to remain impartial to it.

Today, it is generally undisputed that the state should be able to make use of various means of communication in discharging its duties and achieving its goals. In other words, the state is granted a sphere of influence over the people. However, this does not imply that the state is given *carte blanche*: the operative precondition is that the state may attempt to influence the behaviour of the individual by means of information and communication only in those circumscribed domains that are within its purview. Therefore, using the instrument of communication to facilitate the attainment of specific government objectives within these prescribed limits is an acknowledged stratagem that does not, however, extend to its use as an arbitrary means of intervention or coercion. If, for example, a municipality has decreed that energy-conservation measures should be supported or that waste should be sorted according to content, then the authorities are bound in duty to perform the requisite functions. In this, it is feasible that certain communication and diffusion instruments – among them, prohibitions and sanctions – may seem expedient in order to realise these formulated goals. An entirely different situation arises, however, if the municipality has not assumed or been assigned the task in question. In the absence of a foundation for any given task, communication instruments may not be applied to modify behaviour and thereby effect compliance. Whenever an issue does not lie within the purview of the polity, the polity may not communicate pertinent to the issue. This basic principle is unqualifiedly applicable, despite the current trend towards allowing the state greater freedom in the use of communication instruments – including propaganda. This freedom indeed exists, but solely within the framework of predefined and circumscribed state activities.

Should a municipality plan to support GAP activities, the question immediately arises of whether this official promotion of energy conservation or public transport, for instance, lies within the province of the municipality. In Switzerland, depending upon the issue in question, the municipalities may claim the relevant sphere of competency for themselves, or be designated as competent by the cantons (in some cases, by the federation).

Once the primary question of legitimisation has been affirmatively resolved and the municipality has then proceeded to consider various means of communication, two further questions require clarification, which, albeit of lesser overall significance, nonetheless have a practical impact on the daily workings of the municipality. On the one hand, competency for the pending communication activities must be assigned within the municipality. Under certain circumstances, this can lead to difficulties, because informal action on the part of the authorities – as opposed to clearly defined formal action and the rigid assignment of tasks and responsibilities this entails – is not necessarily accommodated by the state organisational structure. And, on the other hand, it must always be taken into account that informal, communication-focused action generates costs, just as do all other state activities. Should a municipality envisage the use of extraordinary communication campaigns that would break the "operational budget" by causing extraordinary expenditure, the relevant financial body must be advised accordingly and, if need be, the allocation of the requisite financial resources must be approved by the competent authorities.

Therefore, municipalities may also attempt to influence the behaviour of the people by means of communication – for example, in the promotion of environmental issues. However, this does not disguise the fact that municipalities, even in the course of "simply" communicating, may not act in a legal vacuum.

Changing Things – Moving People.
Strategies for Promoting Sustainable
Development at the Local Level
R. Kaufmann-Hayoz/H. Gutscher (eds.)

Possibilities and Limits of Environmentally Responsible Action as Perceived by Participants in the Household EcoTeam Programme

Matthias D. Vatter, Wolfgang Gessner, and Urs Wittwer

The acceptance and effectiveness of individual instruments in the typology depend upon the target group's fundamental values, expectations, and behaviour intentions. For this reason it is important to evaluate these factors prior to applying an instrument.

The present contribution starts from the theoretical analysis of some "key concepts" and corresponding empirical findings of a questionnaire survey. Both are then related to the typology of instruments. The survey of participants in the Swiss Household EcoTeam Programme of the Global Action Plan is a study of a special ecologically motivated group. We aim to reach some conclusions about the instruments that should be implemented and the conditions under which they should be applied and to discover what instruments specifically promote environmentally sustainable behaviour. The result is an evaluation of the instruments from both a theoretical and practical perspective.

1. Introduction

This contribution aims to analyse the fundamental values, views on the problem, and behaviour intentions of persons who set demanding goals for behaviour towards the environment. With the aid of the analysis, we hope to arrive at some findings and conclusions with regard to the instruments in the typology. For the analysis, we studied participants in the Swiss Household EcoTeam Programme of the Global Action Plan (GAP), who intend to adopt improved resource-efficient practices in five areas.[1] GAP participants, as persons who are particularly concerned about the environment, should be in a better position to identify the barriers to environmentally friendly action than other, indifferent or less goal-directed, persons. At the same time, they probably show more responsiveness to environmentally friendly structures and take advantage of opportunities to act in an environmentally responsible manner.

The aims of this contribution are threefold. First, concepts that have proven to be fruitful in the research on environmental problems will be theoretically reconstructed (see below, section 1.1.). Based on this theoretical work, we can then survey GAP participants' opinions concerning the conditions of environmental action (see section 1.2. below). The resulting picture represents the second venture of the study, which is to discover the factors that GAP participants think are in fact important for environmentally friendly action. Knowledge of these opinions is necessary for the third objective, which is to estimate the acceptance and effects of the instruments and measures in discussion (see the typology of instruments). With this three-step procedure, the different viewpoints allow us to gain an impression of the likelihood that the measures will succeed in promoting environmentally concerned behaviour.

It is clear that the opinions of GAP participants can deviate considerably from those of the rest of the population. This means that the results of the study are not representative of the entire population, but they show the willingness to act, and possibilities and restrictions thereof, of environmentally concerned actors with paradigmatic clarity. It is also clear that the views of those actors might not (always) correspond to the "actual" or "real" way things work (e.g., causal relations among actions and effects). However, personal opinions lead individuals to action in the first place, and it is to opinions that we must look in order to form hypotheses on the acceptance and the effects of the instruments.

1.1. The theoretical reconstruction of key concepts

So-called key concepts allow access to the greatest possible spectrum of the relevant problems in a field (Flick, 1995:153). The reconstruction of key concepts draws researchers' attention to new aspects of the topic and leads them to new thinking and investigation. Key concepts are like searchlights that illuminate an area from varying angles and in varying colours.

For a theoretical analysis of environmentally responsible behaviour, we find 13 key concepts to be fundamental. These relate to research areas that have already been shown to be central or promise to be fertile in the context of environmentally responsible behaviour. As the key concepts refer to particular scientific disciplines, the findings, results, and theories of those disciplines can be

---

[1] GAP participants are also the subject of the contributions by BRUPPACHER AND ULLI-BEER, who investigated participants' opinions on the basis of interviews, and by MAUCH ET AL. See those contributions for a description of the GAP organisation and its Household EcoTeam Programme.

transposed to the environment context and applied. Theory-based reconstruction of key concepts should, first, allow the development of research instruments and provide meaningful interpretations. Second, it should yield theoretical findings and explanations with regard to environmentally sustainable behaviour.[2]

In this contribution, we will limit ourselves to the presentation and discussion of five key concepts:

- Value orientation
- Attribution of control
- Human error
- Structural constraints
- Co-operative action

The selection is due to limited space, but we also consider these five key concepts the most interesting with regard to both theoretical and empirical findings as well as to the hypotheses concerning the typology of instruments.[3]

1.2. Empirical study

Guided by the key concepts, we designed a questionnaire on everyday actions and the environment. We wanted to discover the personal and structural factors that GAP participants believe to influence their everyday behaviour towards the environment. As we noted above, it is important to know the views of those concerned, for otherwise it is difficult to estimate the impact the various instruments have. The items reflect the content of the key concepts. Besides providing information about the general conditions for environmentally responsible behaviour, the questionnaire responses allow us to evaluate the relevance of the key concepts. The resulting picture reveals the (potential) actors' views about themselves and about the situations in which they act.

The questionnaire survey was conducted from the fall of 1997 to the winter of 1998/1999 with persons who at the time were active participants in the GAP programme.[4] In this contribution we

---

2 This does not mean that the key concepts represent a complete typology of all conceivable theoretical perspectives. Instead, these concepts vary in the level and degree of specificity and are partly interdependent. Certain facts can be viewed under more than one key concept.

3 The key concepts not listed are: perception of side effects, risk perception, forming of intentions, dynamics of prognosis, self-management, perspective taking, prosocial orientation, and habit formation. The reader may refer to the following for a comprehensive treatment of the topic: Vatter (2000) presents a detailed overview of the key concepts, their empirical investigation, and evaluation under special theoretical consideration of the key concept of "co-operative action". Wittwer's (2000) theoretical focus is the key concept of "human error". Gessner et al. (in prep.) discuss the key concepts in detail, their interrelations, and their relevance with regard to environmentally responsible behaviour.

4 A total of approximately 150 questionnaires were distributed; 63 completed or partially completed questionnaires were returned to us by mail. This is a response rate of 42%.

will present a descriptive summary and interpretation of the response tendencies under the various key concepts, so that some conclusions may be drawn with regard to the typology of instruments. More extensive quantitative evaluations of the data, such as relationships among the key concepts, are reported elsewhere.[5]

2. The key concepts and their relation to the typology

This section presents the individual key concepts under three aspects. First, each concept is explicated theoretically (1). While not a detailed theoretical treatment, we hope it is enough of an overview to explain the content of the concept and the connections between the concept and the environment context. In a second step (2), the overall picture that emerges from the questionnaire survey of GAP participants is presented. With these theoretical and empirical results in mind, each concept is then discussed in relation to the typology of instruments under (3). This discussion contains hypotheses and conjectures as to how the various instruments would be accepted and, partly as a consequence, how likely they are to be effective.

2.1. Value orientation

2.1.1. Theoretical explication

The key concept of value orientation refers to fundamental values and attitudes as the background of personal action. Attitude research has shown that while attitudes are not expressed directly in behaviour, there is a moderate correlation between them (see, for example, Ajzen and Fishbein, 1970; 1977). In the area of the environment, there is a large body of findings on "environmental concern". This concept is understood as environmental awareness and experience or expanded to include value orientation related to the environment, knowledge about the environment, environmentally-relevant behaviour intentions, and, more rarely, behaviour itself (Spada, 1996; Urban, 1986). A meta-analysis evaluating various studies on the environment topic found a correlation between attitudes towards the environment and environmental behaviour with a coefficient of $r = .35$ (Hines et al., 1986/87); other investigations have found stronger correlations (e.g., Vatter et al., 1995; see also Kaiser et al., 1999). The differing results on the strength of the correlation between environmental consciousness and behaviour towards the environment is an indication of the problem of defining the concepts (compare Vatter, 1999).

The key concept of value orientation takes reference to this whole research area, but mainly to the aspect of values. Values are understood as an aspect of a mutual social construction of reality that is used in language communication in a group, society, or culture as the reason, justification, or

[5] For evaluations and details on the questionnaire, see Vatter (2000) and Gessner et al. (in prep.).

basis for evaluation of events and actions (Lantermann and Döring-Seipel, 1996: 632f.). Depending on the context, there can be discrepancies between individual and societal value systems. Values with regard to the environment can be regarded as part of these value systems, which can in turn compete with and come into conflict with other values. Their relevance has to be evaluated, and their relative weightings have to be assigned. It is interesting here to investigate what value orientations are found in certain persons and how they are prioritised.

2.1.2. *Empirical findings*

Given this theoretical background, we asked respondents about their motives for acting in an environmentally responsible way, about recognising and weighing societal problems, and about future prospects of environmental development.

The vast majority of GAP participants report that they endeavour to act in environmentally responsible ways because of their ethical principles and personal thinking about the problem of the environment. Other reasons, such as fear of sanctions, model behaviour, social pressure, regulations, and laws are negligible. The respondents to the questionnaire consider the destruction of the environment in Switzerland and in other countries to be the gravest societal problem. Accordingly, the respondents agree that prosperity should not be allowed to have a negative impact on the environment. Environmental protection should have priority over material affluence. The respondents are very worried about the prospects for the future, but they do not agree on the timing of the greatest threat. All concur, however, that time is running out: approximately 80% of the respondents believe that there is little time left to resolve the problem of climate change and that human survival is dependent upon that resolution. The use of nuclear power is seen by 80% to pose an intolerable risk. All in all, the respondents hold a markedly ecological value orientation.

2.1.3. *Relation to the typology of instruments*

The value orientations of the respondents provide valuable basic information that is relevant to possible measures and their expected effectiveness or non-effectiveness. Where people already have very marked environmental value orientations, communication instruments that relate to norms – presenting and persuading about appraisals, goals, and norms – are not necessary. However, some other instruments in this category may be indicated, such as presenting facts, if the ecological facts are unfamiliar. There will certainly be no resistance (conscious or unconscious) on the part of the recipients when taking in this new knowledge; on the contrary, information uptake will be facilitated by the recipients' existing values. The effectiveness of two of the communication instruments with direct requests is dependent upon ecological value orientation: Prompts and reminders work only if they relate to already pre-existing norms, and willingness to make a self-commitment requires at least a moderate level of agreement with personal values. Self-commitment in turn can affect value orientations by reinforcing and stabilising their behaviour-guiding effect. Presenting facts and options and providing experience of reality also probably contain an element of stabilisation if the facts and experience conform to personal values. Even feedback and self-feedback would not make sense unless personal ecological norms set the standards by which the behaviour goals could be defined and feedback evaluated. Conversely, most of the communication instruments can be implemented in order to foster the development of ecological value orientations; this is probably the most common purpose for their use. An effect upon values – similar to that described above – is likely if communication instruments with direct requests can bring people who do not have a strong ecological value orientation to act in an environmentally friendly way. There is a high probability that their personal values will shift to an ecological value orientation. This would pave the way for other interventions and increase environmental action.[6]

Offering or improving services and infrastructure that allow or facilitate ecologically sound action are instruments that typically target persons with environmentally friendly value orientations. This does not mean that no other persons can or will make use of improved services and infrastructures, but the prerequisite willingness to try them out is more likely to be found in people sensitive to ecological interests (compare MAUCH ET AL. and GEHRIG AND NORTH). In other words, the options for action that these instruments make available would more often be used by people with environmentally friendly value orientations than by others.

For the remaining types of instruments, value orientations are less relevant, as they are designed specifically to be effective independently of personal norms. But they are still important insofar as value orientations have a strong influence on the acceptance of such measures (prohibitions, taxes) and, as a result, their effectiveness. The objection can be raised that the implementation of economic incentives for environmentally sound behaviour can undermine pre-existing intrinsic motivation (compare here Deci, 1975). While that is indeed possible, it is more likely in this context that incentives help ecologically-minded persons to put their intentions into practice or to maintain their environmentally friendly behaviour (compare section 2.2.3.).

2.2. *Attribution of control*

2.2.1. *Theoretical explication*

People's subjective perceptions of their own actions and their options to act have a significant influence on behaviour (Flammer, 1990), including behaviour towards the environment (Axelrod and Lehman, 1993; Kaufmann-Hayoz and Peter, 2000). The decision for or against ecologically sound action may depend to a great extent upon whether a person perceives options for action in certain areas (Foppa, 1989). Opinions, experiences, and expectations about the consequences of their own actions also determine the degree to which people behave in an environmentally friendly way. The conviction that through their own behaviour people can influence the problem and contribute to a

---

6 The well-known "foot-in-the-door" technique takes full advantage of this, as a little step is expected to lead to further, bigger steps. Some theories that offer potential explanations of this phenomenon are dissonance theory (Festinger, 1957), the theory of self-perception (Bem, 1972), and impression management theory (Tedeschi, 1981).

solution has a positive effect on their willingness to act in an environmentally sustainable manner (Kaufmann-Hayoz, 1996). Conversely, the lack of a conviction that people have some degree of control can lead to a failure to act (Jaeggi et al., 1996). An adequate attribution of control based on the objective characteristics of the situation is often difficult to achieve in the environment context, as many situations are non-transparent and cause-and-effect relationships are unclear (see also section 2.3.).

For this key concept, then, perceptions of the power and influence relations in the field of action are relevant. Not only are *control beliefs* important (that is, if people feel that they have some control), but also more general *control attributions* (that is, control that is attributed to others). On the basis of personal control beliefs, it should be possible to estimate the potential effectiveness that individuals attribute to themselves in their particular spheres of action. From the more general control attributions, we can find out the degree to which people think that certain individuals, institutions, or the authorities are responsible for and capable of action in particular areas. From this we can generate hypotheses about how various environmental measures and interventions will be accepted.

Attributions of control can also be related to efficiency and sufficiency orientations (compare here MAUCH ET AL.). In our context, an efficiency orientation is the conviction that environmental degradation can be solved by means of technological improvements, while a sufficiency orientation is the conviction that without a reduction of growth, or curbs on affluence and the standard of living, there will be no change to the situation. Efficiency thinking attributes control externally, whereas sufficiency thinking makes internal attributions. It is reasonable to assume that environmentally sustainable action increases if the locus of control is attributed internally.

2.2.2. *Empirical findings*

The theoretical considerations led us to ask questions concerning potential actors' responsibility and possibilities for promoting environmentally friendly behavior, sufficiency and efficiency orientations, and judgements of various regulation options.

GAP participants clearly reject the idea that responsibility for environmental protection should be turned over completely to individuals (90%). They attribute responsibility for the environment to the government, and they therefore approve of governmental measures (taxes, laws, and so on). They are divided in their opinions on the issue of whether maintaining the current standard of living and affluence is compatible or not with environmental protection. Two-thirds of the respondents find these goals mutually exclusive and are convinced that sustainable environmental protection is not possible without cutting back the living standard (sufficiency orientation). At the same time, there is a tendency for the respondents to believe that technology and organisation can compensate and maintain current levels of prosperity (efficiency orientation). There is a moderate negative correlation[7] between the two views: A respondent agreeing with the one statement is likely

[7] r (Pearson correlation coefficient) = .45 ($p<.01$).

to disagree with the other. In the areas for action that the GAP programme outlines, respondents see the following actors as having the greatest influence:

Waste, water:	Mainly individuals, but also industry and the trades as well as the municipal and cantonal authorities
Energy:	Mainly energy suppliers, but also industry and the trades as well as individuals; the federal authorities to a small degree
Transportation:	Mainly individuals, followed by the federal authorities, then industry and the trades
Consumption:	Individuals overwhelmingly appear as responsible (75% of the respondents rank individuals in first place here!), followed by industry and the trades; environmental organisations are also attributed with some influence in this area.

The individual as actor is thus attributed with a great deal of influence on environmentally responsible action; with the exception of the field of energy, respondents rank individual persons in first place.[8] However, they also see possibilities for industry and the trades as well as the authorities to make real contributions to environmental protection. It is astounding how little influence is attributed to environmental organisations; apparently these are seen as admonishing voices rather than do-ers. When asked about the desirability of various control and economic mechanisms, only few respondents reject the idea of more laws and taxes, and usually over half heartily approve of the measures (for example, 80% are in favour of higher taxes in the transport sector!). There is practically no agreement with the idea that control should be left to market supply and demand in a mostly privatised, deregulated economy.

2.2.3. *Relation to the typology of instruments*

Due to the very concept behind the GAP programme, we would expect GAP participants to show a more individual-centred attitude that focuses on personal options for action. The findings confirm that the actions of individual persons are seen as having a great influence. The reigning conviction is that each individual can act in an environmentally friendly manner, and that there are usually ways to do this. However, this attitude points quite naturally to the significance of service and infrastructure instruments, for these serve to create options for action. At the same time, respondents approve of very strong intervention to promote environmentally sustainable practices. Their definite rejection of government withdrawal from the field, as well as their clear rejection of placing responsibility exclusively on individuals, make it likely that this group would show strong acceptance of the entire group of command and control instruments. The same is expected with regard to economic instruments that, as taxes or subsidies, are set by and controlled by the government. As the majority of the respondents explicitly approve of laws and taxes, we expect them as well to accept

8 This is not surprising given the strong emphasis that the GAP programme places on individual responsibility.

governmental and economic instruments and measures. Their refusal to turn all control over to the market, too, allows us to derive, vice versa, acceptance of all types of regulations. Apparently, the actors in the market, industry, and the trades are not perceived as promoters of environmentally friendly action, even though they could be, for respondents attribute quite a lot of influence to them. Industry and the trades could exercise responsibility in the form of collaborative agreements, but they can also introduce ecologically sound products to the market and withdraw environmentally undesirable products (both refer to service and infrastructure instruments).

In sum, in the opinion of the respondents, environmental protection should not be left to individuals in a totally free market system, despite (or perhaps because of) the high rating respondents give to the influence of (the sum of) individuals. They see a clear need for a legal and fiscal framework, a strong government that intervenes in the market and thus influences individual consumption. There does not seem to be a danger of the undermining effect (i.e., decrease of intrinsic motivation induced by extrinsic rewards, compare section 2.1.3.); instead, respondents wish to see additional incentives and regulations as a support to environmentally friendly action. This desire is very likely connected to the hope that more persons would then begin to act in an environmentally sustainable manner (see also 2.5.).

2.3. Human error

2.3.1. Theoretical explication

Actions that have an ecological effect are usually embedded in complex and non-transparent situations. Innumerable influencing factors make it difficult or even impossible for the individual to gain an exact grasp of the action situation. For this reason people often can not anticipate the consequences of their actions and thus adapt these actions to the situation to create successful action patterns. This aspect of environmental problems is very evident when we see that there is frequently a temporal delay between action and consequence, that the consequences often show up at other physical locations, and that some consequences of actions can not be perceived at all (see here Pawlik, 1991). The central issue is the way people react in complex situations, how adequate their perception of such systems is, and what influence this has on behaviour control. These are the conditions that will determine whether ecologically motivated actions will succeed or fail. Human error theory (Reason, 1994; Dörner, 1989) reveals the particular significance of so-called "mistakes"[9] whose origins lie in the complex interaction of situation variables and special, at times sub-optimal, processes in human perception and judgement. The theoretical bases suggest that ecologically relevant, correct, and, most particularly, easily accessible knowledge with regard to successful organi-

[9] "Mistakes" occur in the action planning phase, in contrast to "slips" and "lapses", where an act fails during the execution phase.

sation of actions is most important. This is particularly true for action and planning processes in groups, but it also holds at the individual level.

2.3.2. *Empirical findings*

The theoretical explication showed the importance of asking questions about knowledge of environmental facts (such as consequences of actions), about accessibility of action knowledge, and about perceived complexity of the environment context.

Even though subjective reports by GAP participants state that they have little difficulty in putting their ecological actions into practice successfully, other statements, as expected, indicate that crucial (action) knowledge is frequently lacking or erroneous. Through this, ecologically motivated persons can experience considerable uncertainty about what action alternatives will fulfil their environmentally friendly motives and intentions. Environmental problems are also perceived by many as complex and confusing. On the basis of the findings, it is probable that GAP participants, like others, often underestimate the variety and extent of the actual ecological consequences of their actions. Moreover, they reckon with, in part even explicitly, environmental impacts and resource use in order to achieve important action goals.

2.3.3. *Relation to the typology of instruments*

The theoretical and empirical findings underscore the necessity of the use of communication instruments. Within these, in the main measures that foster knowledge gain and experience gain should be considered, whereby it is important that both knowledge of facts (such as the consequences of a certain act) and knowledge about action (for example, the proper way to build a compost pile) are transmitted. It is therefore not surprising that, when asked about their reasons for participating in the GAP programme, respondents most frequently say it is "because I can learn something". The second most frequent response is that they participate because "GAP gives me practical suggestions". Communication instruments without a direct request can serve as an aid to decrease the complexity of the field of action and thus to make it easier to choose the "right" ecological action alternatives. It is particularly important that adequate diffusion channels are utilised. Knowledge must be transmitted directly to persons who perform the actions or will perform the actions, and in the specific place where the action shall be performed. Through the use of infrastructure instruments we can make sure that no further knowledge is required for choosing the ecological action alternative, for only that alternative is offered.

2.4. *Structural constraints*

2.4.1. *Theoretical explication*

> "What really gets to me is that I feel at the mercy of it all. I mean, that I have to buy so many packaged foods – with too much packaging - that I would prefer to pack myself. I really want to shop carefully, but then I see that I can't manage it; I'd have to go all over town to shop the way I want to. I get really angry that there is so little I can do against the excessive production of waste" (quoted from an interview, in Reichert and Zierhofer, 1993: 24, transl. by ed.).

This is an example of the many constraints that hinder environmentally sustainable action. Constraints can be physical-structural barriers that restrict a person's objective options for action to a degree that the person can act only in the given frame. There are also "soft" constraints that do not control actions absolutely, but instead steer behaviour in a certain direction. The individual may not even be aware of these structural constellations, which lead to "inner constraints" that cannot be overcome without unreasonable efforts. Many individuals may not experience "inner constraints" as constraints at all, but they narrow the subjective scope of action. In this connection, Gibson (1986) speaks of "affordances", by which he means the options or demands for actions that arise from the reciprocity of organism and environment. There is a reciprocity between person and environment in the sense that people do not only perceive the objects in their environment, but also at the same time perceive the meaning of those objects in relation to themselves, or the action possibilities that they offer. In other words, people perceive the options presented by the environment. The options, in turn, are dependent upon action possibilities specific to the species or person. Thus, the options, and the restrictions as well, are simultaneously objective and subjective. Many features of our constructed and designed everyday environments do not foster – as "affordances" or options – environmentally responsible action, but rather actions that waste our resources. It may be that, for example, the negative ecological consequences are explicitly hidden, or products and infrastructures may favour environmentally harmful action alternatives (Kaufmann-Hayoz et al., 1996). For these reasons, the environmentally responsible intentions of individual persons who have no opportunity to change these complex constraints have little chance of being put into successful practice.[10]

2.4.2. *Empirical findings*

The structural constraint questions aim at perceived constraints in the everyday settings of GAP participants and, where constraints are in fact perceived, how participants deal with them.

Surprisingly, in general, GAP participants do not see themselves as subject to many constraints. They usually feel able to put their decisions into practice; their intentions do not seem corrupted through situational factors, although this may conceivably happen in some cases. However, about

---

[10] See Gessner (1996) for a discussion of various types of restrictions.

one-quarter of the respondents believe that opportunities for environmental action are not created by themselves alone, and one-quarter report that they frequently encounter impenetrable barriers in their attempts to act in an environmentally responsible way. As an explanation of the fact that many respondents perceive no constraints on their behaviour, it might not be that they never experience constraints, but rather that subjectively, they feel that there are still plenty of alternatives open to them.[11] The barriers mentioned by respondents are experienced mainly in the areas of energy and transportation and less so in the areas of garbage and water. An interesting question that we can not treat here is the issue of the personality factors and factors in the situation that determine whether or not an obstacle is seen as insurmountable; certainly the limits are not the same for everyone.[12] Most frequent is the temptation to give up a tedious environmentally friendly action alternative and to return to a convenient and simple environmentally unfriendly variant.

Structural constraints thus seem to be perceived at times and pose a barrier to ecological action, but they do not make action impossible. The respondents in general see options for environmentally friendly action, and they put, or at least attempt to put, their intentions into practice by circumventing structural (and perhaps other) constraints.

2.4.3. Relation to the typology of instruments

The GAP participants appear to be people who are particularly clever at circumventing restrictions to action. However, the barriers perceived are of many different kinds, and depending on what type they are, they should be broken down through the implementation of command and control instruments, economic instruments, and service and infrastructure instruments. For example, increasing taxes on energy/resources would help people to overcome the temptation to drive short distances instead of walking or getting on their bicycles (this would break down a "soft" constraint). Similarly, offering or improving services that allow or facilitate ecologically sound action, such as providing new bus connections, would also allow people to forego the use of their cars (this would eliminate a "hard" constraint). The same goal can be achieved through dismantling or degrading infrastructure that hinders or inhibits ecologically sound action, such as constructing attractive bicycle paths instead of four-lane roads. The category of service and infrastructure instruments, which includes the instruments just mentioned, is particularly important with regard to breaking down restrictions and opening up new opportunities for action.

It is important to remember that barriers do not have to be absolute to function as such. GAP participants, who after all make an explicit attempt to act in an environmentally friendly manner, most likely have a higher threshold than the rest of the population when it comes to restrictive effects. Therefore it is extremely important to loosen up both "hard" and "soft" constraints. All command and control instruments and all economic instruments can be useful here, provided that

---

[11] An alternative explanation would be that these respondents just do not have any (or too little) experience in putting intentions into action. However, for GAP participants, this is rather unlikely.

[12] For some answers to this question, see BRUPPACHER AND ULLI-BEER.

they serve to break down individual action restrictions or institute corresponding alternative options for action. Existing infrastructures that are known to pose particular barriers must be eliminated, and lacking, positive infrastructures have to be created.

However, improving options for action and eliminating restrictions will of themselves not alone suffice to promote environmentally friendly behaviour. People need to hear about these changes. Thus, the targeted implementation of communication instruments to inform people about the improvements is indispensable.

2.5. Co-operative action

2.5.1. Theoretical explication

This key concept emphasises the fact that environmentally responsible behaviour is embedded within a social context. Two aspects can be distinguished. On the one hand, (potential) collective action has an effect on actors and on the success of action (this aspect can be called "social motivation and co-ordination"); on the other hand, there is a connection between individual action and the collective (we call this aspect"co-operation").

"Social motivation and co-ordination" refers on the one side to collective (versus individual) action with a view to the social process and how it is perceived by actors. The question is whether actors feel that they are under social pressure and the extent to which this pressure influences their actions. Depending on the social surround and the situation, a person can feel pressured to behave in environmentally friendly or unfriendly ways. Conversely, actors can also receive social support (from a spouse, for example) that strengthens their resolve or that is even the main factor in their resolve. The important role played by such processes (the role of the family, for example) with regard to environmentally relevant action has been pointed out by, among others, Känel et al. (1998). Quite often, goals that are not related to the environment stand in the forefront of actions, even if they have environmental effects (see, for example, Hirsch, 1993). It depends on how compatible social motives are with environmentally friendly action alternatives. On the other hand, it may be that it is easier to achieve a goal together with others not because of social support, but because the goal itself is easier to reach when greater numbers of people participate. For some actions to succeed, the co-ordination of some or many actors sharing the same interest may be necessary. In another case, a person may not have the abilities required to execute a particular action and thus need the help of others. Here collective (versus individual) action will stand at the focus in view of the task to be fulfilled.

"Co-operation" is studied by game theory and in research on "social dilemmas". Game theory investigates (mostly abstract) situations where pursuit of personal interest leads to sub-optimal results, or damages, for the collective. Game theory thus looks at strategies, as individual acts, and their collective consequences. It investigates the emergence of various strategies, their interaction among themselves, and their consequences, primarily in order to discover the "optimal" strategy.

Many game theorists, however, are also interested in descriptive empiricism: They hope to explain action and interactions. Of greater ecological validity is the work conducted in the tradition of the psychological research on social dilemmas, which was inspired in the main by Hardin's (1968) key treatment of the "Tragedy of the Commons". This tradition studies, under experimental conditions, the behaviour of persons who with their individual behaviour influence a collective good (usually a simulated natural resource). The state of the collective good does not, however, depend upon one single act, but rather is the result of the aggregated acts of all participants. There are a wide variety of experimental settings, and, accordingly, a great number of factors influencing behaviour have been found.[13]

Some central questions under the key concept of co-operative action, therefore, are how a collective (usually society in general, but also smaller social systems such as the family) is perceived by the individual, how predominant norms, expectations, aggregated actions, their consequences and effectiveness are perceived and predicted, and how they influence the individual's intentions. Other interesting topics are expectations and attributions with regard to others' behaviour related to the environment and people's beliefs about the influence of others' behaviour on their own actions.

2.5.2. Empirical findings

Following the theoretical explanations, the questions under this key concept concerned the perceived effects of acting together as a team (the GAP team) and the importance of social support. Furthermore, questions included perceptions and predictions of efforts by others to act in an environmentally friendly manner. The respondents were asked what influence these efforts had or potentially have on themselves and their own intentions and actions.

The majority of the surveyed GAP participants state that they gain a lot from collective action (from their GAP team). It seems easier to work out solutions and reach goals together with others, and social control and feedback have a positive effect on the motivation to act as well. The team aspect of the GAP programme is not merely valued by many respondents. For about a third of the respondents, social contact is one of the most important things about GAP. Great value is placed on social co-ordination and co-operation in the environmental sphere. About 90% of the respondents find the support of their spouses important, and 80% believe that more could be achieved in environmental protection if various actors would work together better. In other words, the status quo is perceived to lack co-ordination.

Questionnaire items that state that the sustainable behaviour of others (including politicians and companies) is a prerequisite for one's own environmentally friendly actions are rejected by 90% of the respondents. They do not share the opinion that "others have to start first". On the contrary, 85% say that they try to practice resource-efficient behaviour regardless of others' behaviour. Similarly, the majority of respondents reject the idea that their actions would be more efficient and

---

[13] See, for example, Ernst (1997) for a summary and discussion. For a detailed discussion of the co-operation aspect, see Vatter (2000).

effective if there were a guarantee that the behaviour of others would not undermine their contributions. They feel that each individual contribution is valuable and has an influence. The behaviour of others appears to have some influence, however: 40% of those surveyed feel it would be helpful when deciding what they should do if they could count on others to act responsibly. They are pessimistic in just this regard, for they believe that their fellow citizens are willing neither to pay higher prices and taxes nor to make adjustments to their lifestyles. The majority of respondents find that the dominant interests of certain actors often make successful co-operative environmental action impossible. All in all, then, many of those surveyed see others' particular strivings as a barrier to environmental protection and as a cause of the lack of co-operation – that is, in their perception, the interests of various actors contradict the collective interests in the area of the environment. At the same time, however, the respondents themselves do not feel that they are at the mercy of this dilemma. In other words, the fact that others are not practising sustainable lifestyles is to them no reason not to become actively involved. They are prepared to fight against the deterioration of the environment no matter what others do. But they do feel that being able to have trust in others to do the same would be supportive to their own environmental goals. Reliably co-operative behaviour of others would make it easier to translate their intentions into action and would lead to more environmentally friendly action.

2.5.3. Relation to the typology of instruments

All in all, embeddedness in a social network and, consequently, social relationships are important for environmentally responsible action. Many of the respondents draw the motivation to act in sustainable ways from their social surround. This fact points to the importance of communication instruments and of the design of diffusion instruments in particular. The households that make up a GAP EcoTeam form a network that not only provides team members with relevant information, but also provides social support and inspiration. The EcoTeam could be called an almost ideal form of communication and diffusion, as people in a specific target group are addressed personally by familiar or at least trustworthy communication partners who share a similar background. This guarantees attention, comprehension, acceptance, retention, intention, and behaviour in the best possible way (compare section 2.5., Table 2, in the chapter by KAUFMANN-HAYOZ ET AL. in the Theoretical Framework section). Moreover, one of the biggest advantages of this form of organisation is that it allows members to deepen social contacts to like-minded people, which thus serves a goal that goes beyond the mere exchange of information. We do have to keep in mind that people who are willing to join the GAP programme are probably already actively involved in environmental endeavours and thus represent a special selection. It seems clear that a communication and diffusion strategy like GAP's requires already strong ecological value orientations.

Although the respondents try to act in sustainable ways independently of others' behaviour, it would be easier for them if they could rely on others to do the same, for only commonly shared efforts would produce greater success. This is an aspect that GAP also tries to take into account – by creating social networks within which sustainable practices are self-understood. No longer isolated in single combat against greater forces, people surrounded by supportive, fellow participants more often experience a feeling of success. Unfortunately, this is possible only to a limited degree. GAP participants as well as (and more so) non-participants move within a social surround where unsustainable practices predominate and where individual attempts at responsible action are called a waste of time. On this background, people can hardly expect others to behave in environmentally responsible ways, and this may well weaken personal willingness to act. Here, command and control instruments, economic instruments, and collaborative agreements could create a situation in which a person can expect that a sufficient number of other people will act co-operatively. Once he or she comes to trust in this, the likelihood of environmentally sustainable behaviour increases (see the study by GUTSCHER ET AL.). If "objective" structures cannot be changed in this way, communication instruments could be implemented in an attempt to change expectations and attributions with regard to others' behaviour. This would be important in situations where individual incentives foster unsustainable practices, but a sufficient number of people seem willing to behave in environmentally friendly ways, provided that a certain number of other people would support them (as in some forms of the classic social dilemma). Using instruments like presenting and persuading about facts, options, appraisals, goals, and norms, perhaps the fatal choice of unsustainable practices could be broken and sustainable action established, which, if successful, would automatically spread through self-diffusion. Presenting model behaviour and sending appeals could also be implemented to this purpose. In order to steer expectations and attributions in the desired direction and thus promote sustainable behaviour, very targeted information would have to be presented that focuses on others' commitment and demonstrates the positive effects of personal actions. What is crucial is that a person trust that others are behaving in a similar fashion and therefore expect that a personal contribution to sustainability – together with others' contributions – would be effective.

3. Summary and Conclusions

From the viewpoints of five key concepts, some interesting insights on the possibilities and pitfalls of promoting environmentally responsible behaviour as perceived by GAP participants came into sight. The reason we focus on the views and opinions of the people surveyed is that those very views and opinions influence the effectiveness of promotion instruments to a considerable degree. If we know what people think is a help or hindrance to them in their efforts to behave in an environmentally friendly way, we are better able to estimate the acceptance of various promotion instruments and thus their effectiveness. In reference to the typology of instruments, theoretical analyses and empirical findings yielded some helpful conclusions with regard to application and implementation.

Ecological value orientations facilitate environmentally sustainable action. They are in most cases a necessary but insufficient requirement. Even for measures that do not depend on ecological values for their effectiveness, a certain degree of environmental consciousness is required for political acceptance. Thus, value orientations, even if they are not directly translated into action, form a significant, indirect fundament.

GAP participants believe that it is important that every individual make a contribution to the protection of the environment and adopt resource-efficient lifestyle practices. For this, opportunities for action must be available. This will not be enough, however: To achieve comprehensive, extensive, environmentally responsible action, they believe that people require command and control instruments and economic instruments.

Knowledge of the facts and knowledge on opportunities and execution – together with real options to actually put actions into practice – are also central to environmentally sustainable behaviour. Service and infrastructure instruments are indispensable when it comes to removing structural constraints and opening up alternatives for ecological action.

For people to be convinced of the effectiveness of their efforts to act ecologically and thus to become and remain motivated to act, a social surround must be created in which they can rely on others to do their part. Here we need command and control instruments and economic instruments in the main. Service and infrastructure instruments may also serve this goal. If collaborative agreements can be reached, they are extremely helpful. In favourable social contexts, the implementation of communication and diffusion instruments may suffice.

In this contribution, a number of perspectives under which the various instruments should be viewed were presented. There are no instant recipes for the design of instruments, nor is there any guaranteed prognosis as to their effectiveness. Ultimately, the parties planning to implement an instrument must themselves first evaluate the concrete social and structural context in which the intervention is to take place. The present contribution wishes to aid that endeavour by demonstrating the potential effects of the various factors that are important in environmentally responsible behaviour. The subjective views of potential actors play a major role in their acting in environmentally friendly or unfriendly ways, and knowledge of these specific views helps us to select and tailor the instruments that are the most promising in a given context. Guided by several key concepts, we have identified some factors that may be important for environmentally friendly behaviour. These factors should not escape our attention as we attempt to promote sustainable practices.

References

Ajzen, I. and Fishbein, M. (1970) The prediction of behavior from attitudinal and normative variables. *Journal of Experimental Social Psychology, 6*: 466-487.

Ajzen, I. and Fishbein, M. (1977) Attitude-behavior relations: A theoretical analysis and review of empirical research. *Psychological Bulletin, 84*: 888-918.

Axelrod, L.J. and Lehman, D.R. (1993) Responding to environmental concerns: What factors guide individual action? *Journal of Environmental Psychology, 13*: 149-159.

Bem, D.J. (1972) Self-perception theory. *In*: Advances in experimental social psychology, Berkowitz, L. (ed.), Academic Press, New York.

Deci, E.L. (1975) Intrinsic motivation. Plenum, New York.

Dörner, D. (1989) Die Logik des Misslingens. Rowohlt, Reinbek bei Hamburg.

Ernst, A.M. (1997) Ökologisch-soziale Dilemmata. Psychologische Wirkmechanismen des Umweltverhaltens. Psychologie Verlags Union, Weinheim.

Festinger, L. (1957) A theory of cognitive dissonance. Stanford University Press, Stanford.

Flammer, A. (1990) Erfahrung der eigenen Wirksamkeit: Einführung in die Psychologie der Kontrollmeinung. Huber, Bern.

Flick, U. (1995) Stationen des qualitativen Forschungsprozesses. *In*: Handbuch Qualitative Sozialforschung. Grundlagen, Konzepte, Methoden und Anwendungen, Flick, U., Kardorff, E. v., Keupp, H., Rosenstiel, L. v. and Wolff, S. (eds.), Psychologie Verlags Union, Weinheim.

Foppa, K. (1989) Grundlagen einer ipsativen Theorie des Handelns. Universität Bern; *unveröffentlichtes Manuskript.*

Gessner, W. (1996) Der lange Arm des Fortschritts. *In*: Umweltproblem Mensch: Humanwissenschaftliche Zugänge zu umweltverantwortlichem Handeln, Kaufmann-Hayoz, R. and Di Giulio, A. (eds.), Paul Haupt, Bern.

Gessner, W., Vatter, M.D. and Wittwer, U. Ökologie im Kontext. Barrieren und Chancen des umweltrelevanten Alltagshandelns. IKAÖ, Universität Bern; *in prep.*

Gibson, J.J. (1986) The ecological approach to visual perception. Erlbaum, Hillsdale N.J.

Hardin, G. (1968) The tragedy of the commons. *Science, 162*: 1243-1248.

Hines, J.M., Hungerford, H.R. and Tomera, A.N. (1986/87) Analysis and synthesis of research on responsible environmental behavior: A meta-analysis. *Journal of Environmental Education, 18/2*: 1-8.

Hirsch, G. (1993) Wieso ist ökologisches Handeln mehr als eine Anwendung ökologischen Wissens? *GAIA, 2/3*: 141-151.

Jaeggi, C., Tanner, C., Foppa, K. and Arnold, S. (1996) Was uns vom umweltverantwortlichen Handeln abhält. *In*: Umweltproblem Mensch: Humanwissenschaftliche Zugänge zu umweltverantwortlichem Handeln, Kaufmann-Hayoz, R. and Di Giulio, A. (eds.), Paul Haupt, Bern.

Kaiser, F., Wölfing, S. and Fuhrer, U. (1999) Environmental attitude and ecological behaviour. *Journal of Environmental Psychology, 19*: 1-19.

Känel, E., Magun, B., Öhri, R. and Sanchez, A. (1998) Umweltverantwortliches Alltagshandeln beim Global Action Plan: die Bedeutung sozialer Netze. Schriftenreihe Studentische Arbeiten, Nr. 11. IKAÖ, Universität Bern.

Kaufmann-Hayoz, R. (1996) Förderung umweltverantwortlichen Handelns – Versuch einer Synthese. *In*: Umweltproblem Mensch: Humanwissenschaftliche Zugänge zu umweltverantwortlichem Handeln, Kaufmann-Hayoz, R. and Di Giulio, A. (eds.), Paul Haupt, Bern.

Kaufmann-Hayoz, R., Häuselmann, C. and Gessner, W. (1996) "Eco-Design" – die wahrnehmungspsychologische Erweiterung eines technischen Konzepts. *In*: Naturbilder. Ökologische Kommunikation zwischen Ästhetik und Moral, Lesch, W. (ed.), Birkhäuser, Basel.

Kaufmann-Hayoz, R. and Peter, U. (2000) The concept of control: A key concept in understanding and overcoming barriers to responsible environmental behavior. *In*: Control of human behavior, mental processes and consciousness: Essays in honor of the 60th birthday of August Flammer, Perrig, W.J. and Grob, A. (eds.), Erlbaum, Mahwah.

Lantermann, E.-D. and Döring-Seipel, E. (1996) Umwelt und Werte. *In*: Ökologische Psychologie. Ein Handbuch in Schlüsselbegriffen, Kruse, L., Graumann, C.-F. and Lantermann, E.-D. (eds.), Psychologie Verlags Union, Weinheim.

Pawlik, K. (1991) The psychology of global environmental change: Some basic data and an agenda for cooperative international research. *International Journal of Psychology, 26/5*: 547-563.

Reason, J. (1994) Menschliches Versagen. Psychologische Risikofaktoren und moderne Technologien. Spektrum, Heidelberg.

Reichert, D. and Zierhofer, W. (1993) Umwelt zur Sprache bringen. Westdeutscher Verlag, Opladen.

Spada, H. (1996) Umweltbewusstsein: Einstellung und Verhalten. *In*: Ökologische Psychologie. Ein Handbuch in Schlüsselbegriffen, Kruse, L., Graumann, C.-F. and Lantermann, E.-D. (eds.), Psychologie Verlags Union, Weinheim.

Tedeschi, J.T. (ed.) (1981) Impression management theory and social psychological research. Academic Press, New York.

Urban, D. (1986) Was ist Umweltbewusstsein? Exploration eines mehrdimensionalen Einstellungskonstruktes. *Zeitschrift für Soziologie, 15/5*: 363-377.

Vatter, M.D. (1999) Moral, Ökonomie, Kollektivität: ihre Bedeutung für umweltverantwortliches Handeln. Diskussion einiger Anwendungsprobleme. Unveröffentlichter Arbeitsbericht Nr. 13 aus dem Projekt "Veränderungshindernisse". IKAÖ, Universität Bern.

Vatter, M.D. (2000) Einstellungen, Handlungserfahrung und Kooperation im Umweltbereich. Lizentiatsarbeit. Institut für Psychologie, Universität Bern.

Vatter, M.D., Wittwer, U. and Aragon, M. (1995) Problembewusstsein und Umweltverhalten. Vordiplomarbeit. Institut für Psychologie, Universität Bern.

Wittwer, U. (2000) Menschliches Versagen im Kontext individuellen umweltverantwortlichen Handelns. - Schriftenreihe Studentische Arbeiten, Nr. 18. IKAÖ, Universität Bern.

Conclusions

Changing Things – Moving People.
Strategies for Promoting Sustainable
Development at the Local Level
R. Kaufmann-Hayoz/H. Gutscher (eds.)

Suggestions for Designing Policy Strategies for Sustainable Development

Ruth Kaufmann-Hayoz, Silvia Ulli-Beer, Myriam Garbely, and Heinz Gutscher

In this chapter we discuss the main findings of the studies presented in the previous sections with respect to their usefulness for designing policy strategies for sustainable development. The discussion is centred on the question of how optimal combinations of instruments can be designed by careful consideration of their respective modes of operation in time. We will also discuss the question of transferring the findings to socio-economic contexts different from the one encountered in Switzerland.

Up to the present, the discourse on environmental policy instruments and on instruments for promoting sustainable development has mainly taken place within the economic and political sciences and jurisprudence. In these discussions the focus is usually on the state as an actor and on its capacity to steer the behaviour of various social actors in order to achieve the policy goals (an extensive review of this perspective has been given, e.g., by König and Dose, 1993).

Such an approach does not typically look at the various policy instruments from the perspective of the target groups, i.e., of those whose behaviour is to be influenced and who influence each other and the state by using certain instruments as well. In this book, we have made an attempt to integrate both perspectives. We have discovered that this requires an integrated theoretical framework that relies on theoretical elements developed in different disciplines:

- theories of individual action, developed mainly in psychology and philosophy
- theories of organisational and collective action, developed mainly in political, economic, and business sciences
- theories of societal change, developed mainly in history and sociology.

The typology and the studies that we have presented are the result of interdisciplinary co-operation among scholars representing most of these disciplines, who brought in their respective theoretical backgrounds. This helped us to construct the theoretical framework that is outlined in the Introduction and in section 4 of the typology chapter by KAUFMANN-HAYOZ ET AL. We did not attempt, however, to construct an overall "super theory" of human individual, collective, and societal action. This would probably not be possible or even desirable. What is needed is sound, integrative, problem-oriented work that takes advantage of the differential strengths of the various disciplines for understanding and explaining the complexities of human-environment interactions. We hope to have given an example of how this can be achieved.

One conclusion we can draw clearly from the theoretical as well as empirical work presented in the previous chapters is the following: There is no single "best instrument" or general "best policy mix" for improvements of ecological sustainability. Instead, the "appropriate mix" has to be adapted to the specific problem to be tackled and to its specific context. Since the problems and contexts are always new and unique, it does not make sense to look for a universally applicable "recipe". Rather, we would call for unlimited imagination and creativity when designing policies: The typology that we have presented can be modified and extended, new instruments can be invented, and unconventional combinations can be tried. After all, society is no machine, applying policy instruments is no technical procedure, and sustainability is no static final state. We must rather strive for initiating a complex steering process towards sustainability as a dynamic balance of ecological, economic, and social concerns and for continuous learning on the way.

In the following sections, we will point out what we think is important to know about combining policy instruments. It is knowledge of principles that should be helpful when it comes to choosing sets of measures that are optimally adapted to the concrete local situation.

1. The significance of communication and diffusion instruments

Although "communication" has always been regarded in a general way as an important part of policy making, usually only relatively simple measures – mostly the presenting of facts in a verbal form by media – are subsumed under this term. We have for the first time systematically described communication and diffusion instruments as an instrument type in its own right, with reference to theoretical bases, and we have related them to more conventional policy instruments that have been treated extensively in the literature. By doing so, we have put in more concrete terms the proposition made by (Kaufmann-Hayoz, 1996: 534, transl. by ed.) in a review of the many obstacles to environmentally responsible action and possibilities to reduce them: "Key terms [for finding solutions] are reflection, communication, and organisation."

Communication and diffusion instruments as they are described in this book are means to organise individual and collective learning and communication processes. Two main forms of application of these instruments can be distinguished:

- Communication instruments may be applied by themselves – perhaps replacing conventional instruments – because they create preconditions for voluntary action based on insight, in the absence of a special economic benefit and of control and sanctions. The studies by GUTSCHER ET AL. and by BRUPPACHER AND ULLI-BEER are examples of this mode of application. As these studies show, carefully selected sets of communication instruments undoubtedly have an impact on the behaviour of the target persons. This is not surprising, given that these instruments are based on well-known socio-psychological mechanisms. The study by MAUCH ET AL., however, has shown that they have a weak point: their ecological outcome is often minimal unless their full "market potential" can be reaped. Therefore, when communication instruments are applied in this function as parts of action campaigns, a major challenge is to reach a sufficiently large number of participants, i.e., to apply not only an effective set of communication instruments, but also powerful diffusion instruments. The study by GEHRIG AND NORTH shows that the same applies for some service and infrastructure instruments, where marketing aspects seem to be of utmost importance if substantial ecological outcomes are to be achieved. The second important limitation to the effectiveness of communication instruments applied alone is that they cannot work against hard constraints resulting from the socio-economic, legal-administrative, or physical conditions of action (for practical examples, see the chapter by BRUPPACHER AND ULLI-BEER; for theoretical considerations, the study by VATTER ET AL.).

- Communication and diffusion instruments can also be applied in various ways for the preparation of conventional policy instruments and for improving the implementation and enforcement of these instruments. By this we mean more than the rather trivial task of disseminating information about the implementation of new command and control or economic instruments, although – as FLURY-KLEUBLER AND GUTSCHER point out – this is, of course, a necessary condition of their implementation. The studies by GARBELY AND MCFARLANE, MÜLLER AND KRAMER, BRUPPACHER AND ULLI-BEER, and BÄTTIG AND BALTHASAR contain a great number of suggestions about how communication and diffusion instruments could reinforce the acceptance and the effectiveness of conventional instruments. In democratic societies this is a very important function of those instruments, and the target groups are political decision-makers as well as the broad public. Illustrative examples of promising combinations are given in section 3 below.

2. Finding optimal combinations of instruments

2.1. Why are combinations necessary?

JEGEN reports that most decision-makers in the Swiss energy policy domain are in favour of a combination of two or three energy policy instruments for achieving better energy efficiency in the construction sector. This shows that the idea of trying to reach policy objectives by combining different types of instruments is already quite well rooted in common sense.[1] But **why** are combinations – especially of instruments belonging to different types – superior? To answer this question, we can draw on the theoretical analyses as well as on a number of the empirical findings presented in the previous chapters.

From the **theoretical analyses** we have learnt that potentials and resources for responsible environmental action and sustainable development emerge from multiple aspects of the external situation as well as of the action conditions within the actors. Any specific instrument will primarily modify only a few of these factors, which will in most cases not be enough to reorient societal development as a whole towards sustainability. Combinations of instruments are therefore needed in order to create **constellations of favourable conditions** by modifying as many factors as possible and thus to produce enough momentum for change (see section 4 of the typology chapter by KAUFMANN-HAYOZ ET AL., and the chapters by FLURY-KLEUBLER AND GUTSCHER, VATTER ET AL., BRUPPACHER AND ULLI-BEER; also Loessøe, 1996).

A second insight gained from the theoretical analyses is that the different types of instruments have different rationales or "modes of operation", i.e., their effectiveness is based on different mechanisms (see Table 3, page 90-91 in the typology chapter by KAUFMANN-HAYOZ ET AL.). These mechanisms are not only related to economic processes and legal constraints, but also have to do with personal and social goals, norms, and values. Of course, goal hierarchies vary both within and across individuals, organisations, and social groups. From the statement by FLURY-KLEUBLER AND GUTSCHER that "the better an instrument fits the motivational preferences of the target persons, the more efficient it will be" (section 4.2, page 84), we conclude that different instruments will not be equally effective with all persons or organisations (see also VATTER ET AL.). Working with combinations of instruments will therefore be more effective on the whole, because the potential to influence a large number of actors increases.

In addition to these basic theoretical reasons, we have presented a number of **empirical findings** pertaining to the above question:

- GARBELY AND MCFARLANE have shown by means of their equilibrium models that homeowners will be more willing to invest in energy efficient technology of buildings, if certain combinations of economic instruments (specific forms of incentive taxes and subsidies) are applied. However, they also state that despite this positive effect, "rebound effects" resulting from user behaviour can outweigh the technical energy conservation effects, and they hypothesize that influencing people's goals and knowledge by communication and diffusion instruments will be necessary in order to avoid such undesired side-effects. They also state that the creation of labels that signal the high energy standard of a building would enhance the effectiveness of the economic instruments.

[1] This finding pertains to conventional instruments and collaborative agreements; communication and diffusion instruments were not included in JEGEN's survey.

- MÜLLER AND KRAMER have shown how Switzerland's policy on water pollution control, which is based almost exclusively on command and control instruments, could be improved significantly if it were complemented by specifically designed economic instruments. They also suggest that campaigns using communication and diffusion instruments in certain phases of communal decision-making on technical renovations of wastewater treatment plants could have important economic and ecological long-term effects and thus increase the efficiency of the classic instruments.
- GEHRIG AND NORTH have highlighted the fact that the service and infrastructure instrument of solar power supply could have considerably more impact if it were combined with a label, with more deliberate use of communication and diffusion instruments to encourage demand, and if the price difference between alternative and conventional power were reduced by economic instruments.
- BÄTTIG AND BALTHASAR'S analysis of the Energy 2000 Action Programme of the Swiss federal government has shown that combining moderate subsidies and the use of communication and diffusion instruments within a set-up of network management has played an essential role in municipalities' decisions on concrete energy policy measures.
- BRUPPACHER AND ULLI-BEER have found that identical clusters of factors influence environmentally responsible action at the household level in different domains, and that municipalities must use combinations of measures in order to create and modify structures in such a way that these factors will mobilise resources and act as catalysts for more sustainable lifestyles.

2.2. *What instrument combinations are particularly promising?*

Although it is always necessary to carefully adjust the selection of instruments to the problem to be tackled and to the specific political and institutional context, a number of principles when combining instruments seem to hold quite generally, again for theoretical reasons as well as in the light of the empirical evidence.

- In many domains it might be promising to set a **minimal standard** by command and control instruments and to **motivate** target groups to invest or to make efforts to go beyond this minimum by economic instruments, by communication and diffusion instruments, and/or by service and infrastructure instruments (see the studies by GARBELY AND MCFARLANE and by MÜLLER AND KRAMER).
- Combining economic instruments with communication and diffusion instruments is very promising, because it seems to result in **mutual reinforcement** of the instruments (see the studies by BÄTTIG AND BALTHASAR, BRUPPACHER AND ULLI-BEER, and GARBELY AND MACFARLANE).
- Public-private agreements seem to be effective only when they are **legally binding** and **embedded in a strategy** that provides that economic or command and control instruments will be

applied if the targets are not achieved on the basis of agreements (see Example 8 on page 67 of the typology chapter by KAUFMANN-HAYOZ ET AL.).

- Labels have the potential to **reinforce** economic as well as service and infrastructure instruments, because they reduce the costs and effort of target groups to gather relevant information (see the studies by GARBELY AND MCFARLANE and GEHRIG AND NORTH).
- Within the category of service and infrastructure instruments, repulsion based and attraction based measures should be carefully combined. Repulsion based instruments alone are often not well accepted; when combined with an attraction based instrument they will not only gain higher acceptance, but will also enhance the salience and attractiveness of attraction based measures. For example, in order to make people refrain from private car use for attending a large-scale sports event, one could severely limit parking space at the site of the event (a repulsion based measure) and at the same time offer free public transportation to the event (an attraction based measure).

2.3. *What has to be avoided when combining instruments?*

It is very important that the instruments selected do not contradict each other and that newly applied instruments are not inconsistent with existing regulations and structures. This is especially important with communication instruments, because they will have a minimal effect or even undesired consequences, if they contradict the powerful messages conveyed by economic, legal, or physical structures. For example, an inconsistency is created by propagating very low speed limits or by calling on people to refrain from car use, given a well developed net of roads that encourages private car use and fast driving and given sophisticated advertising for cars. BÄTTIG AND BALTHASAR have pointed out the risk that the Energy 2000 Action Programme will interfere with cantonal energy policies. MÜLLER AND KRAMER have described the difficult situation of municipalities with respect to water saving campaigns: If such campaigns are well done and have the desired effect of reducing the amount of wastewater, the municipality will have to increase waste water treatment charges (and thus punish people for saving water!), because the total costs of treatment cannot be reduced in a short period of time. Therefore, such a campaign would only make sense if decisions on technical renovation or on capacity increase of a treatment plant are on the agenda. Otherwise it could be counterproductive.

2.4. *What aspects of timing have to be considered?*

Timing is an important element of strategy formation. Aspects of timing have to be considered at the level of individual instruments, but they are of special importance when combining instruments. Among the questions to be answered are the following: What preparation times do the various instruments require? For how long will an instrument be applied? In the case of limited application of an instrument or a set of instruments (such as communication and diffusion instruments implemented within an action campaign): How is the campaign ended? What comes after the campaign, i.e., how does the campaign fit into a long-term strategy?

2.4.1. Preparation times

As WICHTERMANN points out, all public action needs to be legitimised by a nation's constitutional and legal framework. If the legal basis for an instrument does not exist, the political and legislative process to create it may require several years or even decades before a new instrument can be applied. This is well known for the classic instruments. For agreements and for communication and diffusion instruments, a general legitimation of the state to be active in a certain domain is sufficient, so that these more informal instruments do not require as much time for legislation. Nevertheless, a political decision is usually required, at least for approving the costs of these instruments.

As we have seen, some instruments (service and infrastructure instruments, agreements, and communication and diffusion instruments) are not only available to public authorities, but also to private organisations and firms. These actors can of course act much more flexibly and try out new approaches in experimental ways. We would like to point out that we consider such experimentation to be very important for setting new trends and for maintaining societal progress and innovative force. An innovation-promoting pattern of public policy should therefore be regarded an essential strategy element.[2] For this reason, communication and diffusion instruments should be used deliberately in the early stages of strategy implementation. Through these instruments, pioneers are found who pave the way and test the practicability of solutions that may later be enforced by conventional instruments. In that case, the pioneers will have an advantage due to their longer experience.

2.4.2. Windows of opportunity

Clever politicians and business leaders have always been able to chose the right moments in time to optimise their chances of success. This is of course also important for a policy strategy for sustainable development. Windows of opportunities may arise when specific acute problems or even catastrophical events create heightened public awareness and sensibility. For example, the frequent floods in the late nineteenth century were an important success factor for the rapid implementation of new forest regulations in Switzerland (Brändli, 1998). Likewise, the vehement public discussion on "Dying Forests" in Europe in the 1980s, the poisoning of the Rhine following a fire at the Sandoz chemical plant in Basel in 1986, and the incident at the nuclear plant of Tchernobyl in 1986 were important facilitators of progress in environmental policy.

Windows of opportunity may also open when external events lead to new coalitions within the network of political actors, as is presently the case in Switzerland's energy policy domain due to

---

2 This idea has been worked out in some detail by Blazeicjak et al. (1999).

the liberalisation in the European power sector (see the contributions by JEGEN and GEHRIG AND NORTH). Finally, windows of opportunity are frequently associated with the appearance of new and charismatic key players in any of the relevant actor groups, or to changes in the political leadership following elections. Needless to say, such events may also close existing windows of opportunity and create inopportune conditions.

For communication and diffusion instruments, windows of opportunity are of particular importance. As FLURY-KLEUBLER AND GUTSCHER point out, one of the primary functions of these instruments is to arouse attention and emotions:

> "The **arousal** caused by attention and emotions is a necessary prerequisite if people are to learn the contents of certain messages. Only if messages trigger a certain level of arousal will their content be remembered and become able to influence later behaviour" (page 122).

An illustrative example is reported by BRUPPACHER AND ULLI-BEER: They have found that a window of opportunity for presenting facts and options, for persuading about appraisals and goals, and for showing model behaviour on energy efficient construction is when people are about to construct or renovate a house and therefore apply for a construction permit at the municipality. They suggest that public information policy should much more deliberately look for such "moments of readiness" when delivering information to specific target groups. Disruption of daily routines caused by life events (moving, starting a new job, onset of an illness, birth of a child, and others) can also constitute windows of opportunity, because new routines have to be established, and the well-planned use of communication instruments at such times may be very effective.

2.4.3. Duration of application

Each instrument has its own characteristics with respect to the optimal duration of application. Some instruments – services and infrastructures that provide necessary action opportunities (such as facilities for collecting waste for recycling), zoning regulations, charges, and so on – have to persist, because they are necessary in order to maintain the desired behaviour. Other instruments are primarily intended to stimulate change, and once this is achieved, the instruments are no longer necessary or may even become counterproductive, i.e., inhibit further change and innovation. This might be the case with subsidies: in order to keep their potential to stimulate innovations they have to be limited in time (which is usually the case with subsidies as environmental policy instruments), or they have to be adapted in concordance with the process of change. Dynamic economic instruments, e.g., increasing the tax rates of incentive taxes over time or tightening the criteria for receiving subsidies, will be much more efficient than static instruments, provided that this dynamism is announced at the time of implementation and thus predictable for the target groups when they decide on investments (Müller, 1995).

When applying communication and diffusion instruments, special attention has to be given to effects of saturation: the same message delivered again and again over long times will no longer arouse attention and thus lose its efficacy. Prompts, for example, are usually effective for only a few days or weeks if left unchanged. On the other hand, repetitions over a long period and reminders from time to time are necessary in order to stabilise new behaviours and prevent forgetting. The challenge consists in finding a good balance between stability and change, i.e., finding ever novel forms to deliver the basic key messages (Zimbardo and Leippe, 1991: 175 ff.).

3. Applicability of the findings in other contexts

In examining the various instruments of environmental policy, we have strongly accented their concrete applicability. The context, at least implicitly, has been Switzerland. With the exception of a few studies that deal specifically with political decision-makers (JEGEN) and the decision process (MÜLLER AND KRAMER, GEHRIG AND NORTH) in Switzerland, however, the findings can be applied to other countries. This holds particularly for countries with a comparable level of socio-economic development.

For instruments implemented by the state on a legal basis, the state, as a prerequisite, must have the required means at its disposal: The state must be in a position to enforce laws, prescriptions, and prohibitions; to administer taxes and charges efficiently; and to finance subsidies and infrastructure. A further prerequisite is the existence of general political consensus on the necessity for environmental policy, and individual and corporate actors must be in a position to act upon options.

These conditions are not always satisfied to a sufficient degree, particularly in developing and transition countries. Even though there is no question that environmental policy is equally urgent in these countries – in view of the often massive consequences of environmental degradation on public health and the direct dependency of large sections of the population on intact natural resources (especially water and soil) – its priority status is not quite the same because of the acute social and economic problems. The state is not always in a position to implement environmental measures efficiently and often does not have the required means. Moreover, individuals and companies have a very limited scope of action due to both a lack of financial means and insufficient infrastructures.

In forming environmental strategies, extra special consideration in developing countries has to be given to social and economic aspects. This means that environmental policy **must be integrated** in a policy towards sustainable development. This holds not only for the implementation of measures, but for the setting of the actual goals. For instance, reducing the use of fossil energies in developed countries is an undisputed goal in view of both the greenhouse effect and air pollution. In a country like Algeria, however, where petroleum makes up 50% of the Gross National Product (GNP) and the export of fossil energy generates approximately 80% of the national income, progress towards the goal can only be made if at the same time a decisive policy on economy and development makes radical restructuring of the economy possible.

It is also not possible to simply apply the results of our evaluation of measures and the empirical studies to the context of the developing world. In particular, possible ecological side effects have to be evaluated anew. In highly industrialised countries, we assume that with a CO_2 tax, consumers can respond immediately by curbing consumption (e.g., not driving during leisure time) or through substitute behaviours (e.g., shifting to public transportation), and that they will pay the tax for the "indispensable" part of their energy use. In the longer term, consumers will make investments for energy saving and the use of clean, renewable energy. In very poor, rural areas of Africa, however, where petroleum is used mainly as a cooking fuel, an increase in the energy price will make it unaffordable for a large segment of the population. The only available alternative fuel is then traditional, self-collected fuel (mainly wood). This in turn increases the danger of continued deforesting, which has devastating consequences for the land and agriculture. The example shows how crucial the socio-economic context is to intelligent implementation of instruments.[3]

Specific regional problems as well as the state's limited scope of action in third world countries demand new methods. Innovative instruments are needed that can be implemented also by non-governmental actors, in particular by NGOs and local communities. Accordingly, communication and diffusion instruments take on significance, especially if they are implemented in a framework of participatory problem solving and planning methods.[4] The findings gained through the studies presented here are therefore thoroughly valuable for the situation in developing countries as well. However, for the concrete formulation of environmental and sustainability strategies in that context in particular, it will be important to investigate in detail (a) what instruments described here can be implemented exclusively for environmental goals and what instruments are also suitable for serving social and economic goals, and (b) whether the typology of instruments should be complemented by specific instruments (classic or new) in order to comprehensively represent the instruments for sustainable development.

References

Brändli, D. (1998) Mit Bäumen gegen Fluten: Überschwemmungsrisiko und Forstwesen während des 18. und 19. Jahrhunderts. Lizenziatsarbeit. Historisches Institut der Universität Bern.

Blazeicjak, J., Edler, D., Hemmelskamp, J. and Jänicke M. (1999) Umweltpolitik und Innovation: Politikmuster und Innovationswirkungen im internationalen Vergleich. *Zeitschrift für Umweltpolitik und Umweltrecht 22/1*: 1-32.

Kaufmann-Hayoz, R. (1996) Förderung umweltverantwortlichen Handelns – Versuch einer Synthese. *In*: Umweltproblem Mensch. Humanwissenschaftliche Zugänge zu umweltverantwortlichem Handeln, Kaufmann-Hayoz, R. and Di Giulio, A. (eds.), Haupt, Bern.

König, K. and Dose, N. (1993) Instrumente und Formen staatlichen Handelns. Carl Heymanns Verlag, Köln.

Loessøe, J. (1996) Consumption, lifestyles and constellatoric strategies. Nytenkning omkring effektiv energibruk og bërekraftig forbrul i husholdinger, Norges forskingsgråd.

Müller, T. (1995) Effets de la migration et de la taxation der l'énergie sur l'économie suisse, Thèse No 417 de la Faculté des sciences économiques et sociales. Université de Genève.

---

3 An overview of the problems of applying "traditional" instruments of environmental policy in the Third World is provided by O'Connor (1996).

4 See here O'Connor and Turnham (1992) as well as publications by the Discussion Forum North-South of the SPP Environment.

O'Connor, D. (1996) Applying economic instruments in developing countries: From theory to implementation. OECD Development Centre, Paris.

O'Connor, D. and Turnham, D. (1992) Gestion de l'environnement dans les pays en voie de développement. *Cahier de politique économique, 2.* Centre de développement de l'OCDE, Paris.

Zimbardo, P.G. and Leippe, M,R. (1991) The psychology of attitude change and social influence. McGraw-Hill, New York.

Changing Things – Moving People.
Strategies for Promoting Sustainable
Development at the Local Level
R. Kaufmann-Hayoz/H. Gutscher (eds.)

Inter and Transdisciplinary Processes – Experience and Lessons Learnt

Rico Defila and Antonietta Di Giulio

The aim of this contribution is not to present a theoretical discussion but to report on the procedures of and experience with inter and transdisciplinary processes in a project group, and recommendations resulting from these.[1] *The Integrated Project, "Strategies and instruments for sustainable development: Bases and evaluation of applications, with special regard to the municipality level" (IP), was a group of nine research projects (cf. List of Projects at the end of this book). Four of these sub-projects were located at the Swiss universities of Bern, Geneva, and Zurich; five were carried out by private institutions. The IP was managed by the Interdisciplinary Centre for General Ecology (IKAÖ) of the University of Bern. The project group was formed in the context of the 2nd stage of the Priority Programme "Environment" of the Swiss National Science Foundation, and funded from 1997 to 2000. The terms of this programme demanded proposals by project groups, in which individual sub-projects were to make a contribution to the objectives of the project group while pursuing their own research objectives.*

The introduction of this contribution (section 1) will briefly summarize the theoretical requirements for assessing the experience and making recommendations. Section 2 will outline the way in which the IP was organised in the given context of the programme, followed by a debate of whether this structure can meet the requirements indicated in section 1. The next two parts will focus on co-operation within the project group as it occurred in the IP (section 3), and with end-users (section 4). Section 5, finally, contains recommendations on the design of such project

---

1 The authors supported the IP director as scientific collaborators. They are also participants in an international co-operation that may interest the reader as a complement to the present report (see http://www.ikaoe.unibe.ch). In the framework of the DACH co-operation, a survey was conducted of all researchers participating in four inter and transdisciplinary environmental research programmes in Switzerland, Germany, and in Austria. The goal was to examine researchers' experiences over the past years and to prepare recommendations in the form of a manual for the design of inter and trandisciplinary processes.

groups, based on experience, as regards management and project organisation, integration and consensus, the participation of end-users, and internal and external communication.

1. Theoretical requirements

"Interdisciplinarity" is defined as the integration-oriented co-operation of individuals from at least two disciplines, towards common objectives, i.e., the synthesis of various disciplinary points of view, while "transdisciplinarity" refers to a special form of interdisciplinarity involving extra-scientific practitioners in research. In particular, based upon theoretical considerations and case-studies, the following requirements must be met by inter and transdisciplinary project groups (compare also Gibbons et al., 1994; Defila and Di Giulio, 1998; 1999):

- *Consensus:* By means of suitable procedures and methods, the participants need to arrive at a shared view of problems, at common objectives, shared questions and a common approach to dealing with them, and they need to develop a common language. Here, consensus does not mean "agreement" or "authorisation" in an everyday meaning, but the development of shared methods, models, and theories that integrate the various disciplinary viewpoints so that what is shared, is shared by all. For example, a shared description of the research subject needs to be developed. Of course, this requires of researchers the ability to overcome their disciplinary perspectives. These prerequisites are not discussed here in detail (see Defila and Di Giulio, 1996).
- *Integration (Synthesis):* From the beginning, suitable methods and processes need to be combined in such a way that the results of individual sub-projects form a whole, which is more than the simple addition of individual results, and which aims at answering the common questions. Common results and products therefore need to be developed.
- *Management:* Processes of consensus building and integration need to be initiated and monitored; co-operation within such a research group needs to be structured.
- *Diffusion:* The results need to be adequately disseminated, and their reception with the target audience needs to be promoted. As a rule, this audience is neither disciplinary nor purely scientific, just as the channels of dissemination are often different from the disciplinary ones. The knowledge gained needs to be useful to the target audience and their activities.
- In transdisciplinary research especially, it is necessary to ensure *from the outset the adequate participation of end-users* in the research. End-users are those at whom the research products and results are targeted, i.e., who are intended to use the products and results in their professional or everyday activities. They may be equal members of the research team – both at project group and sub-project level – or may be involved as "external participants" who do not belong to the research team, but are nevertheless substantially involved in a project, e.g., as members of a monitoring group (the term "external participants" does not apply to individuals who are the object of studies in the context of surveys, observations, etc.).

2. The Integrated Project "Strategies and Instruments" – Structure

2.1. Defining the project structure

One of the first steps following the inception of research work was to define the internal structure of the Integrated Project (IP) in order to clarify the rights, duties, and areas of authority of all the researchers involved in the IP. It was important to ensure that the overall interests of the IP as well as the interests of the individual sub-projects were taken into account. In other words, a structure was required that would ensure the development of a synthesis without putting the participating researchers at risk of neglecting their research work in the sub-projects. The IP's structure was therefore defined in a process of consensus building: The IP Management drafted a first outline of the IP's structure in the research proposal to the Swiss National Science Foundation. A "Co-ordination Group", consisting of three to five persons, and therefore not representing every sub-project, was foreseen. This group was chiefly intended to support joint planning and consensus building within the IP. It was also to assist development of the synthesis, which was predominantly to evolve in the course of various workshops. All researchers involved in the IP were to participate in these workshops. During a kick-off meeting, the researchers were asked to outline their ideas on the IP's structure, including the Co-ordination Group (its composition and tasks, and expected performance). It became clear that they did not deem such a group to be suitable for developing the synthesis, suggesting instead to set up a synthesis group representing all sub-projects, and bearing greater responsibility for the contents of the synthesis. Following this debate, a new proposal concerning the structure of the IP was drafted and accepted subsequent to a consultation process in writing.

A similar process attended the establishment of the Monitoring Group. The research proposal already foresaw this approach to ensure a continuous exchange with the target audience of the IP, and to promote the diffusion of its results. The initial plan was to create various monitoring groups for different target audiences. However, discussion with the researchers led to establishing one single monitoring group at IP level, which was considered more appropriate and more efficient. In parallel, some sub-projects created their own monitoring groups with practitioners to support their work (e.g., the sub-project of H. Gutscher).

Finally, it was decided to carry out as many studies as possible in the same municipalities, so-called "core municipalities", in order to promote the synthesis. These core municipalities should also be represented in the Monitoring Group.

2.2. *Organisation chart and task specifications*

The IP's structure foresaw clearly defined organs, such as the Synthesis Group, or the IP Management, as well as organs whose composition was flexible and whose tasks were circumscribed in a more general manner (e.g., working groups, sub-groups). The structure of the IP was intended not only to take into account the needs of the participants but also to remain sufficiently flexible and adaptable to new requirements.

IP Management: IP Management consists of the IP director and two scientific collaborators. It is responsible for the IP, which it represents both vis-à-vis the Swiss National Science Foundation and the public. One of the IP director's main tasks concerns synthesis, its responsibility being the *process* of consensus building and integration, rather than contents and products of the synthesis. The IP director therefore has to initiate and monitor the required processes, rather than determining their contents. It is supported by the scientific collaborators who, on the one hand, are in operative charge of the IP organs, of documenting processes, and of administration. On the other hand, these collaborators also carry out conceptual work on behalf of the IP director regarding measures and instruments, especially to promote consensus building and integration.

Sub-projects: The sub-projects are in charge of their own studies, of co-operation with other sub-projects, and of diffusion of their research results among their respective scientific communities. Their directors represent them vis-à-vis the Swiss National Science Foundation and the public. However, they are committed to the IP as a whole, including co-operation in synthesis processes and making available materials for synthesis building and products.

Synthesis Group: In the Synthesis Group, each sub-project is represented by one person, who is not necessarily the director of the sub-project. These delegates are nominated by the sub-projects and remain members of the Synthesis Group for the entire duration of the IP; substitution is foreseen only in the event of illness. The Synthesis Group is in charge of the contents of the synthesis; it therefore decides about synthesis instruments and products, and elaborates both the synthesis and its products. If required, it may form and call upon the support of special working groups with particular assignments.

Monitoring Group: The Monitoring Group ensures the continuous exchange between the IP and its target audience. In particular, it monitors practical, actual intelligibility and applicability of research results, and introduces stimuli, questions, and practical experience into the research process. Moreover, it supports the IP in diffusing the IP results.

Working groups, sub-groups: If required, working groups may be established to deal with particular issues or tasks. Such working groups support the Synthesis Group, by whom they are convened and which determines their objectives and tasks. Sub-groups may be formed for co-operation between two or more sub-projects, in particular joint data collection, exchange of data, joint development of methods, joint publications, etc. Their targets and tasks are determined by the sub-groups themselves.

The scope within which the researchers intended to operate was outlined by the organisation chart and task specifications. Does this seem adequate to meet the requirements of inter and transdisciplinary research as drawn up in section 1 above?

- *Consensus and integration:* The IP explicitly created an organ to elaborate the synthesis, i.e., the Synthesis Group. Its composition is representative so that, provided the delegates have been well chosen, a joint result can be achieved integrating viewpoints, know-how and results from all the sub-projects. Certain concessions have to be made as regards the range of the processes of consensus building and integration. Contrary to what would ideally have been desirable and was foreseen by the original IP Management concept, not every IP researcher participates equally in these processes. Rather, consensus building and integration are concentrated among the members of the Synthesis Group. At any rate, the present division of tasks was jointly elaborated and agreed following a debate at the inception of the IP.
- *Management*: IP Management has been charged with those tasks which, considering the theoretical requirements, it needs to fulfil if inter and transdisciplinary research is to be successful.
- *Diffusion*: The tasks resulting from this requirement at various levels have been identified and are reflected in the task specifications. The IP Management is in charge of general public relations work, while the sub-projects are responsible for diffusion of their results among their respective scientific communities. The Monitoring Group, finally, is a body which, if the delegates have been well chosen, should be in a good position to promote diffusion among the IP's non-scientific target communities.
- *Participation of end-users:* This requirement has been taken into account in two respects: the Monitoring Group consists of members of the IP's target audience, representatives from core municipalities among them. They are to be integrated into IP research activities as external participants. Also among the IP's target audience are certain parts of the private sector, such as consulting companies of municipalities. As sub-projects, such private companies, which often do both consultancy and research, are equal partners in the IP.

3. Interdisciplinarity – Co-operation within the project group

3.1. Management – Principles and perception of role

Below, the management principles of the IP are presented, as well as the way in which IP Management carried out its tasks. This presentation will be followed by an evaluation of experience gained in the IP. In its work, IP Management was guided by the following principles:

- *Create simple structures and clear responsibilities.* Simple structures are flexible and can be adapted to new situations and conditions. Clear responsibilities avoid uncertainties, as well as time and energy-consuming negotiations as to who is in charge of which task. Keeping a record of responsibilities and mutual commitments allows timely identification and discussion of conflicting interests and incongruous expectations between the IP and its sub-projects.

- *No delegation of responsibility for content to IP Management.* While IP Management is in charge of the process of creating the synthesis, it is not responsible for the results of the synthesis. IP Management, therefore, does not prescribe contents or form of the synthesis; it simply takes care that consensus building and integration are set in motion and that things move in the direction which was agreed upon consensually. This ensures that responsibility for content is jointly borne by the IP Management and the sub-projects.
- *Whatever is being done has everyone's support.* Although consensus-based decisions are time consuming, they guarantee the sub-projects' identification with processes and results at IP level. If consensual decisions are not being implemented, this will be discussed; decisions may need to be revised.
- *Concentration rather than excessive diversification.* IP Management takes care that a manageable amount of synthesis-related work is being undertaken, and that synthesis-related tasks do not suddenly drift apart. This will help avoid fragmenting the IP, and ensures continuous mutual communication and increasing understanding. Concentrating on central issues avoids energies being scattered. However, fewer tasks can be undertaken and fewer products will result. Accordingly, prior to submitting any measures to the researchers, IP Management will examine each one as to whether it is appropriate and justifies the effort.
- *Measures must be useful to everyone concerned.* It is the researchers themselves who decide whether and how any measure is implemented in synthesis building or diffusion. Only those measures will be optimally implemented whose meaning and purpose they can appreciate, and which will benefit not only the IP, but also their respective sub-projects.
- *The researchers set the pace.* With regard to consensus building and integration, IP Management maintains the dynamism of the process, but it does not set the rate at which this process evolves, nor does it set deliverables for tasks. In this way, the researchers will not be overtaxed, nor does IP Management forge ahead, thereby determining content.

Although, initially, it was indeed IP Management that made some quite prescriptive decisions, such as on preliminary internal and external communication structures. Even in this respect, however, the researchers moulded IP structures according to their own ideas, for example by re-defining the composition and tasks of the Synthesis Group (cf. section 2, above), thereby deciding that all sub-projects but not all researchers of the IP should participate in synthesis work. Therefore, this group was the "core" of the IP, assembling approximately every five months for extended meetings, and making all the decisions pertinent to synthesis in consensus.

The actual process can be described as follows: IP Management either proposed measures and instruments on consensus building, integration and diffusion, or developed them during talks with researchers (for example during Synthesis Group meetings, or workshops). When measures were agreed, IP Management ensured their implementation. The processes were largely guided bottom-up, by means of negotiations or in consultations. What set the direction was the common objectives and questions of the IP, which had also been determined in a procedure of consensus building, having been discussed and formulated jointly prior to submitting the proposal, and having been discussed again and modified about a year later on the occasion of the kick-off meeting. In the second half of the project, IP level work was further guided by joint products, i.e., this book, and a jointly arranged symposium.

The principles guiding IP Management in its tasks, and the role IP Management played, proved to be useful. IP Management also succeeded in setting in motion consensus building and integration and in keeping them going. The chief challenges and difficulties facing IP Management can be described as follows:

- *Withdrawal into sub-projects:* The question presented itself how the demands resulting from sub-project-related targets could be met alongside those resulting from common IP objectives. IP Management had to pay close attention to common objectives and questions not being lost from view, that is, to research work in the sub-projects not gaining predominance over synthesis-building in the course of the project. The problem revealed itself in that sub-project contributions to synthesis were occasionally tardy, delaying work at IP level. IP Management had no means to influence these processes by means of incentives or sanctions.
- *Lack of resources:* The issue here was to harmonise temporal requirements at the sub-project and IP levels. Although attempts were made to achieve this, for example by jointly assessing temporal demands for the IP, and by jointly defining a time budget for each task, it was ultimately impossible to arrive at a satisfactory solution. The fact that the Swiss National Science Foundation remunerated research work for sub-projects but left researchers with no remuneration for the time and effort spent on IP-level work cut into the time-budgets of the sub-projects.
- *Promoting and sustaining identification with the IP:* The challenge here was to sustain researchers' motivation to actively participate in processes of consensus building, integration, and diffusion. One difficulty IP Management was faced with was the uncertainty of knowing whether people were still motivated and still identified with the IP, and whether the pace was appropriate. IP Management had to develop specific indicators for this purpose as well as appropriate measures. For example, if there was no response to a proposal, this was not interpreted as a sign of tacit consent but as a "warning signal". In consequence, IP Management required that everyone had to express their explicit agreement with every suggestion submitted in writing. Also, if there was no reaction to written requests for contributions from members of the Synthesis Group, this was interpreted as indicating that this procedure was not considered purpose-oriented and that further action needed to be discussed in the Synthesis Group. To encourage debate, IP Management only rarely presented the researchers with finished proposals, attempting to initiate debate by making suggestions that occasionally required a lot of further thought.

3.2. Internal communication

Internal communication corresponded to the researchers' needs as they were ascertained in the context of the kick-off workshop.

- As far as possible, and in principle, communication exchanges took place electronically. E-mail distributors and an ftp-server were installed for this purpose.
- The primary purpose of meetings in person was not just for mutual exchanges; such meetings were always dedicated to joint work. Accordingly, the Synthesis Group did not meet at regular intervals but as and when the workload required. Once a year all researchers met for a two-day workshop, during which research results were exchanged and joint work was undertaken.
- Information and materials were distributed on demand. Documentation, papers, and publications from the sub-projects were not simply sent to everyone in circular mailings, but offered and mailed to whomever requested them. This prevented the researchers from being flooded with unnecessary information.
- It was up to IP Management to evaluate and select information for the attention of the researchers. It made a pre-selection, only passing on what was deemed necessary and meaningful to the researchers. This also meant that not every step and decision taken by the IP Management was communicated, provided they were in agreement with decisions taken jointly beforehand.

Communicative organisation stood the test of time, even though mastering all the technical problems with electronic communication media was very time-consuming. Owing to the principle of information on demand, however, scarcely any exchange took place between sub-projects outside the workshops. Finally, the sub-groups were seen to be superfluous; in such a small project group, it is not necessary to formally regulate co-operation between sub-projects. On the whole, the style of co-operation within the IP was collegial, frank, and informal, which both considerably enhanced synthesis-building and made it possible, towards completion of the work, to integrate an additional project as an equal partner in the IP.

3.3. Procedures of synthesis development

According to Rossini and Porter (1978) as well as Krott (1994; 1996), there are four different procedures according to which a synthesis can be developed. Each of these procedures was implemented in the IP:

"System": *Results from the sub-projects are continuously integrated according to a given theory or method (e.g., system analysis or model).* In an initial step, using the approach of policy evaluation (Bussmann et al., 1997), the Synthesis Group formulated key questions regarding policy instruments, which were answered from the perspective of the sub-projects. This enabled a more precise location and correlation of hypotheses, approaches, and questions from all the sub-projects with their widely differing contents on the background of one single theory. Most particularly, it helped identify those questions that required consensus building. The framework of policy evaluation was kept, albeit in a slightly modified form: it informs the description of types of instruments from the typology, the most important part of the synthesis results (see section 4 of the typology chapter by KAUFMANN-HAYOZ ET AL.), likewise the empirical studies of the sub-projects.

"Group": *Integration of results from sub-projects is achieved by the researchers as a group. They act as a group, acquiring joint group knowledge – the integrated results – shared by all participants.* The Synthesis Group evolved a common point of view and joint knowledge, albeit not on every detail of the synthesis, but only with regard to principles and criteria according to which the instruments for sustainable development were categorised, described, and compared, as well as the knowledge pertinent to these issues. It was decided not to focus further on the acquisition of more in-depth (disciplinary) group knowledge concerning all these instruments. What was jointly acquired and defined, however, was the basis of synthesis work (e.g., the model of human action described in the chapter by KAUFMANN-HAYOZ AND GUTSCHER in the Introduction of this book) and the guidelines informing further work. It enabled participants to select and search for specialised knowledge as well as results from sub-projects. This knowledge was then reformulated and combined with that basis.

"Negotiation": *Integration of results from sub-projects occurs by gradually enmeshing the sub-projects themselves. The integrated result of the entire project group then grows from such partial syntheses.* Members of the Synthesis Group occasionally worked in sub-teams that pursued particular questions on behalf of the Group, adducing specific disciplinary knowledge and results from sub-projects. The Synthesis Group also convened two working groups to carry out specific work concerning the further development of instrument typology. In other words, a smaller group initially acquired a common point of view and integrated knowledge prior to submitting the results of this process to the Synthesis Group. It was therefore possible to divide the workload. Moreover, disciplinary knowledge and results from the sub-projects were reformulated in pre-consolidated partial syntheses, which were nevertheless still informed by joint work (to be) done within the Synthesis Group.

"Project Management": *Nearing or following completion of the project, the managers of a project group integrate the results of not previously integrated sub-projects.* This type of procedure was not applied in pure form, although the Synthesis Group did delegate certain tasks to individuals, and to IP Management: individuals were asked to evaluate and reformulate insights from their sub-projects as well as further specialised knowledge concerning specific questions related to synthesis building. Such commissions, and work on them, were always based on what had jointly been developed and agreed on in the Synthesis Group. Moreover, IP Management from the start took charge of coordinating and closely monitoring synthesis work, making sure that the sub-groups, working groups, and individuals were making their contributions and that their contents and format corresponded to the Synthesis Group's stipulations. IP Management also submitted suggestions to the Synthesis Group regarding further steps and traced open questions and contradictions. Towards completion of the project, IP Management also took charge of producing this book as a joint product.

On the whole, the strategy of elaborating a synthesis in the Synthesis Group, the procedures employed, as well as the distribution of roles among this group and IP Management proved to be appropriate. Likewise, it made sense that the Synthesis Group had a consistent composition – ensuring continuity and avoiding repeat debates owing to changes in membership. Beyond these aspects, experience at the IP level also covers the following points:

- *Various synthesis procedures are not mutually exclusive:* Synthesis procedures as distinguished by Rossini and Porter (1978) as well as Krott (1994; 1996) are not mutually exclusive alternatives. In practice, as in the IP, a mixture of procedures is usually applied, depending on what is perceived to be sensibly done jointly, by all participants, or to be delegated to subgroups, and on what issues the specific knowledge of individuals will have to be consulted. It was impossible for everyone to do everything, but neither was this necessary to arrive at a synthesis. However, it was essential to arrive at a joint definition of certain "cornerstones", both with regard to contents and to methodology. It also proved necessary and useful for IP Management to make its own content-related contributions to synthesis building, and, from a certain distance, to locate differences in opinion and contradictions, which were then presented for debate.
- *It is difficult to involve researchers who are not part of the Synthesis Group:* It proved impossible to involve all researchers in synthesis building. The Synthesis Group was perceived as an increasingly inaccessible, closed circle. Whenever the Synthesis Group presented its results to other researchers in workshops, their advance knowledge, their having debated the issues and already having an opinion often impeded a true debate because the other researchers felt left out. Only rarely were any of their suggestions adopted by the Synthesis Group, by way of the representative of their sub-project. This was partly due to a clear division of tasks within the sub-project teams in order to economise resources. Neither could researchers who were not members of the Synthesis Group be integrated into working groups, which might have been another avenue for more active participation in the work of the Synthesis Group. On the other hand, it is debatable whether it would have been possible and have made sense to more deeply involve all researchers in synthesis building.
- *Synthesis is more than and different from the mere integration of the results of the sub-projects:* It was not possible to deal with the common objectives and questions of the IP on the basis of research work carried out in the sub-projects alone. On the one hand, not every one of their results could be integrated into the synthesis; on the other hand, the synthesis also depended largely on the researchers' further background, on their knowledge and experience gained outside the IP.

- *Linear planning is not possible for the processes of consensus building and integration:* Elaborating the synthesis took longer and proceeded more slowly than intended. It was necessary to present many issues for debate again and again, and to adapt them until all participants felt that they were in and of themselves correct and appropriate. In part, this was also due to the fact that contributions not made during Synthesis Group meetings were not always in synch; in other words, the various studies were not always at the same stage, nor were the participants always able to deal with synthesis work at the same time and to the same degree of intensity. Not to take this into account, that is, not to return to controversial issues, would at first sight have sped up the process, but it would have jeopardised synthesis building. To harmonise participants' work schedules was only possible up to a point.
- *A common language and theoretical basis are ideally developed by working towards a common product:* A common language was developed while working on a specific case, i.e., concerning those terms and concepts required at the time. An attempt to produce an independent glossary met with failure, as the researchers saw no need to agree on common terminology. The same is true for a common theoretical basis. This, too, was always developed in the context of needs emerging from work being done.

4. Transdisciplinarity – Co-operation with end-users

Some of the sub-projects of the IP were handled by private consulting companies who also do research. The question needs to be asked whether and to what extent researchers from these companies can be viewed as "practitioners". IP experience showed that the integration of individuals from such companies as equal partners in research was of particular benefit in guaranteeing the transdisciplinary nature of the IP because they customarily move back and forth between science and practical application. At the same time, it was found that the integration of such individuals cannot be a substitute for the integration of end-users who are not themselves involved in research work. This is probably true in general whenever end-users are integrated as equal partners in research: while those who do research will remain in touch with "their" fields of application, their position relative to those fields will shift; in other words, they will likely distance themselves from them, to a certain extent.

The following recommendation therefore follows from IP experience: to ensure the transdisciplinary nature of a project, it is necessary to integrate end-users both as equal partners in research and as external participants, for example as members of a monitoring group. The following paragraphs present and assess the way in which co-operation occurred with external participants, and how external communication was achieved in the IP.

4.1. Co-operation with core municipalities

Based on preliminary contacts, certain municipalities were selected as common study areas ("core municipalities"): as far as possible, the sub-projects focused their research on these municipalities, for example by conducting empirical studies or campaigns there, which were evaluated later. Agreements with these municipalities defined the contribution they were to make towards the IP, such as making available extensive documentation about themselves, or financially supporting the campaign of a sub-project (cf. study by GUTSCHER ET AL.). One council member and two members of the administration represented the core municipalities in the IP Monitoring Group. The following summary can be made of IP experience as regards co-operation with these core municipalities:

- *The creation of core municipalities is no guarantee of synthesis:* The fact that most of the sub-projects were focused on common study areas did not lead to the projects being more closely harmonised, nor to an enhanced integration of their results. A common study area – a "playing field" – is not identical with a common object of study, and common objectives and questions; hence, no synthesis is guaranteed to emerge. What is more, for some particular sub-projects, the efforts necessary to conduct studies in municipalities with which they were not familiar, or which were not located in their vicinity, were greater than the benefits. Contrary to original intentions, core municipalities did not catalyse synthesis.
- *Everyone needs to benefit from and desire co-operation:* One basic difficulty lay in the fact that it was not the municipalities that approached the IP with specific needs, questions, or problems. Quite the reverse was the case, with the IP searching for a common study area, and approaching three municipalities, requesting co-operation. The benefits to the municipalities were not evident, and were rated differently. This ultimately led to the municipalities identifying with the IP to varying, though not very great, extents. Hence, their willingness to make active contributions to the IP also differed. As a consequence, co-operation had to be cancelled with one of the municipalities because its contribution was not forthcoming. Also, on the whole, co-operation with the municipalities was less close than had originally been intended.
- *The varying cycles in research and practice should not be underrated:* One of the more challenging aspects of co-operation was the different timeframes for activities in the municipalities and in the IP and its sub-projects. A municipality usually needs to plan for campaigns within a different time span than research projects need to plan for their studies. As a consequence, work on one of the sub-projects in one municipality began with considerable delay. Moreover, municipalities could not always commit to activities that still required political sanction at a time when they would have been desirable for both sub-projects and IP. For example, in one municipality political considerations caused one of the planned campaigns to fall through altogether. In other words, co-operation is also impeded by the fact that the political situation of a municipality can shift greatly within two years (different majorities, different political agenda, etc.), including a complete change of conditions for projects.

- *Divergent priorities lead to uncertainties:* To a municipality, co-operation with a research project is a matter that, understandably, does not enjoy top priority, especially if no investments of their own are implicated. If other, more urgent business crops up, co-operation with research may have to be suspended, as it were. To commit fully to co-operation without any guarantees that they would be able to do what was needed to achieve the objectives of their sub-projects, was not always easy for the researchers – it was in fact rather risky, especially given the short duration of the projects.
- *Lack of special funding for co-operation leads to incomplete integration of practitioners:* To a municipality, co-operation with the IP meant that it had to make an additional effort, besides daily business, which did not seem to be of immediate benefit. To the researchers, on the other hand, it meant that they invested a lot of time at the expense of their research work, for example into creating and defining their co-operation with a municipality, which was neither credited to them as part of that work, nor funded separately. Hence, either side attempted to keep its efforts as low as possible and to invest as little time in co-operation as necessary. A similar problem already occurred during the planning stage of the IP and its sub-projects: any one municipality (as well as other end-users) can only invest limited resources in the conception of a (research) project that eventually may not be supported by the funding institution. In the case of the IP, actual work with the municipalities could therefore only begin when the project had been approved, which meant that the objectives and questions of the IP were formulated without the participation of what were later to be the core municipalities. This also meant that the point of view and the specific needs of these municipalities could not be taken into consideration at the planning stage.

4.2. *Working with the Monitoring Group*

The Monitoring Group of the IP was composed according to the researchers' needs and comprised 15 individuals from politics, the administration (federal, cantonal, municipal), as well as NGOs. The group held a total of four meetings lasting between three and four hours each, discussing the sub-projects' preliminary results, the current state of synthesis work, proposals for synthesis products and implementation projects, as well as preliminary ideas for the public symposium at the conclusion of the IP. At each meeting, attempts were made to set the Monitoring Group a specific task regarding IP synthesis work, such as the development of what they deemed to be a meaningful structure for a "tool kit" of policy instruments.

Fruitful and stimulating debates took place both with and within the Monitoring Group which also permitted the taking into account of practitioners' needs in the IP's further activities. However, the Monitoring Group was unable to play the expected role of facilitator in accessing specific practical ideas and questions that could then have been integrated into the IP's work. The following points particularly impeded cooperation:

- *The Monitoring Group had little influence on the direction of the IP's work:* Since the Monitoring Group was only established after the inception of the project, its scope of influence on objectives and questions of the IP and its sub-projects was very limited. Therefore, the Monitoring Group's identification with the IP was relatively weak, and its members placed low priority on participation in the group.
- *The members of the Monitoring Group did not feel personally committed:* Since the Monitoring Group was rather large, individual members did not feel personally committed to participate in meetings. On the one hand, this led to Monitoring Group meetings usually being attended by only about half the members, and on the other hand there was a large fluctuation in the attendees from one meeting to the next. Neither a group spirit nor continuously evolving discussions could emerge from this situation.
- *The Monitoring Group was a purely reactive body:* Since the Monitoring Group had neither an autonomous objective nor its own specific tasks, all that was left was for this group to react to issues presented by the researchers. Hence, it could not launch its own activities. This situation was exacerbated by the fact that the members of the Monitoring Group were active in an honorary capacity and the fact that direct benefits to their practical, daily work were insufficiently thematised. To the researchers, the meetings of the Monitoring Group represented a noticeable additional effort (preparation, participation, follow-up), and the relation between effort and return was not always satisfactory.

4.3. External communication

Alongside publications and presentations of papers by the sub-projects and IP Management, the following elements ensured external communication:

- An IP logo, which was consistently used in both internal and external communication.
- Joint web sites describing the IP and all sub-projects.
- The IP Bulletin, which was published approximately every six months, and in which the sub-projects reported on preliminary results, and the members of the Monitoring Group gave interviews on these results, as well as on the IP in general, for example discussing their expectations and their assessment of the applicability of preliminary results. The Bulletin was targeted at a wide audience of scientists and practitioners rather than at a specialised audience.
- Towards completion of the research work, a public symposium was held for the target audience of the IP. Besides the IP researchers, the members of the Monitoring Group were also involved in planning this symposium. Moreover, this event was arranged jointly with one of the core municipalities of the IP and also offered a platform to selected representatives of its target audience.

One of the main challenges in external communication was to adapt research results for a wider public. It was not only difficult to find a language that corresponded to scientific standards while being intelligible to lay people; it was also a challenge to get to the heart of scientific results and to present them in such a way that their practical consequences became evident. Without such adaptation, however, the significance and use of scientific results will remain opaque and ultimately useless to lay people (scientists from other disciplines, practitioners). Yet, the researchers were often unable to achieve these adaptations on their own. They were supported by IP Management which formulated some guidelines for writing contributions to the IP Bulletin and both edited the texts and played the role of the lay person.

5. Conclusions – Lessons learnt

A synthesis was developed in the IP, in spite of the fact that the sub-projects were not completely in synch with each other and despite the additional effort synthesis required of the researchers. Thus, on the whole, the IP can be considered a success. Its structure, distribution of tasks, procedures, and principles stood the test of time. What is the conclusion to be drawn from this experience? What are the consequences as regards the arrangement of processes in inter and transdisciplinary project groups?

5.1. Management and project organisation

The managers of the project group monitor and animate synthesis building without, however, defining the contents of the synthesis results: This ensures that the points of view, results and knowledge of all participants can be made part of the discourse, enabling consensus building and the development of a synthesis which does not rely on the managers alone. It is, however, essential for the management to contribute towards synthesis building as regards both content and methodology, in the sense of making suggestions, and coordinating and closely monitoring synthesis work. In other words, stimulating and monitoring work by the managers is "rich in content". Accordingly, the team of managers should consist of people with specific knowledge of the subject of the project. At the same time, steps need to be taken to ensure that content is not dominated by the managers, and that the researchers need not fear that it will appropriate their results of synthesis work.

The team of managers needs to represent more than just specialised knowledge of the project's research object: It is preferable not to designate one single manager of the project group, because it is highly unlikely that one individual will have all the necessary abilities. The team of managers should consist of individuals who have skills in communication and monitoring as well as systematic and analytical abilities, which play a crucial role when it comes to supporting the development of the synthesis.

Ensure true researcher participation at all levels: This ensures that the measures agreed on by the project group have everyone's support and will be implemented by all. It also enables the researchers to voice any misgivings about and objections to the managers' proposals, and to present counter-proposals of their own. It should be remembered that unease and misgivings may not always be expressed – the managers will therefore always have to watch out for signals of non-verbal and indirect communication.

Jointly define project structure in the project group: This ensures transparency of decision pathways, creating clarity about everyone's rights and duties, and, in the event of conflict, allowing to take recourse to a clearly defined consensus on rights and duties.

Jointly establish important stages, milestones, and timetables in the project group: This ensures that planning at the project group level stays transparent and is harmonised with work schedules within the sub-projects. It should be remembered that internal deadlines usually have lower priority than external ones and therefore are often not respected, so that it may be advisable to allow sufficient leeway and to discuss how to deal with internal deadlines.

Concentrate strengths and define how much effort is required for individual tasks at the project group level: This ensures that researchers do not lose focus and spread their resources too thinly; also, it enables the managers and the researchers to have tight control, and to continuously adjust resources for tasks, at both the sub-project and project group levels.

5.2. Integration and consensus

Establish a synthesis group that is responsible for developing the synthesis, and with representatives from each sub-project.

Advantages:

- If it is the synthesis group that decides on how to proceed in synthesis building as well as on form and content of the synthesis, the result will be true consensus and identification with the outcome.
- Placing responsibility for the integration of results with the synthesis group forces it to make decisions, and impedes its withdrawing into a purely consultative or affirmative role.
- Placing key decisions with the synthesis group avoids disregarding researchers' interests and conditions, and setting a pace that is too high for the researchers, for example owing to limited resources.

Bear in mind:

- It is essential for participants to the synthesis group to remain constant.
- Synthesis building remains restricted to members of the synthesis group – the other researchers are not involved in this process to the same degree and will therefore not identify with results to the same extent. This may entail friction and/or conflict.
- Relationships between the synthesis group and the other researchers concerning duties and areas of authority need to be clearly defined.
- Reciprocal consequences for sub-projects deriving from integrated results, and for integrated results emerging from sub-projects need to be assessed and monitored continuously.

Align synthesis work with joint products.

Advantages:

- If taken seriously, a joint product (such as a publication or presentation) will serve as a catalyst for developing a common language, theoretical basis, and view of problems, leading to an integration of results.
- Debates arising from work on a specific product more clearly reveal deficits in consensus building than abstract discussions.
- Completing joint products enhances the feeling of belonging to a group, improves identification with, as well as willingness to make a contribution to the project group as a whole.

Bear in mind:

- It may take a long time before joint products can be identified. It is therefore advisable to start the search for potential, specific products early.
- Searching for potential, specific products may involve errors and detours. It is essential to allow these to occur.
- Completing a joint product will sooner or later require an individual, or a small team, to take charge of actual production. This cannot be done by a large group, and will require planning in good time.

5.3. Participation of end-users

To ensure exchange between researchers and end-users, establish a monitoring group with practitioner experts.

Advantages:

- If researchers and monitoring group meet at regular intervals, a continuous exchange between researchers and selected end-users will take place, and research will continuously be fed input from practitioners, while they will receive new research results.

- Since feedback can be obtained from end-users on first ideas and preliminary results rather than on well-rounded ideas and final results only, work can be continuously oriented to the needs of practice.

Bear in mind:

- The monitoring group needs to strive towards an objective of its own; it needs to have a mission to avoid being a mere "echoing chamber" to research – resources need to be earmarked accordingly.
- To ensure that each member feels personally committed to work towards the common objective, the monitoring group must not be too large.
- From the outset, the monitoring group should have adequate influence on the design and direction of research.

Identify a common, real-life study area ("playing field") to carry out as much work there as possible while integrating appropriate end-users, at least as "external participants".

Advantages:

- Greater ease of comparability and exchange of results.
- End-users in the common study area identify with "their" research, "their" project group.
- If required, researchers can approach clearly identified contacts in "their" study area.

Bear in mind:

- A common, real-life study area does not of and by itself lead to a common research object, nor to a synthesis.
- End-users in the common study area need to be able to draw specific benefits from their co-operation with the project group, resulting in their own interest in shaping research.
- Co-operation can only be successful if these end-users are prepared to invest resources (manpower, funding) of their own.
- It is time-consuming but necessary to harmonise the different timeframes and conditions of work in research and in the common study area.

5.4. Internal and external communication

Cultivate an informal atmosphere: Informal relationships make it easier for people to express discontent and conflicting interests and improve constructive participation and co-operation, thereby contributing towards synthesis building.

Plan for multi-day meetings: Extended personal contacts enable the kind of intensive exchanges required by consensus and synthesis building. Electronic communication by e-mail or via ftp servers, for example, can only partly replace personal contact.

Establish simple communication structures and media: Clear and technically simple lines of communication and media help participants keep an overview and not waste time and help avoid technical problems and incompatibilities.

Present a corporate identity (e.g., logo, clearly identified contact persons): A corporate identity immediately identifies the project group as a whole to the outside world.

Publish a newsletter, bulletin, etc.: Regular publication of information, reporting on results from the project group and the sub-projects, can promote awareness of the target audience even during the course of the studies and can enhance reception of the research results. It should be remembered, however, that this effect cannot be assessed without considerable effort, and that wide diffusion does not ensure actual reception.

Plan for specific activities for and with the target audience: Integrating the target audience into the planning of an event ensures that it will meet the needs of that audience. Integrating representatives of the audience into the event itself will make it more attractive to the audience.

Lend support to researchers to translate their research results: If a lay person from the team of managers asks specific questions and makes suggestions for formulations, etc., this will help researchers to get to the point and to adapt research results for the target audience.

To conclude, it has to be pointed out that these suggestions and recommendations apply primarily to project groups with a size and structure similar to that of the IP, consisting of eight to ten sub-projects or research groups, and where there exist objectives and questions at the project group level as well as distinctive objectives and questions at the sub-project level. In practice, the latter will probably always be the case, even if sub-projects or research groups are more strongly aligned with the project group as a whole than was the case in the IP. The more closely the sub-projects are oriented to each other and to common goals, the finer the netting of the connections among the sub-projects, and the more intensively that a synthesis can be developed. Ensuring this, however, is not only a task for the managers and researchers of inter and transdisciplinary projects, but also for the funding institutions that must define the framework conditions of such research endeavours accordingly.

References

Bussmann, W., Klöti, U. and Knoepfel, P. (eds.) (1997) Einführung in die Politikevaluation. Helbing und Lichtenhahn, Basel, Frankfurt am Main.

Defila, R. and Di Giulio, A. (1996) Voraussetzungen zu interdisziplinärem Arbeiten und Grundlagen ihrer Vermittlung. *In*: Ökologie und Interdisziplinarität – eine Beziehung mit Zukunft? Wissenschaftsforschung zur Verbesserung der fachübergreifenden Zusammenarbeit, Balsiger, P., Defila, R. and Di Giulio, A. (eds.), Birkhäuser, Basel, Boston, Berlin.

Defila, R. and Di Giulio, A. (1998) Interdisziplinarität und Disziplinarität. *In*: Zwischen den Fächern – über den Dingen? Universalisierung versus Spezialisierung akademischer Bildung, Olbertz, J.-H. (ed.), Leske+Budrich, Opladen.

Defila, R. and Di Giulio, A. (1999) Evaluating transdisciplinary research. Evaluation criteria for inter and transdisciplinary research. *Panorama, Special Issue 1/99*. SPP Environment, Bern.

Gibbons, M., Limoges, C., Nowotny, H., Schwartzman, S., Scott, P. and Trow, M. (1994) The new production of knowledge. The dynamics of science and research in contemporary societies. Sage Publications, London, Thousand Oaks, New Delhi.

Krott, M. (1994) Management vernetzter Umweltforschung. Wissenschaftspolitisches Lehrstück Waldsterben. Böhlau Verlag, Wien, Köln, Graz.

Krott, M. (1996) Interdisziplinarität im Netz der Disziplinen. *In*: Ökologie und Interdisziplinarität – eine Beziehung mit Zukunft? Wissenschaftsforschung zur Verbesserung der fachübergreifenden Zusammenarbeit, Balsiger, P., Defila, R. and Di Giulio, A. (eds.), Birkhäuser, Basel, Boston, Berlin.

Rossini, F.A. and Porter, A.L. (1978) The management of interdisciplinary, policy-related research. *In*: Management handbook for public administrators, Sutherland, J.W. and Legasto, A.J. (eds.), Van Nostrand Reinhold Company, New York, Cincinatti, Atlanta, Dallas, San Francisco.

Annex

List of Authors

Artho Jürg, lic. phil. I
Psychologisches Institut, Abteilung Sozialpsychologie, Universität Zürich,
Plattenstr. 14, CH-8032 Zürich
Tel: ++41 (1) 634 21 23; Fax: ++41 (1) 634 49 31
eMail: artho@sozpsy.unizh.ch

Bättig Christoph, dipl. geogr.
Interface, Institut für Politikstudien, Kapellgasse 1, CH-6004 Luzern
Tel: ++41 (41) 412 07 12; Fax: ++41 (41) 410 51 82
eMail: baettig@interface-politikstudien.ch

Balthasar Andreas, Dr. phil.
Interface, Institut für Politikstudien, Kapellgasse 1, CH-6004 Luzern
Tel: ++41 (41) 412 07 12; Fax: ++41 (41) 410 51 82
eMail: balthasar@interface-politikstudien.ch

Bruppacher Susanne, Dr. phil.
Wissenschaftliche Mitarbeiterin, Eidgenössische Anstalt für Wasserversorgung, Abwasserreinigung und Gewässerschutz (EAWAG), Seestr. 79, CH-6047 Kastanienbaum
Tel: ++41 (41) 349 21 64; Fax: ++41 (41) 349 21 62
eMail: susanne.bruppacher@eawag.ch

Defila Rico, Fürspr.
Interfakultäre Koordinationsstelle für Allgemeine Ökologie (IKAÖ), Universität Bern,
Falkenplatz 16, CH-3012 Bern
Tel: ++41 (31) 631 33 62; Fax: ++41 (31) 631 87 33
eMail: defila@ikaoe.unibe.ch

Di Giulio Antonietta, lic. phil. hist.
Interfakultäre Koordinationsstelle für Allgemeine Ökologie (IKAÖ), Universität Bern,
Falkenplatz 16, CH-3012 Bern
Tel: ++41 (31) 631 39 56; Fax: ++41 (31) 631 87 33
eMail: digiulio@ikaoe.unibe.ch

McFarlane Alastair, Ph.D. in Economics
U.S. Department of Housing and Urban Development, Office of Policy Development and Research, 451 7th Street, Washington DC, USA-20410
eMail: Alastair_McFarlane@HUD.GOV

Flury-Kleubler Peter, Dr. phil.
Psychologisches Institut, Abteilung Sozialpsychologie, Universität Zürich,
Plattenstr. 14, CH-8032 Zürich
eMail pflury@sozpsy.unizh.ch

Friederich Ueli, Dr. iur., Fürspr.
Advokatur Arn und Friederich, Kramgasse 70, CH-3000 Bern 8
Tel: ++41 (31) 312 33 30; Fax: ++41 (31) 312 24 64
eMail: ufriederich@swissonline.ch

Garbely Myriam, Dr. ès sciences économiques et sociales
Centre universitaire d'étude des problèmes de l'énergie (CUEPE), Université de Genève,
7, route de Drize, CH-1227 Carouge
Tel: ++41 (22) 705 96 42; Fax: ++41 (22) 705 96 39
eMail: myriam.garbely@cuepe.unige.ch

Gehrig Sonja, dipl. geogr.
INFRAS, Gerechtigkeitsgasse 20, CH-8039 Zürich
Tel: ++41 (1) 205 95 23; Fax: ++41 (1) 205 95 99
eMail: sgehrig@infras.ch

Gessner Wolfgang, Prof. Dr. des.
Hochschule für Wirtschaft Olten, Riggenbachstr. 16, CH-4601 Olten
Tel: ++41 (62) 286 04 82
eMail: wolfgang.gessner@fhso.ch

Gutscher Heinz, Prof. Dr. phil.
Psychologisches Institut, Abteilung Sozialpsychologie, Universität Zürich,
Plattenstr. 14, CH-8032 Zürich
Tel: ++41 (1) 634 21 13; Fax: ++41 (1) 634 49 31
eMail: gutscher@sozpsy.unizh.ch

Jegen Maya, DES en science politique
Energy and Resource Group, University of California, Berkeley. Barrows Hall,
Berkeley, CA, USA-94720
eMail: maya.jegen@uclink.berkeley.edu

Kaufmann-Hayoz Ruth, Prof. Dr. phil.
Interfakultäre Koordinationsstelle für Allgemeine Ökologie (IKAÖ), Universität Bern,
Falkenplatz 16, CH-3012 Bern
Tel: ++41 (31) 631 39 54; Fax: ++41 (31) 631 87 33
eMail: rkaufmann@ikaoe.unibe.ch

Kramer David, lic. rer. pol.
Ecoplan, Wirtschafts- und Umweltstudien, Thunstr. 22, CH-3005 Bern
Tel: ++41 (31) 356 61 61; Fax: ++41 (31) 356 61 60
eMail:kramer@ecoplan.ch

Mauch Ursula, lic. phil. nat.
Mauch Consulting, Ruchweid 23, CH-3917 Oberlunkhofen
Tel: ++41 (56) 634 15 27; Fax: ++41 (56) 634 31 40
eMail: umauch@pop.agri.ch

Mosler Hans-Joachim, PD Dr. phil. et dipl. zool.
Psychologisches Institut, Abteilung Sozialpsychologie, Universität Zürich,
Plattenstr. 14, CH-8032 Zürich
Tel: ++41 (1) 634 21 18; Fax: ++41 (1) 634 49 31
eMail: mosler@sozpsy.unizh.ch

Müller André, lic. rer. pol., Dipl. Ing. ETH
Ecoplan, Wirtschafts- und Umweltstudien, Thunstr. 22, CH-3005 Bern
Tel: ++41 (31) 356 61 61; Fax: ++41 (31) 356 61 60
eMail: mueller@ecoplan.ch

North Nicole, dipl. geogr.
INFRAS, Gerechtigkeitsgasse 20, CH-8039 Zürich
Tel: ++41 (1) 205 95 95; Fax: ++41 (1) 205 95 99
eMail: nnorth@infras.ch

Raffael Pulli, dipl. natw. ETH
INFRAS, Gerechtigkeitsgasse 20, CH-8039 Zürich
Tel: ++41 (1) 205 95 21; Fax: ++41 (1) 205 95 99

Stern Paul C., PhD in Psychology
U.S. National Research Council, 2101 Constitution Ave. N.W., HA-178,
Washington DC, USA-20418
Tel: ++1 (202) 334 30 05; Fax: ++1 (202) 334 35 84
eMail: PStern@nas.edu

Ulli-Beer Silvia, lic. rer. pol.
Interfakultäre Koordinationsstelle für Allgemeine Ökologie (IKAÖ), Universität Bern,
Falkenplatz 16, CH-3012 Bern
Tel: ++41 (31) 631 39 40; Fax: ++41 (31) 631 87 33
eMail: ulli@ikaoe.unibe.ch

Vatter Matthias D., lic. phil. hist.
Länggassstr. 97, CH-3012 Bern
Tel: ++41 (31) 301 77 63
eMail: thys.vatter@gmx.ch

Wichtermann Jürg, Fürspr.
Blumensteinstr. 17, CH-3012 Bern
Tel: ++41 (31) 301 79 60
eMail: juerg.wichtermann@ziv.unibe.ch

Wittwer Urs, lic. phil. hist.
Aids-Hilfe Schweiz, Konradstr. 20, CH-8031-Zürich
Tel: ++41 (1) 447 11 11; Fax: ++41 (1) 447 11 12
eMail: urs.wittwer@aids.ch

Translators

Russon Ellen, lic.phil.
Übersetzerin
24 Putting Green Circle East Sandwich, Massachusetts, USA-02537
Tel: ++1 (508) 362 10 41; Fax: ++1 (508) 428 07 01
eMail: erusson@tiac.net
Translation of the chapters by GUTSCHER ET AL., VATTER ET AL., MAUCH ET AL.; correction and language editing of the entire volume.

Powell Joss Margret, lic. phil. I
Übersetzerin SUEDTV, Sprachlehrerin CTEFLA-RSA,
Muesmattstr. 28, CH-3012 Bern
Tel: ++41 (31) 302 61 62, Fax: ++41 (31) 302 61 62
eMail: mpowell@access.ch
Translation of the chapter by DEFILA AND DI GIULIO.

Zach-Miller Barbara
Z.a.ch. GmbH, Freudenreichstr. 31, CH-3047 Bremgarten
Tel: ++41 (31) 305 60 03
Translation of the chapters by FRIEDERICH and WICHTERMANN.

Zupnickel Ulrike
German Language Service, The Meadows, Lyndhurst, GB-5043 7EJ
Tel: ++41 (1) 703 283 984; Fax: ++41 (1) 703 283 887
eMail: uli@gerlan.co.uk
Translation of the chapters by BÄTTIG AND BALTHASAR, MÜLLER AND KRAMER.

List of Projects

Integrated Project

Strategies and instruments for sustainable development: Bases and evaluation of applications, with special regard to the municipality level
Strategien und Instrumente für eine nachhaltige Entwicklung: Grundlagen und Evaluation von Anwendungen, mit besonderer Berücksichtigung der Gemeindeebene
Kaufmann-Hayoz Ruth, IKAÖ, Universität Bern, CH-Bern
Carlevaro Fabrizio, CUEPE, Université de Genève, CH-Genève
Gutscher Heinz, Institut für Psychologie, Universität Zürich, CH-Zürich
SNF no. 5001-48826

Sub-Projects

Evaluation of water and waste water taxes and charges of cantons and municipalities
Evaluation kantonaler und kommunaler Abgaben für Abwasser und Wasser
ECOPLAN, CH-Bern
Müller André
Sommer Heini
Walter Felix
SNF no. 5001-48836

World views and sustainable action – religious and cutural values in ecological strategies
Weltanschauung und nachhaltiges Handeln - religiöse und kulturelle Grundwerte bei Umweltstrategien
IKF, Institut für Kommunikationsforschung, CH-Meggen
Jäggi Christian J.
Krieger David J.
SNF no. 5001-48831

Towards sustainable development by improving interactions between governmental environmental instruments, voluntary agreements and actions at the grass-root level
Nachhaltiges Handeln durch optimales Zusammenwirken von Aktionen an der Basis und staatlichen Rahmenbedingungen
INFRAS, CH-Zürich
Mauch Ursula
Von Stokar Thomas
North Nicole
SNF no. 5001-48835-1

Network management: A strategy for successful implementation of environmental policy?
Netzwerkmanagement als neues Instrument der Umweltpolitik?
Interface, Institut für Politikstudien, CH-Luzern
Balthasar Andreas
Bättig Christoph
SNF no. 5001-048827

Environmentally responsible behavior in community settings: Theoretical analysis and empirical investigation of overcoming barriers to change
Umweltverantwortliches Alltagshandeln in kommunalen Umfeldern: Theoretische Analyse, empirische Untersuchung und Überwindung von Veränderungshindernissen
Kaufmann-Hayoz Ruth
Gessner Wolfgang
Interfakultäre Koordinationsstelle für Allgemeine Ökologie (IKAÖ) der Universität Bern, CH-Bern
SNF no. 5001-48832

Ecological taxation of equipments and sustainable development. Evaluation of the environmental effectiveness and the economic efficiency of an eclogical taxation of buildings
Contribution de la taxation écologique des équipements au développement durable
Garbeley Myriam
Carlevaro Fabrizio
Weber Willi
CUEPE, Université de Genève, CH-Genève
SNF no. 5001-48829

Energy policy and Swiss elites: The acceptability of energy measures
Acceptabilité des mesures en matière de politique énergétique au sein de l'élite politique suisse
Kriesi Hanspeter
Jegen Maya
Département de Science politique, Unversité de Genève, CH-Genève
SNF no. 5001-48833

Contributions to sustainable development in municipalities: Simulation-aided testing and diffusion of psychological forms of interventions
Beiträge zur Nachhaltigkeit in Gemeinden: Simulationsgestützte Erprobung und Diffusion psychologischer Interventionsformen
Gutscher Heinz
Mosler Hans-Joachim
Psychologisches Institut, Abteilung Sozialpsychologie, Universität Zürich, CH-Zürich
SNF no. 5001-35273

Municipalities' legal room for manoeuvre in the domain of environmentally-relevant taxation
Rechtlicher Handlungsspielraum der Gemeinden im Bereich umweltrelevanter Abgaben
Advokatur Arn und Friederich, CH-Bern
Friederich Ueli
Keller Peter M.
Wichtermann Jürg
SNF no. 5001-58414

Discussion Forum North-South
Diskussionsforum Nord-Süd
Ruth Kaufmann-Hayoz, Interfakultäre Koordinationsstelle für Allgemeine Ökologie (IKAÖ) der Universität Bern, CH-Bern
Flury Manuel, IKAÖ, Interfakultäre Koordinationsstelle für Allgemeine Ökologie (IKAÖ) der Universität Bern, CH-Bern
Hurni Hans, Geographisches Institut (GIUB) der Universität Bern, CH-Bern
Mauch Samuel, INFRAS, CH-Zürich
SNF no. 5001-48826

☞ **More information is available online in the WWW at <http://www.ikaoe.unibe.ch> and <http://www.snf.ch/SPP_Umwelt/Overview.html>**

Index

This index is limited to the names of the instrument types and sub-types, in accordance with Table 1A (pages 40-41).

GPSR Compliance

The European Union's (EU) General Product Safety Regulation (GPSR) is a set of rules that requires consumer products to be safe and our obligations to ensure this.

If you have any concerns about our products, you can contact us on ProductSafety@springernature.com

In case Publisher is established outside the EU, the EU authorized representative is:

Springer Nature Customer Service Center GmbH
Europaplatz 3
69115 Heidelberg, Germany

Batch number: 10732714

Printed by Printforce, the Netherlands